Infosys Science Foundation Series

Infosys Science Foundation Series in Mathematical Sciences

The **Infosys Science Foundation Series in Mathematical Sciences**, a Scopus-indexed book series, focuses on high-quality content in the domain of pure and applied mathematics, biomathematics, financial mathematics, operations research and theoretical computer science. With this series, Springer hopes to provide readers with monographs, textbooks, edited volume, handbooks and professional books of the highest academic quality on current topics in relevant disciplines. Literature in this series will appeal to a wide audience of researchers, students, educators and professionals across the world.

Arpita Mal · Kallol Paul · Debmalya Sain

Birkhoff–James Orthogonality and Geometry of Operator Spaces

Springer

Arpita Mal
Department of Mathematics
Indian Institute of Science Bangalore
Bengaluru, Karnataka, India

Kallol Paul
Department of Mathematics
Jadavpur University
Kolkata, West Bengal, India

Debmalya Sain
Indian Institute of Information Technology
Raichur, Karnataka, India

ISSN 2363-6149 ISSN 2363-6157 (electronic)
Infosys Science Foundation Series
ISSN 2364-4036 ISSN 2364-4044 (electronic)
Infosys Science Foundation Series in Mathematical Sciences
ISBN 978-981-99-7110-7 ISBN 978-981-99-7111-4 (eBook)
https://doi.org/10.1007/978-981-99-7111-4

Mathematics Subject Classification: 46B20, 47L05, 46B28, 47L25, 46A32, 47A58, 46C50

This Springer imprint is published by the registered company Springer Nature Singapore Pte Ltd.
The registered company address is: 152 Beach Road, #21-01/04 Gateway East, Singapore 189721, Singapore

Arpita Mal would like to dedicate the book to her parents Bidyut Kumar Mal and Kalpana Mal, Kallol Paul dedicates the book to his mother Smt. Hira Paul and Debmalya Sain dedicates the book to his father Dr. Dwijendra Nath Sain

Preface

This monograph is intended to be a gentle guide to some of the latest developments taking shape in the broad area of geometry of Banach spaces. In particular, it provides insights into the geometric aspects of the spaces of bounded linear operators between Banach spaces, in light of the notion of Birkhoff–James (B-J) orthogonality. The notion of B-J orthogonality was introduced in 1935 by Birkhoff to generalize the notion of usual orthogonality on Euclidean spaces. A little less than a decade later, James elegantly demonstrated the applications of the said notion of orthogonality in the study of the geometric and the analytic properties of Banach space.

We wish to illustrate through this monograph that B-J orthogonality and related topics remain a fruitful area of research even after so many years. Indeed, a survey of the relevant literature should convince the reader that applications of B-J orthogonality techniques in the Banach space theory have picked up pace at an unprecedented order, especially in the last decade. Our goal, as far as this monograph is concerned, is to convince the reader of the wide applicability of this notion towards a better understanding of some long-standing problems related to the geometry of Banach spaces, including smoothness, convexity, extreme contraction, etc. It further carries forward this spirit to the study of the geometric properties of bounded linear operators de-fined between Hilbert spaces as well as Banach spaces.

In Chap. 3, we have studied the norm attainment set of an operator and its properties which play a very important role in the characterization of B-J orthogonality of operators. The possible structures of the operator norm attainment sets for Hilbert space operators are rather easy to understand. For their counterparts in the Banach setting, the situation is far more non-trivial. Although an analytic characterization of the operator norm attainment set has recently been obtained for bounded linear operators between Banach spaces, several basic questions remain unanswered. It might be surprising to the readers that the problem in its fullest generality remains open even for linear operators on real ℓ_p^2 spaces! Some of the aspects central to this problem have been presented in a carefully crafted manner, with the aim of presenting the state of the art in this direction.

In Chap. 4, we discuss a step-by-step development of the characterization of B-J orthogonality of bounded linear operators between Banach spaces. We explore the interrelation between B-J orthogonality in the ground space and in the space of operators, using the role of norm attainment set. As we will observe, this approach allows us to obtain a meaningful extension of some of the features of Hilbert space operators to the realm of Banach spaces.

Chapter 5 deals with two generalizations of B-J orthogonality, known as approximate B-J orthogonality. We investigate the geometries of both the ground space and the space of operators, from the perspective of approximate orthogonality. We present a geometric characterization of approximate B-J orthogonality sets in terms of two-dimensional normal cones. We also obtain characterizations of the approximate B-J orthogonality of operators between Hilbert spaces as well as Banach spaces.

Chapter 6 concerns the deeply studied and classical notion of smoothness of bounded linear operators between Hilbert spaces and Banach spaces. In fact, we complete the long-pursued program of characterizing smoothness in the space of bounded linear operators between given Banach spaces. Continuing the study further, we also investigate the k-smoothness of operators. We obtain a complete characterization of the k-smoothness of operators between Hilbert spaces and observe that the problem remains open for operators between Banach spaces.

In Chap. 7, we explore the notion of symmetricity of B-J orthogonality of operators both in the setting of Hilbert spaces and Banach spaces. The asymmetric nature of B-J orthogonality ensures that it is an interesting problem to study symmetricity of B-J orthogonality. A geometric characterization of Radon planes (two-dimensional Minkowski planes where B-J orthogonality is symmetric) is provided. We also investigate left-symmetric and right-symmetric operators between Hilbert spaces and Banach spaces. In particular, such operators on $\ell_1^n(\mathbb{R})$ and $\ell_\infty^n(\mathbb{R})$ have been studied in detail.

Chapter 8 is devoted to the study of extreme contractions, i.e., the extreme points of the unit ball of the space of operators between Banach spaces, endowed with the usual supremum norm. We introduce two geometric notions to deal with this problem. Using them we study extreme contraction on strictly convex and smooth Banach spaces. We also study extreme contractions on polyhedral Banach spaces. We will see that the notion of k-smoothness of operators plays an important role in studying extreme contractions on polyhedral Banach spaces.

As mentioned earlier, B-J orthogonality techniques have been firmly established, even more so in the recent years, as an indispensable resource in the theory of Banach spaces. All of the authors (along with several other collaborators around the world) have been directly associated with this particular direction of research. The present monograph consists of some of the results proved by us in the last decade, which seem particularly relevant towards demonstrating the applicability of B-J orthogonality in different areas of geometry of Banach spaces, and more generally speaking, functional analysis and operator theory. In our humble opinion, two particular features of this monograph best describe its significance and importance among the already existing literature on Banach spaces which are both profound and vast. First of all, we present the current state of the art of the various applications of B-J orthogonality and this is perhaps the first such monograph dealing exclusively with the concerned topic. In addition to that, the wide accessibility of the present work is another strong feature of our project. A thorough understanding of basic functional analysis (including the foundational theorems) and the theory of linear operators and matrices are mostly enough to understand and appreciate the main content of this monograph. Keeping in mind these dual benefits, we wish the readers a happy expedition in the fascinating realm of Banach spaces and bounded linear operators between them.

Bengaluru, India Arpita Mal
Kolkata, India Kallol Paul
Raichur, India Debmalya Sain
April 2022

Contents

Symbols

$\mathbb{C}$	The field of complex numbers
$\mathbb{R}$	The field of real numbers
$\mathbb{X}$, $\mathbb{Y}$	Banach spaces
$\mathbb{L}(\mathbb{X}, \mathbb{Y})$	The space of bounded linear operators between $\mathbb{X}$ and $\mathbb{Y}$
$\mathbb{K}(\mathbb{X}, \mathbb{Y})$	The space of compact linear operators between $\mathbb{X}$ and $\mathbb{Y}$
$\mathbb{L}(\mathbb{X})$	The space of bounded linear operators on $\mathbb{X}$
$\mathbb{K}(\mathbb{X})$	The space of compact linear operators on $\mathbb{X}$
$\mathbb{X}^*$	The dual space of $\mathbb{X}$
$\mathbb{H}$	Hilbert space
$\mathbb{L}(\mathbb{H})$	The space of bounded linear operators on $\mathbb{H}$
$\mathbb{K}(\mathbb{H})$	The space of compact linear operators on $\mathbb{H}$
T, A	Bounded linear operators
x, y	Elements of the spaces $\mathbb{X}$, $\mathbb{Y}$, $\mathbb{H}$
f, g	Bounded linear functionals
$\Re z$	The real part of the complex number z
$\Im z$	The imaginary part of the complex number z

Chapter 1
Notations and Terminologies

The notations and terminologies to be used throughout this monograph are mentioned in this chapter.

Let $\mathbb{R}$, $\mathbb{C}$ denote the fields of real numbers and complex numbers, respectively. The letter $\mathbb{H}$ stands for a Hilbert space and the letters $\mathbb{X}$, $\mathbb{Y}$ stand for Banach spaces over the field $\mathbb{R}$ or $\mathbb{C}$, unless otherwise specified explicitly. Let $B_{\mathbb{X}} = \{x \in \mathbb{X} \colon \|x\| \leq 1\}$ and $S_{\mathbb{X}} = \{x \in \mathbb{X} \colon \|x\| = 1\}$ be the unit ball and the unit sphere of $\mathbb{X}$, respectively. The unit ball and the unit sphere of a Hilbert space are denoted by $B_{\mathbb{H}}$ and $S_{\mathbb{H}}$, respectively. Let $\mathbb{L}(\mathbb{X}, \mathbb{Y})$ ($\mathbb{K}(\mathbb{X}, \mathbb{Y})$) denote the space of all bounded (compact) linear operators from $\mathbb{X}$ to $\mathbb{Y}$, endowed with the usual operator norm. We write $\mathbb{L}(\mathbb{X}, \mathbb{Y}) = \mathbb{L}(\mathbb{X})$ ($\mathbb{K}(\mathbb{X}, \mathbb{Y}) = \mathbb{K}(\mathbb{X})$), if $\mathbb{X} = \mathbb{Y}$. Thus, $\mathbb{L}(\mathbb{H})(\mathbb{K}(\mathbb{H}))$ is the space of all bounded (compact) linear operators on a Hilbert space $\mathbb{H}$. One of the main aims of this monograph is to explore the geometry of the space $\mathbb{L}(\mathbb{X}, \mathbb{Y})$, and to illustrate the far more general behavior of the norm function, in contrast to several niceties of the inner product which essentially ensures that the structure of $\mathbb{L}(\mathbb{H})$ is better understood. We do sincerely hope that the reader would appreciate this while gradually progressing through the subject matter developed in this monograph.

For a Banach space $\mathbb{X}$, $\mathbb{X}^*$ denotes the dual space of $\mathbb{X}$. For $x \in \mathbb{X}$, $y^* \in \mathbb{Y}^*$ and $x^{**} \in \mathbb{X}^{**}$, we use the symbols $y^* \otimes x$ and $x^{**} \otimes y^*$ to denote the linear functionals on $\mathbb{L}(\mathbb{X}, \mathbb{Y})$, defined, respectively, as : $y^* \otimes x(T) = y^*(Tx)$ and $x^{**} \otimes y^*(T) = x^{**}(T^* y^*)$ for all $T \in \mathbb{L}(\mathbb{X}, \mathbb{Y})$.

The following definitions will be used repeatedly in our work, and therefore, it makes sense to introduce them here.

Definition 1.0.1 (*Extreme point*) Let S be a non-empty convex subset of a Banach space $\mathbb{X}$. An element $z \in S$ is said to be an extreme point of S if $z = (1-t)x + ty$, for some $x, y \in S$ and for some $t \in (0, 1)$ implies that $z = x = y$. The collection of all extreme points of S is denoted by $Ext(S)$.

A. Mal et al., *Birkhoff–James Orthogonality and Geometry of Operator Spaces*, Infosys Science Foundation Series, https://doi.org/10.1007/978-981-99-7111-4_1

Definition 1.0.2 (*Extreme contraction*) An operator $T \in \mathbb{L}(\mathbb{X}, \mathbb{Y})$ is said to be an extreme contraction if T is an extreme point of the unit ball of $\mathbb{L}(\mathbb{X}, \mathbb{Y})$.

Definition 1.0.3 (*Strictly convex space*) A Banach space $\mathbb{X}$ is said to be strictly convex if the unit sphere does not contain a non-trivial line segment. Equivalently, $\mathbb{X}$ is strictly convex if $Ext(B_{\mathbb{X}}) = S_{\mathbb{X}}$.

Definition 1.0.4 (*Supporting functional*) Let $x \in S_{\mathbb{X}}$. A linear functional $f \in S_{\mathbb{X}^*}$ is said to be a supporting functional at x if $f(x) = 1$.

By the famous Hahn–Banach Theorem, for a non-zero element $x \in \mathbb{X}$, there always exists a supporting functional at x. The collection of all supporting functionals at x is denoted by $J(x)$, i.e.,

$$J(x) = \{f \in S_{\mathbb{X}^*} : f(x) = \|x\|\}.$$

It is easy to observe that $J(x)$ is a non-empty weak* compact convex subset of $\mathbb{X}^*$. Moreover, $J(x)$ is an extremal subset of $B_{\mathbb{X}^*}$ in the sense that if $f \in J(x)$ and there exist $g, h \in B_{\mathbb{X}^*}$ such that $f = (1-t)g + th$, for some $t \in (0, 1)$ then $g, h \in J(x)$.

Definition 1.0.5 (*Smooth point*) An element $x \in S_{\mathbb{X}}$ is said to be a smooth point of $\mathbb{X}$ if there exists a unique supporting functional $f \in S_{\mathbb{X}^*}$ at x, i.e., there exists a unique bounded linear functional f on $\mathbb{X}$ such that $f(x) = 1$ and $\|f\| = 1$.

The space $\mathbb{X}$ is said to be a smooth space if every element of $S_{\mathbb{X}}$ is smooth.

We next mention the following concepts which are essential to describe the geometry of normed spaces, and in particular, operator spaces.

Definition 1.0.6 (*Norm attainment set*) Let T be a bounded linear operator from $\mathbb{X}$ to $\mathbb{Y}$. Then the norm attainment set of T, denoted as M_T, is defined as the collection of all unit vectors x for which $\|Tx\| = \|T\|$, i.e.,

$$M_T = \{x \in S_{\mathbb{X}} : \|Tx\| = \|T\|\}.$$

Definition 1.0.7 (*L-ideal and M-ideal*) A linear projection P on $\mathbb{X}$ is called an L-projection if $\|x\| = \|Px\| + \|x - Px\|$ for all $x \in \mathbb{X}$. A closed subspace $J \subseteq \mathbb{X}$ is called an L-summand (or L-ideal) if it is the range of an L-projection. A closed subspace $J \subseteq \mathbb{X}$ is called an M-ideal if J^0 is an L-summand of $\mathbb{X}^*$, where $J^0 = \{f \in \mathbb{X}^* : f(x) = 0 \; \forall \; x \in J\}$, is the annihilator of J.

Definition 1.0.8 Let $\mathbb{X}$ be a Banach space. A function $[\, , \,] : \mathbb{X} \times \mathbb{X} \longrightarrow \mathbb{K}(= \mathbb{R}, \mathbb{C})$ is a semi-inner-product (s.i.p.) if for any scalars $\alpha, \; \beta$ and for any $x, \; y, \; z \in \mathbb{X}$, it satisfies the following:
(a) $[\alpha x + \beta y, z] = \alpha[x, z] + \beta[y, z]$,
(b) $[x, x] > 0$, whenever $x \neq 0$,
(c) $|[x, y]|^2 \leq [x, x][y, y]$,
(d) $[x, \alpha y] = \overline{\alpha}[x, y]$.

Before we end this section, we would like to draw the attention of the readers to the fact that although the definitions and the results have been discussed in the setting of Banach spaces, almost all of them work equally well in normed linear spaces. For the sake of uniformity and also to maintain focus on the main topics of our interest, we have worked exclusively with the completeness assumption. The only instances when we make an exception to this rule are due to historical reasons, as mentioned in the appropriate places. The reader is encouraged to keep this in mind while reading this monograph. It should also be noted that nowhere in this article we will be considering unbounded linear operators. In particular, the terms "bounded linear operators", "linear operators", and "operators" carry the same meaning, and have been used interchangeably in order to break the monotony of repetitions.

Chapter 2
Basic Theory of B–J Orthogonality in Banach Space

2.1 Introduction

The Hilbert space theory has an inherent orthogonality notion that comes easily through inner product and that differentiates the geometry of the Hilbert space from the Banach space. Two elements x and y in a Hilbert space are said to be orthogonal if $\langle x, y\rangle = 0$. This orthogonality notion is a natural generalization of perpendicularity of two vectors $\vec{x}$ and $\vec{y}$ in the real plane $\mathbb{R}^2$, note that two vectors $\vec{x}$ and $\vec{y}$ are perpendicular if $\vec{x} \cdot \vec{y} = 0$. This orthogonality notion, in general, is not available in a Banach space. It is easy to see that orthogonality of two elements x and y in a real Hilbert space is equivalent to each of the following:

(i) $\|x + \lambda y\| \geq \|x\|$ for all scalars λ

(ii) $\|x - y\|^2 = \|x\|^2 + \|y\|^2$.

(iii) $\|x + y\| = \|x - y\|$.

Each of the above equivalent conditions is described in terms of the norm induced by the inner product. Motivated by the above equivalent conditions, the following notions of orthogonality are introduced in a Banach space.

Definition 2.1.1 ([20, 59] *Birkhoff–James orthogonality*) Let $\mathbb{X}$ be a Banach space and $x, y \in \mathbb{X}$. Then x is said to be Birkhoff–James orthogonal to y, written as $x \perp_B y$, if for any scalar λ, $\|x + \lambda y\| \geq \|x\|$.

Definition 2.1.2 ([61] *Isosceles orthogonality*) Let $\mathbb{X}$ be a Banach space and $x, y \in \mathbb{X}$. Then x is said to be isosceles orthogonal to y, written as $x \perp_I y$, if $\|x + y\| = \|x - y\|$.

A. Mal et al., *Birkhoff–James Orthogonality and Geometry of Operator Spaces*, Infosys Science Foundation Series, https://doi.org/10.1007/978-981-99-7111-4_2

Definition 2.1.3 ([61] *Pythagorean orthogonality*) Let $\mathbb{X}$ be a Banach space and $x, y \in \mathbb{X}$. Then x is said to be orthogonal to y in the sense of Pythagorean or Pythagorean orthogonal to y, written as $x \perp_P y$, if $\|x - y\|^2 = \|x\|^2 + \|y\|^2$.

The geometrical interpretation of Birkhoff–James orthogonality is as follows: Given two elements x and y in a Banach space $\mathbb{X}$ with $\|x\| = 1$, we see that $x \perp_B y$ if and only if the line passing through x parallel to y is outside the open unit ball.

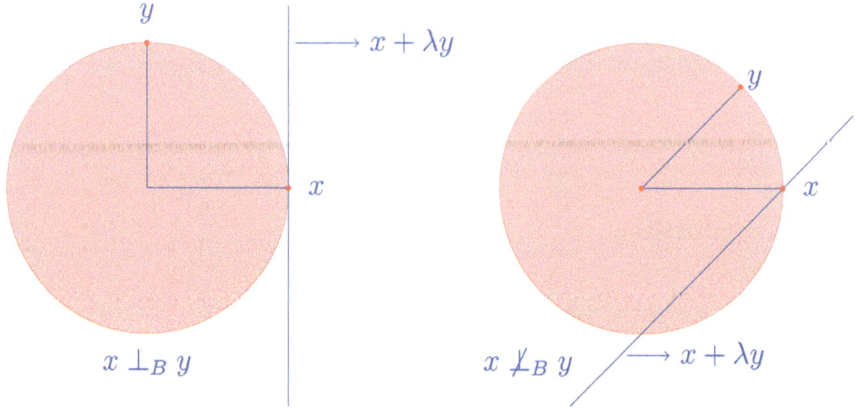

We note that in a Hilbert space, Birkhoff–James orthogonality is equivalent to the usual inner product orthogonality even if the space is complex but the case is not so for isosceles orthogonality and Pythagorean orthogonality. We first note that Birkhoff–James orthogonality is homogenous, i.e., $x \perp_B y$ implies $\alpha x \perp_B \beta y$ for all scalars α and β. However, the same is not true for Pythagorean orthogonality and isosceles orthogonality. Although all of the above three notions of orthogonality are equivalent in a real Hilbert space, they are not so in a Banach space. We give examples to show that Birkhoff–James orthogonality is independent of both Pythagorean orthogonality and isosceles orthogonality.

Example 2.1.4 (i) (Birkhoff–James orthogonality does not imply Pythagorean orthogonality)
Consider the real Banach space ℓ_1^2. Let $x = (1, 0), y = (-\frac{1}{2}, -\frac{1}{2}) \in \ell_1^2$. Then $\|x + \lambda y\| = \|(1 - \frac{1}{2}\lambda, -\frac{1}{2}\lambda)\| = |1 - \frac{1}{2}\lambda| + |\frac{1}{2}\lambda| \geq 1$ and so $x \perp_B y$. But $\|x - y\|^2 = \|(1, 0) - (-\frac{1}{2}, -\frac{1}{2})\|^2 = 4 \neq \|(1, 0)\|^2 + \|(-\frac{1}{2}, -\frac{1}{2})\|^2 = \|x\|^2 + \|y\|^2$. Thus, $x \not\perp_P y$.

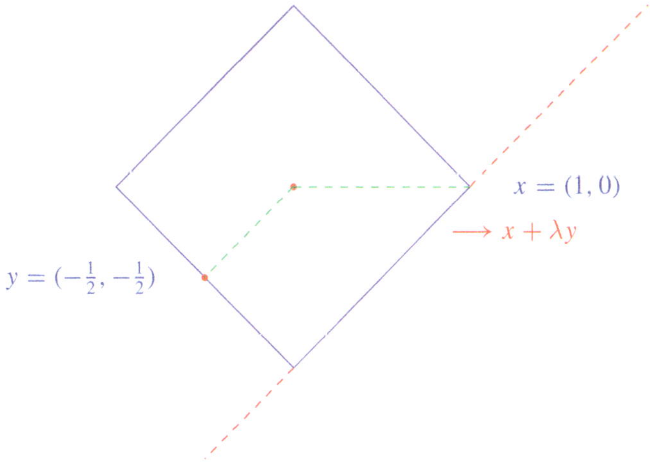

Unit sphere of ℓ_1^2 : $x \perp_B y$ but $x \not\perp_P y$

(ii) (Pythagorean orthogonality does not imply Birkhoff–James orthogonality) Consider the real Banach space ℓ_1^2. Let $x = (2, -1)$, $y = (-1, -3) \in \ell_1^2$. Then $\|x - y\|^2 = \|(2, -1) - (-1, -3)\|^2 = 25 = \|(2, -1)\|^2 + \|(-1, -3)\|^2 = \|x\|^2 + \|y\|^2$. So $x \perp_P y$. But $\|x - \frac{1}{3}y\| = \|(2, -1) - \frac{1}{3}(-1, -3)\| = \|(2 + \frac{1}{3}, 0)\| = 2 + \frac{1}{3} < 3 = \|(2, -1)\|$. So $x \not\perp_B y$.

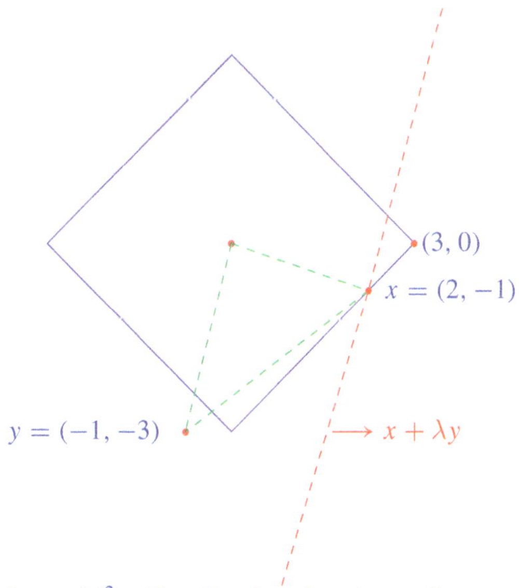

Sphere of ℓ_1^2 with radius 3 : $x \perp_P y$ but $x \not\perp_B y$

(iii) (Birkhoff–James orthogonality does not imply Isosceles orthogonality) Consider the real Banach space ℓ_1^2. Let $x = (1, 0)$, $y = (\frac{1}{2}, \frac{1}{2}) \in \ell_1^2$. Then $\|x + \lambda y\| = \|(1 + \frac{1}{2}\lambda, \frac{1}{2}\lambda)\| = |1 + \frac{1}{2}\lambda| + |\frac{1}{2}\lambda| \geq 1$ and so $x \perp_B y$. But $\|x - y\| = \|(1, 0)$

$-(\frac{1}{2},\frac{1}{2})\| = \|(\frac{1}{2},-\frac{1}{2})\| = 1 \neq 2 = \|(1+\frac{1}{2},\frac{1}{2})\| = \|(1,0)+(\frac{1}{2},\frac{1}{2})\| = \|x+y\|$. Thus, $x \not\perp_I y$.

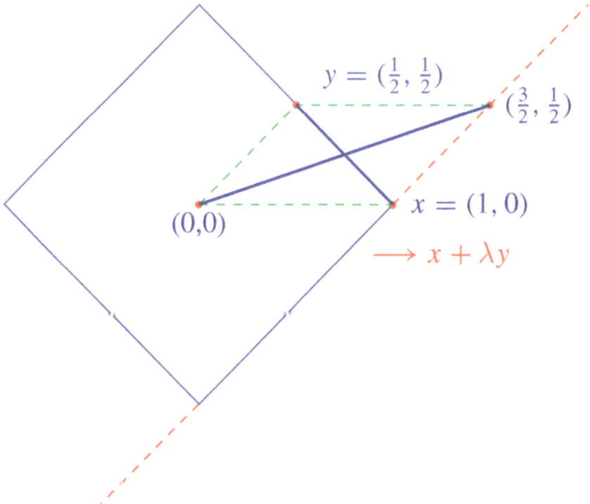

Unit sphere of ℓ_1^2 : $x \perp_B y$ but $x \not\perp_I y$

(iv) (Isosceles orthogonality does not imply Birkhoff–James orthogonality)
Consider the real Banach space ℓ_1^2. Let $x = (2,-1), y = (-1,-3) \in \ell_1^2$. Then $\|x-y\| = \|(2,-1)-(-1,-3)\| = \|(3,2)\| = 5 = \|(1,-4)\| = \|(2,-1)+(-1,-3)\| = \|x+y\|$. So $x \perp_I y$. But $\|x-\frac{1}{3}y\| = \|(2,-1)-\frac{1}{3}(-1,-3)\| = \|(2+\frac{1}{3},0)\| = 2+\frac{1}{3} < 3 = \|(2,-1)\|$. So $x \not\perp_B y$.

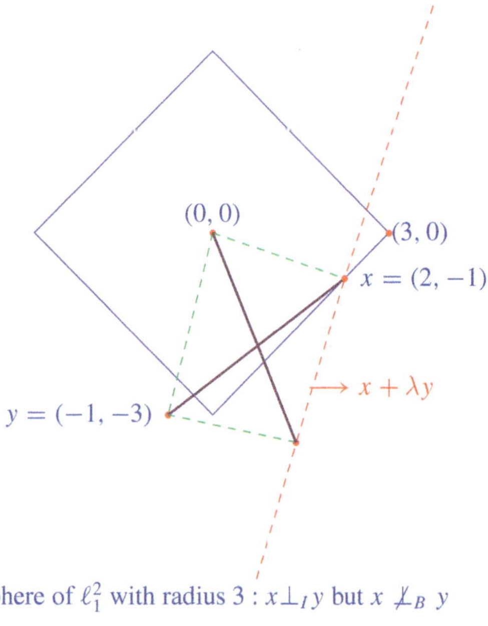

Sphere of ℓ_1^2 with radius 3 : $x \perp_I y$ but $x \not\perp_B y$

We now focus our attention on the Birkhoff–James orthogonality because of the fact that it is going to be the guiding light for us, for the purpose of understanding the normed geometry and the geometry of operator spaces, throughout this monograph. As mentioned in the Preface, the study of orthogonality in the setting of metric linear spaces (in contrast to the more restrictive study in the setting of Hilbert spaces), was initiated by Birkhoff [20] and developed significantly by James [59]. It can be safely asserted that far-reaching applications of this abstract concept of orthogonality have been found, in several areas of the geometric theory of Banach spaces. While this should be fairly obvious from the rapidly growing literature on this, it is our intention to further emphasize the importance of the concept of orthogonality in understanding the structure of Banach spaces through the results presented in the subsequent chapters. The reader will find that the concept of orthogonality can be applied successfully to tackle geometric and analytic problems of a surprisingly large spectrum. In view of the authors, this is indeed a natural consequence of the rich analytic structure associated with the Birkhoff–James orthogonality via support functionals. The reader is strongly advised to consult the pioneering works by James [59–61] to gain more insight into this context. Indeed, in our view, the wide applicability of the said concept is the primary motivating factor behind the proposal for writing this monograph. We have tried to present some of the latest developments taking place in this direction, in order to highlight the importance of the Birkhoff–James orthogonality. For the sake of brevity, we will refer to it as B–J orthogonality, every time fondly remembering the pioneers associated with this nice concept.

Let us now mention some of the possible special attributes of the B–J orthogonality in a Banach space, which are essentially equivalent to several nice properties on the underlying space.

Definition 2.1.5 (*Left unique*) Given x and y in $\mathbb{X}$, there always exists a scalar α such that $\alpha x + y \perp_B x$. The B–J orthogonality is said to be left unique if and only if for no elements $x \neq 0$ and y there are more than one scalar α for which $\alpha x + y \perp_B x$.

Definition 2.1.6 (*Right unique*) Given x and y in $\mathbb{X}$, there always exists a scalar β such that $x \perp_B \beta x + y$. The B–J orthogonality is said to be right unique if and only if for no elements $x \neq 0$ and y there are more than one scalar β for which $x \perp_B \beta x + y$.

Definition 2.1.7 (*Left additive*) The B–J orthogonality is said to be left additive if for any $x, y, z \in \mathbb{X}$, $x \perp_B z$ and $y \perp_B z$ imply that $(x + y) \perp_B z$.

Definition 2.1.8 (*Right additive*) The B–J orthogonality is said to be right additive if for any $x, y, z \in \mathbb{X}$, $x \perp_B y$ and $x \perp_B z$ imply that $x \perp_B (y + z)$.

It is elementary to observe that the B–J orthogonality relation is not symmetric, in general. In other words, for two elements x, y in a Banach space $\mathbb{X}$, $x \perp_B y$ does not necessarily imply that $y \perp_B x$. Observe that in the real Banach space ℓ_∞^2, for $x = (1, 1)$ and $y = (0, 1)$, $x \perp_B y$. However, $y \not\perp_B x$.

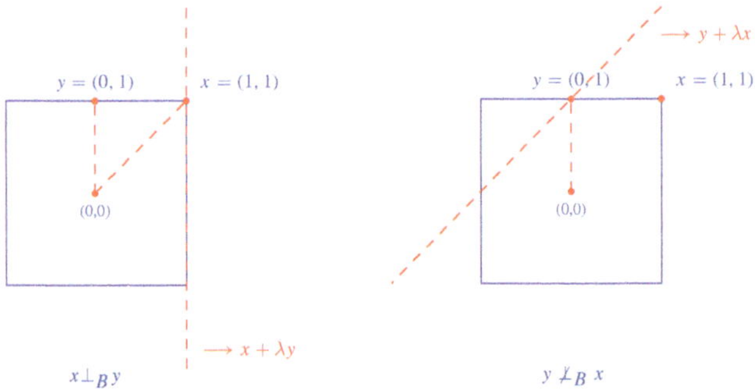

The global symmetry of the B–J orthogonality had once been a topic of much discussion, and the readers should consult [29, 60, 63] for more information in this regard. In fact, we will take up the symmetry issue in this monograph itself, from a local perspective. We intend to illustrate that the local approach has its own rewards, resulting in several useful information regarding the structure of the underlying Banach space. In view of this, we present the following definitions of local symmetry of the B–J orthogonality.

Definition 2.1.9 (*Left symmetric element*) An element $x \in \mathbb{X}$ is said to be a left symmetric element if $x \perp_B y \Rightarrow y \perp_B x, \ \forall y \in \mathbb{X}$.

Definition 2.1.10 (*Right symmetric element*) An element $x \in \mathbb{X}$ is said to be a right symmetric element if $y \perp_B x \Rightarrow x \perp_B y, \ \forall y \in \mathbb{X}$.

Definition 2.1.11 (*Symmetric element*) An element $x \in \mathbb{X}$ is said to be a symmetric element if it is both left symmetric and right symmetric.

The symmetricity of B–J orthogonality is a very strong property and it induces an inner product on the space if the dimension of the space is greater or equal to 3. However, there are two-dimensional Banach spaces, where B–J orthogonality is symmetric but the norm is not induced by an inner product. Such two-dimensional Banach spaces are known as Radon planes. See [14, 30, 73, 90] and the references therein for construction and characterizations of Radon planes. In [59], James provided a way to construct Radon planes. The following figure gives an example of the unit sphere of a Radon plane $\mathbb{X}$, where for $(a, b) \in \mathbb{X}$, $\|(a, b)\| = \max\{|a|, |b|\}$ if $ab \geq 0$ and $\|(a, b)\| = |a| + |b|$ if $ab < 0$.

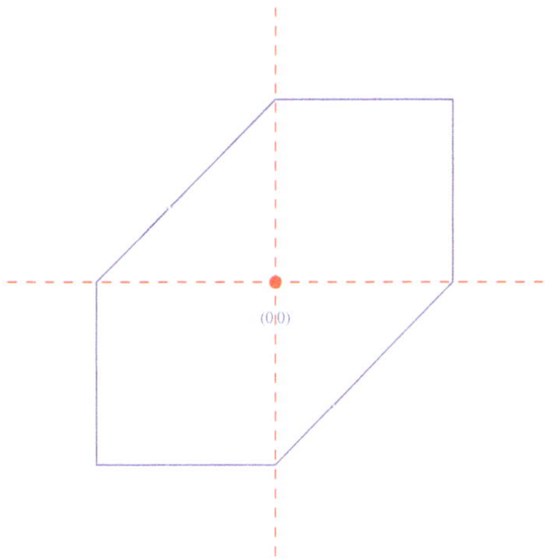

Unit sphere of a Radon plane constructed from $\ell_\infty(\mathbb{R}^2)$ and its dual

Observe that the Hahn–Banach theorem guarantees the existence of orthogonal elements in the sense of Birkhoff–James in a Banach space $\mathbb{X}$. Given $(0 \neq)x \in \mathbb{X}$, consider $M = span\{x\}$. Define $f : M \to \mathbb{R}$ by $f(ax) = a\|x\|$ for all scalars a. Then f is a bounded linear functional on M and $\|f\| = 1$. By the Hahn–Banach theorem, we can find a bounded linear functional $\widetilde{f}$ on $\mathbb{X}$ such that $\widetilde{f}|_M = f$, $\|\widetilde{f}\| = \|f\| = 1$. Then $\widetilde{f}(x) = \|x\|$. Let $H = \ker(\widetilde{f})$ and $y \in H$. Then, for any scalar λ, $\|x + \lambda y\| \geq |\widetilde{f}(x + \lambda y)| = |\widetilde{f}(x)| = \|x\|$, so that $x \perp_B y$. Moreover, for any $z \in \mathbb{X}$, if $a = -\frac{\widetilde{f}(z)}{\widetilde{f}(x)}$, then $\widetilde{f}(ax + z) = 0$ and so $x \perp_B (ax + z)$. Thus, we have the following theorem.

Theorem 2.1.12 ([59, Theorem 2.2, Corollary 2.2]) *Let $\mathbb{X}$ be a Banach space. Then, for every non-zero $x \in \mathbb{X}$, there exists a closed hyperspace H of $\mathbb{X}$ such that $x \perp_B H$.*

Moreover, for any $z \in \mathbb{X}$, there exists $a \in \mathbb{R}$ such that $x \perp_B (ax + z)$.

On the other hand, we have the following theorem.

Theorem 2.1.13 ([59, Theorem 2.3]) *Let $\mathbb{X}$ be a Banach space. Given $x, y \in \mathbb{X}$, there exists a scalar a such that $ax + y \perp_B x$.*

Proof Let $x, y \in \mathbb{X}$, we consider the set $S = \{\|\lambda x + y\| : \lambda \text{ is a scalar}\}$. Then S is a non-empty subset of real numbers bounded below by 0 and so the infimum of S exists. By continuity of the norm function, it is easy to see that there exists a scalar a such that $\|ax + y\| = \inf_\lambda \|\lambda x + y\|$. Then $\|\lambda x + y\| \geq \|ax + y\|$ for all scalars λ, and so $\|(ax + y) + \lambda x\| \geq \|ax + y\|$ so that $ax + y \perp_B x$. This completes the proof. □

In the following theorem, we provide a nice characterization of B–J orthogonality in terms of linear functionals.

Theorem 2.1.14 ([59, Theorem 2.1]) *Let $\mathbb{X}$ be a Banach space. Then, for $(0 \neq)x \in \mathbb{X}$ and a closed hyperspace H of $\mathbb{X}$, $x \perp_B H$ if and only if there exists a bounded linear functional f such that $f(x) = \|x\|$, $\|f\| = 1$ and $f(h) = 0$ for all $h \in H$.*

Proof We first prove the necessary part. Let $z \in \mathbb{X}$. Then $z = \alpha x + h$, where α is a scalar and $h \in H$. Define a functional f on $\mathbb{X}$ as follows: $f(z) = \alpha\|x\|$. If $\|z\| = 1$, then $1 = \|\alpha x + h\| \geq |\alpha|\|x\|$ and so $|f(z)| = |\alpha|\|x\| \leq \|z\|$. Moreover, $f(x) = \|x\|$ implies that $\|f\| = 1$. Thus, f is a bounded linear functional on $\mathbb{X}$ such that $f(x) = \|x\|$, $\|f\| = 1$ and $f(h) = 0$ for all $h \in H$.

Next, we prove the sufficient part. Consider $H = \ker(f)$. Then H is a closed hyperspace of $\mathbb{X}$. Observe that, for any scalar λ and for each $h \in H$, we have

$$\|x\| = \|f\|\|x\| = |f(x)| = |f(x + \lambda h)| \leq \|f\|\|x + \lambda h\| = \|x + \lambda h\|.$$

This shows that $x \perp_B H$. □

In particular, for $x, y \in \mathbb{X}$, the characterization of $x \perp_B y$, in terms of linear functionals will be used time and again in this monograph and so we state the result in the form of the following theorem.

Theorem 2.1.15 ([59]) *Let $\mathbb{X}$ be a Banach space and let $x, y \in \mathbb{X}$. Then $x \perp_B y$ if and only if there exists $f \in \mathbb{X}^*$ such that $f(x) = \|f\|\|x\|$ and $f(y) = 0$.*

The following two definitions of possible approximate versions of the B–J orthogonality are useful in understanding the stability of the B–J orthogonality relation, and provide us with important information on the geometry of the space.

Definition 2.1.16 (*Approximate B–J orthogonality $x \perp_D^{\epsilon} y$*) Let $\epsilon \in [0, 1)$. Then, for $x, y \in \mathbb{X}$, x is said to be approximate $\epsilon-$ Birkhoff–James orthogonal to y if

$$\|x + \lambda y\| \geq \sqrt{1 - \epsilon^2}\|x\| \quad \text{for all scalars } \lambda.$$

In this case, we write $x \perp_D^{\epsilon} y$.

Definition 2.1.17 (*Approximate B–J orthogonality $x \perp_B^{\epsilon} y$*) Let $\epsilon \in [0, 1)$. Then, for $x, y \in \mathbb{X}$, x is said to be approximate $\epsilon-$ Birkhoff–James orthogonal to y if

$$\|x + \lambda y\|^2 \geq \|x\|^2 - 2\epsilon\|x\|\|\lambda y\| \quad \text{for all scalars } \lambda.$$

In this case, we write $x \perp_B^{\epsilon} y$.

The notion of strict convexity, uniform convexity, smoothness, and reflexivity in a Banach space is directly related to B–J orthogonality, see the remarkable article by James [59]. In recent times, many researchers are actively engaged in studying the geometry of spaces of bounded linear operators defined between Hilbert as well as Banach spaces using the notion of B–J orthogonality (see [17, 84, 87, 88, 107, 108, 116]). We here demonstrate the role of B–J orthogonality in the geometry of Banach space, which might be useful for the students and first-time readers.

2.2 B–J Orthogonality and Strict Convexity

We begin with the theorem that gives a characterization of strictly convex Banach space in terms of left uniqueness of B–J orthogonality.

Theorem 2.2.1 ([59, 101]) *A Banach space $\mathbb{X}$ is strictly convex if and only if B–J orthogonality is left unique.*

Proof We first prove the sufficient part. Let B–J orthogonality be left unique. If possible, let $\mathbb{X}$ be not strictly convex. Then there exist $u, v \in S_{\mathbb{X}}$ such that $u \neq v$ and $(1-t)u + tv \in S_{\mathbb{X}}$ for all $0 \leq t \leq 1$. Thus, $\|u + \lambda(v-u)\| = \|(1-\lambda)u + \lambda v\| = 1$, for all $0 \leq \lambda \leq 1$. Using the convexity of norm, it can be shown that $\|u + \lambda(v - u)\| \geq 1$ for all $\lambda \geq 0$. Now, let $\lambda < 0$. If possible, let $\|u + \lambda(v-u)\| < 1$. Then $u = (1 - \frac{\lambda}{\lambda-1})(u + \lambda(v-u)) + \frac{\lambda}{\lambda-1}v$ and $0 \leq \frac{\lambda}{\lambda-1} \leq 1$. So,

$$\|u\| \leq \left(1 - \frac{\lambda}{\lambda-1}\right)\|(u + \lambda(v-u))\| + \frac{\lambda}{\lambda-1}\|v\| < \left(1 - \frac{\lambda}{\lambda-1}\right) + \frac{\lambda}{\lambda-1} = 1,$$

a contradiction. Therefore, $\|u + \lambda(v-u)\| \geq 1 = \|u\|$ for all $\lambda < 0$. Hence, $u \perp_B (v-u)$. Similarly, it can be shown that $v \perp_B (v-u)$, i.e., $u + (v-u) \perp_B (v-u)$. This shows that B–J orthogonality is not left unique, a contradiction. Therefore, $\mathbb{X}$ is strictly convex.

Next, we prove the necessary part. Let $\mathbb{X}$ be strictly convex. If possible, let B–J orthogonality be not left unique. Then there exist $x, y \in \mathbb{X}$ and $a, b \in \mathbb{R}$ such that $ax + y \perp_B x$ and $bx + y \perp_B x$, where $a \neq b$. Let $a < b$. Now, $\|ax + y + \lambda x\| \geq \|ax + y\|$, for all scalars λ. Putting $\lambda = b - a$, we have, $\|bx + y\| \geq \|ax + y\|$. Again, $\|bx + y + \lambda x\| \geq \|bx + y\|$, for all scalars λ. Putting $\lambda = a - b$, we have, $\|ax + y\| \geq \|bx + y\|$. Thus, $\|ax + y\| = \|bx + y\|$. So, $\|kx + y\| = \|ax + y\|$ for all $a \leq k \leq b$. This contradicts that $\mathbb{X}$ is strictly convex. This completes the proof of necessary part. □

B–J orthogonality is not necessarily left additive. For example, consider the real Banach space ℓ^2_∞. Let $x = (1, 0)$, $y = (-1, 1)$, $z = (0, 1)$. Then $x \perp_B z$, $y \perp_B z$ but $x + y \not\perp_B z$.

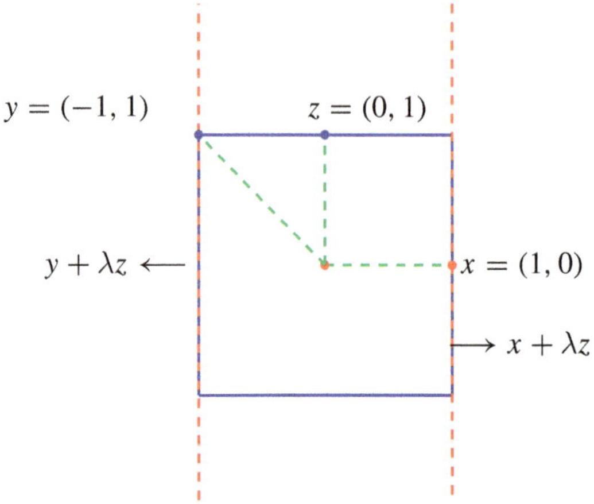

$x \perp_B z, y \perp_B z$ but $x + y \not\perp_B z$

Unit sphere of ℓ^2_∞

Next, we show that left additivity of B–J orthogonality characterizes strictly convex spaces if the dimension of the space is two.

Theorem 2.2.2 ([60, 101]) *A two-dimensional Banach space $\mathbb{X}$ is strictly convex if and only if B–J orthogonality is left additive.*

Proof Let $\mathbb{X}$ be strictly convex. Then B–J orthogonality is left unique. Let $(0 \neq)u, v, w \in \mathbb{X}$ be such that $v \perp_B u$ and $w \perp_B u$. Clearly, the sets $\{u, v\}$ and $\{u, w\}$ are linearly independent. Therefore, $w = \alpha u + \beta v$ for some scalars α, β, where $\beta \neq 0$. Now, from $\alpha u + \beta v \perp_B u$, it follows that $\frac{\alpha}{\beta} u + v \perp_B u$. Since, B–J orthogonality is left unique and $v \perp_B u$, we get $\frac{\alpha}{\beta} = 0$. Thus, $\alpha = 0$, which gives $w = \beta v$. Then $v + w \perp_B u$. Thus, B–J orthogonality is left additive.

Conversely, let B–J orthogonality be left additive. If possible, let $\mathbb{X}$ be not strictly convex. Then there exist $u, v \in S_{\mathbb{X}}$ such that $u \neq v$ and $(1-t)u + tv \in S_{\mathbb{X}}$ for all $0 \leq t \leq 1$. Hence, $\|u + \lambda(v-u)\| = \|(1-\lambda)u + \lambda v\| = 1$, for all $0 \leq \lambda \leq 1$. Now, let $\lambda < 0$. If possible, let $\|u + \lambda(v-u)\| < \|u\|$. Then $u = (1 - \frac{\lambda}{\lambda-1})(u + \lambda(v - u)) + \frac{\lambda}{\lambda-1} v$ and $0 \leq \frac{\lambda}{\lambda-1} \leq 1$. So, $\|u\| \leq (1 - \frac{\lambda}{\lambda-1})\|(u + \lambda(v-u))\| + \frac{\lambda}{\lambda-1}\|v\| < (1 - \frac{\lambda}{\lambda-1})\|u\| + \frac{\lambda}{\lambda-1}\|u\| = \|u\|$, a contradiction. Therefore, $\|u + \lambda(v-u)\| \geq \|u\|$ for all $\lambda < 0$. So, $u \perp_B (v-u)$. Similarly, it can be shown that $v \perp_B (v-u)$. Since B–J orthogonality is left additive, $(v-u) \perp_B (v-u)$. Therefore, $v - u = 0$, i.e., $v = u$, a contradiction. Thus, $\mathbb{X}$ is strictly convex. □

The following theorem gives a characterization of Hilbert space if B–J orthogonality is left additive when the dimension of the space is greater than or equal to three.

Theorem 2.2.3 ([60, Theorem 2]) *Let $\mathbb{X}$ be a Banach space such that* $\dim(\mathbb{X}) \geq 3$. *Then $\mathbb{X}$ is a Hilbert space if and only if B–J orthogonality is left additive.*

Proof We first prove the necessary part. Since B–J orthogonality coincides with usual inner product orthogonality in a Hilbert space, it is easy to see that Birkhoff–James orthogonality is left additive in a Hilbert space.

Next, we prove the sufficient part. We claim that there exists a norm one projection on every two-dimensional subspace of $\mathbb{X}$. Let V be a two-dimensional subspace spanned by x, y. Then there exist hyperspaces H_x and H_y in $\mathbb{X}$ such that $x \perp_B H_x$ and $y \perp_B H_y$. Consider, $H = H_x \bigcap H_y$. Then $x, y \notin H$ and $codim H = 2$. Since B–J orthogonality is left additive, we have, $ax + by \perp_B H$, for all $a, b \in \mathbb{R}$. It is easy to see that any $z \in \mathbb{X}$ can be written as $z = p(z) + w$, where $p(z) = ax + by \in V$ and $w \in H$. Observe that $\|p(z)\| = \|ax + by\| \leq \|p(z) + w\| = \|z\|$, since $p(z) \perp_B w$. Now, from $\|p(x)\| = \|x\|$, it follows that $\|p\| = 1$. So, there exists a projection of norm 1 on V, which establishes our claim. Since there is a projection of norm 1 on any given two-dimensional subspace of $\mathbb{X}$, by Kakutani's theorem [63], $\mathbb{X}$ is a Hilbert space. □

2.3 B–J Orthogonality and Smoothness

One of the most important geometric notions in a Banach space is smoothness. The following figure illustrates geometrically the difference between a smooth element and a non-smooth element.

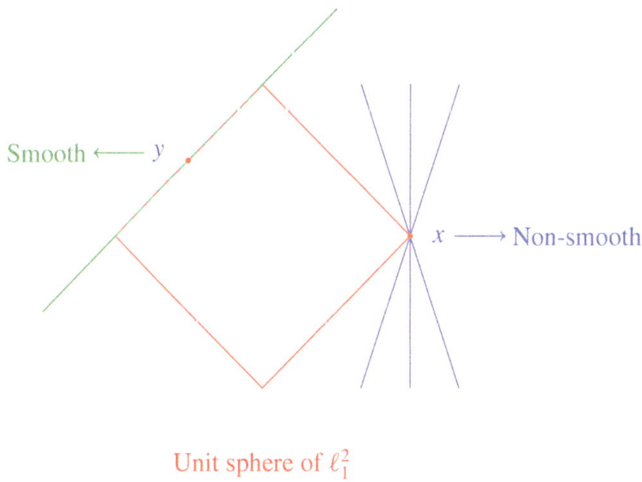

Let us recall from Theorem 2.1.12 that given $(0 \neq) x \in \mathbb{X}$, there exists a hyperspace H such that $x \perp_B H$. The following theorem shows that x is smooth if and only if $x \perp_B y$ implies that $y \in H$.

Theorem 2.3.1 ([101]) *A Banach space $\mathbb{X}$ is smooth if and only if for any non-zero $x \in \mathbb{X}$, $H_x = \{y \in \mathbb{X} : x \perp_B y\}$ is a subspace.*

Proof We first prove the necessary part. Let $0 \neq x \in \mathbb{X}$. Then there exists a unique $g \in S_{\mathbb{X}^*}$ such that $g(x) = \|x\|$. We show that $\ker(g) = H_x$, where $H_x = \{y \in \mathbb{X} : x \perp_B y\}$. Let $y \in \ker(g)$. Then, for all scalars λ, $\|x + \lambda y\| = \|g\|\|x + \lambda y\| \geq |g(x + \lambda y)| = |g(x)| = \|x\|$. So, $x \perp_B y$. Thus, $y \in H_x$. Hence, $\ker(g) \subseteq H_x$. Now, let $0 \neq y \in H_x$. Then $x \perp_B y$. Let $Z = span\{x, y\}$. Define $f : Z \to \mathbb{R}$ by $f(ax + by) = a\|x\|$. Then clearly, $\|f\| = 1$, $f(x) = \|x\|$ and $f(y) = 0$. By the Hahn Banach theorem, there exists $h \in \mathbb{X}^*$ such that $\|h\| = \|f\| = 1$, $h(x) = f(x) = \|x\|$ and $h(y) = f(y) = 0$. Thus, h is a norming linear functional of x. Since $\mathbb{X}$ is smooth, we get $g = h$. Thus, $y \in \ker(h) = \ker(g)$. Hence, $H_x \subseteq \ker(g)$. Therefore, $H_x = \ker(g)$. Thus, H_x is a subspace.

For the sufficient part, let us assume that H_x be a subspace for each $0 \neq x \in \mathbb{X}$. If possible let $\mathbb{X}$ be not smooth. Then there exists $0 \neq x \in \mathbb{X}$ such that x has at least two norming linear functionals f, g. Clearly, $\ker(f) \neq \ker(g)$. Proceeding as above, it can be shown that $\ker(f) \subseteq H_x$ and $\ker(g) \subseteq H_x$. Since $x \notin H_x$, H_x is a proper subspace of $\mathbb{X}$. Therefore, $\ker(f) = H_x = \ker(g)$. This gives that $f = \alpha g$ for some scalar α. Again, $f(x) = \|x\| = g(x)$ gives that $\alpha = 1$. Thus, $f = g$, a contradiction. Therefore, $\mathbb{X}$ is smooth. □

We note that B–J orthogonality is not necessarily right additive. For example, consider the real Banach space ℓ_∞^2. Let $x = (1, 1)$, $y = (1, 0)$, $z = (0, 1)$. Then $x \perp_B y$, $x \perp_B z$ but $x \not\perp_B y + z (= x)$.

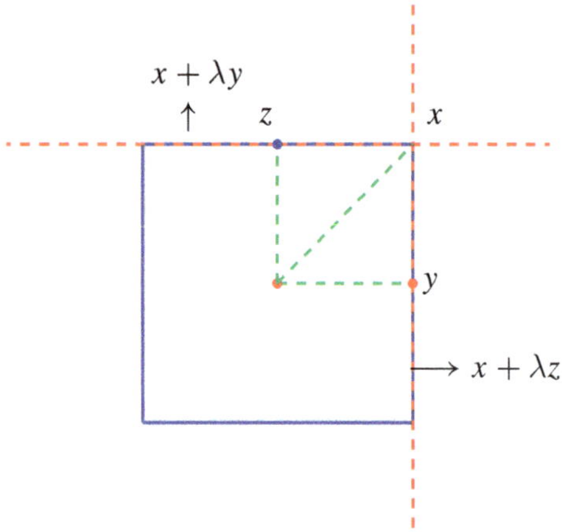

$x \perp_B y$, $x \perp_B z$ but $x \not\perp_B y + z(= x)$

Unit sphere of ℓ_∞^2

We now explore the relation between right additivity of B–J orthogonality and smoothness in a Banach space (see [59]).

Theorem 2.3.2 ([59, 101]) *A Banach space $\mathbb{X}$ is smooth if and only if B–J orthogonality is right additive.*

Proof Let $\mathbb{X}$ be smooth. Let $x, y, z \in \mathbb{X}$ be such that $x \perp_B y$ and $x \perp_B z$. Then $y, z \in H_x$, where $H_x = \{u \in \mathbb{X} : x \perp_B u\}$. Since $\mathbb{X}$ is smooth, from Theorem 2.3.1, it follows that H_x is a subspace. So, $y + z \in H_x$. Hence, $x \perp_B (y + z)$. Thus, B–J orthogonality is right additive.

Conversely, suppose B–J orthogonality is right additive. Let $0 \neq x \in \mathbb{X}$. Let $H_x = \{u \in \mathbb{X} : x \perp_B u\}$. We show that H_x is a subspace. Let $y, z \in H_x$. Then $x \perp_B y$ and $x \perp_B z$. Since B–J orthogonality is right additive, $x \perp_B (y + z)$. Thus, $y + z \in H_x$. Again, since B–J orthogonality is homogeneous, $\alpha y \in H_x$, for each scalar α. Thus, H_x is a subspace and, so from Theorem 2.3.1, it follows that $\mathbb{X}$ is smooth. □

We have already noted that left uniqueness of B–J orthogonality is related to strict convexity of the space. We now explore the relationship between smoothness and right uniqueness of B–J orthogonality.

Theorem 2.3.3 ([59, 101]) *A Banach space $\mathbb{X}$ is smooth if and only if B–J orthogonality is right unique.*

Proof We first prove the sufficient part. Let $0 \neq x \in \mathbb{X}$. Let g, h be two norming linear functionals of x. We show that $g = h$. Let $y \in \ker(g)$. Then, for any scalar λ, $\|x + \lambda y\| = \|g\|\|x + \lambda y\| \geq |g(x + \lambda y)| = \|x\|$. Thus, $x \perp_B y$. If possible, let $y \notin \ker(h)$. Then $h(y) = r \neq 0$. Let $z = x - \frac{\|x\|}{r} y$. Then $h(z) = h(x) - \frac{\|x\|}{r} h(y) = \|x\| - \frac{\|x\|}{r} r = 0$. So, $x \perp_B z \Rightarrow x \perp_B (x - \frac{\|x\|}{r} y) \Rightarrow x \perp_B (-\frac{r}{\|x\|} x + y)$, contradicting that B–J orthogonality is right unique. Therefore, $y \in \ker(h)$. So, $\ker(g) \subseteq \ker(h)$. Again, $\ker(h) \neq \mathbb{X}$, since $x \notin \ker(h)$. Hence, $\ker(g) = \ker(h)$. Moreover, $g(x) = h(x) = \|x\|$. So, $g = h$. Thus, x has unique norming linear functional and so $\mathbb{X}$ is smooth.

We next prove the necessary part. Let $\mathbb{X}$ be smooth. Then B–J orthogonality is right additive. Let $(0 \neq) x, y \in \mathbb{X}$ be such that $x \perp_B (ax + y)$ and $x \perp_B (bx + y)$, where $a, b \in \mathbb{R}$. Then $x \perp_B (ax + y - bx - y)$, i.e., $x \perp_B (ax - bx)$. Hence, $a - b = 0 \Rightarrow a = b$. So, B–J orthogonality is right unique. □

Remark 2.3.4 We note that Theorem 2.3.2 is also locally true. More precisely, an element $x \in \mathbb{X}$ is smooth if and only if B–J orthogonality is right additive at x, i.e., x is smooth if and only if for $y, z \in \mathbb{X}$, $x \perp_B y, x \perp_B z$ implies that $x \perp_B (y + z)$. The same holds for Theorem 2.3.3.

Remark 2.3.5 It is worth mentioning here that B–J orthogonality is related to Gateaux differentiability. The norm of the space $\mathbb{X}$ is said to be Gateaux differentiable at $x_0 \in S_{\mathbb{X}}$ in the direction y_0 if and only if

$$\lim_{t \to 0} \frac{\|x_0 + t y_0\| - \|x_0\|}{t}$$

exists, and in this case, it is written as $G(x_0, y_0)$. The norm of the space is said to be Gateaux differentiable if and only if the limit exists for each $x_0 \in S_{\mathbb{X}}$ and each $y_0 \in \mathbb{X}$. It is well known that the norm is Gateaux differentiable at $x_0 \in S_{\mathbb{X}}$ if and only if x_0 is a smooth element. Thus, from Theorem 2.3.2, it follows that the norm is

Gateaux differentiable (at a point $x_0 \in S_{\mathbb{X}}$) if and only if B–J orthogonality is right additive (at a point $x_0 \in S_{\mathbb{X}}$). For more on Gateaux differentiability, we refer the readers to [34, 94].

2.4 B–J Orthogonality and Reflexivity

From Theorem 2.1.12, it follows that given $x \in \mathbb{X}$, there exists a hyperspace H such that $x \perp_B H$. A natural question in this connection is as follows: Given a hyperspace H, does there exist x such that $x \perp_B H$. The answer is, in general, no. Consider the Banach space c_0 of all sequences of scalars converging to 0. Let $H = \left\{(x_1, x_2, x_3, \ldots, x_n, \ldots) \in c_0 : \sum_{n=1}^{\infty} \frac{x_n}{n^2} = 0\right\}$. Then H is a hyperspace but there exists no $x \in c_0$ such that $x \perp_B H$. Note that c_0 is not reflexive. The next theorem shows that the result is a characteristic property of reflexive Banach spaces.

Theorem 2.4.1 ([59, 101]) *A Banach space $\mathbb{X}$ is reflexive if and only if for any closed hyperspace $H \subseteq \mathbb{X}$, there exists $x(\neq 0) \in \mathbb{X}$ such that $x \perp_B H$.*

Proof We first prove the necessary part. Suppose that $\mathbb{X}$ is reflexive. Let H be a closed hyperspace of $\mathbb{X}$ and f be a bounded linear functional on $\mathbb{X}$ such that $H = \ker(f)$. Since $\mathbb{X}$ is reflexive, f attains its supremum at $x \in S_{\mathbb{X}}$, say. Let $y \in H$. Then, for all $\lambda \in \mathbb{R}$, $\|f\|\|x + \lambda y\| \geq |f(x + \lambda y)| = |f(x)| = \|f\|\|x\|$. So, $\|x + \lambda y\| \geq \|x\|$ for all scalars λ. Thus, $x \perp_B y$. Hence, $x \perp_B H$. We next prove the sufficient part. We note from [57, Theorem 2] that a Banach space $\mathbb{X}$ is reflexive if and only if every bounded linear functional on $\mathbb{X}$ attains its norm. Let f be a non-zero bounded linear functional on $\mathbb{X}$ and $H = \ker(f)$. Clearly, H is a closed hyperspace of $\mathbb{X}$. By hypothesis, there exists $x(\neq 0) \in \mathbb{X}$ such that $x \perp_B H$. Clearly, $f(x) \neq 0$, since $f \neq 0$. Let $f(x) = r \neq 0$. Now, every element z of $\mathbb{X}$ can be uniquely written as $z = ax + h$, where a is a scalar and $h \in H$. So, $\|z\| = \|ax + h\| \geq |a|\|x\|$. Now, $|f(z)| = |af(x)| = |r||a| \leq |r|\frac{\|z\|}{\|x\|}$. Thus, $\|f\| \leq \frac{|r|}{\|x\|}$. Again, $|f(\frac{x}{\|x\|})| = \frac{|r|}{\|x\|}$. Therefore, $\|f\| = \frac{|r|}{\|x\|}$ and f attains its supremum at $\frac{x}{\|x\|}$. Thus, every bounded linear functional attains its supremum on $\mathbb{X}$ and so $\mathbb{X}$ is reflexive. □

2.5 B–J Orthogonality and Inner Product

Now, we raise the question: Given a hyperspace H in a Banach space $\mathbb{X}$, does there exist $x \in \mathbb{X}$ such that $H \perp_B x$? This property characterizes Hilbert space if the dimension of the space is ≥ 3, which is our next result.

Theorem 2.5.1 ([60, Theorem 4]) *Let $\mathbb{X}$ be a Banach space such that* $\dim(\mathbb{X}) \geq 3$. *Then $\mathbb{X}$ is a Hilbert space if and only if for any closed hyperspace H of $\mathbb{X}$, there exists an element $x \in \mathbb{X}$, such that $H \perp_B x$.*

Proof We only need to prove the sufficient part as the necessary part is immediate. Let $\mathbb{V}$ be a two-dimensional subspace of $\mathbb{X}$. Suppose that S is the set of all subspaces $\mathbb{M}$ of $\mathbb{X}$ such that $\mathbb{V} \perp_B \mathbb{M}$. From the hypothesis, it is clear that S is non-empty. By well ordering the set, S has a maximal element $\mathbb{M}_0$. We claim that $span\{\mathbb{V}, \mathbb{M}_0\} = \mathbb{X}$. If not, then we can find a hyperspace H of $\mathbb{X}$ such that $span\{\mathbb{V}, \mathbb{M}_0\} \subseteq H$. From hypothesis, there exists $x \in \mathbb{X}$ such that $H \perp_B x$. Clearly, for any $v \in \mathbb{V}$, $m \in \mathbb{M}_0$, we have

$$\|v + m + \lambda x\| \geq \|v + m\| \geq \|v\|,$$

for any scalar λ. This implies $\mathbb{V} \perp_B span\{\mathbb{M}_0, x\}$, which contradicts the maximality of $\mathbb{M}_0$. Thus, $span\{\mathbb{V}, \mathbb{M}_0\} = \mathbb{X}$. This ensures the existence of a norm one projection P from $\mathbb{X}$ onto $\mathbb{V}$, which is given by $P(v + m) = v$, for all $v \in \mathbb{V}$, $m \in \mathbb{M}_0$. Hence, by Kakutani's theorem [63], it follows that $\mathbb{X}$ is a Hilbert space. □

Remark 2.5.2 The geometry of a Banach space can be so wild that there may be no hyperspace H for which there exists an element x such that $H \perp_B x$. Consider the space $C[0, 1]$ of all continuous real-valued functions on $[0, 1]$. Then there exists no hyperspace H such that $H \perp_B x$. Again there exist spaces like $\ell_p (p \geq 1, p \neq 2)$ in which there are infinitely many hyperspaces H_α for which there exist elements x_α such that $H_\alpha \perp_B x_\alpha$ and, at the same time, there exists hyperspace H for which no element x can be found with $H \perp_B x$. If the space is strictly convex then for no hyperspace H, two elements x_1, x_2 can be found such that $H \perp_B x_1$, $H \perp_B x_2$.

Our last question in this direction is as: Given an element $x \in \mathbb{X}$, does there exist hyperspace H such that $H \perp_B x$? This is again a characteristic property of a Hilbert space if the dimension of the space is ≥ 3.

Theorem 2.5.3 ([60, Theorem 5]) *Let $\mathbb{X}$ be a Banach space such that* $\dim(\mathbb{X}) \geq 3$. *Then $\mathbb{X}$ is a Hilbert space if and only if for any element $x \in \mathbb{X}$, there exists a hyperspace H such that $H \perp_B x$.*

Proof The necessary part follows easily and so we prove the sufficient part. We complete the proof into two cases, in the first case, we assume the space to be strictly convex. Let $x, y, z \in \mathbb{X}$ be such that $x \perp_B z$, $y \perp_B z$. By hypothesis, there exists a hyperspace H such that $H \perp_B z$. Since the space is strictly convex, it follows from Theorem 2.2.1 that the orthogonality is left unique and so $x \in H$, $y \in H$. Then $x + y \in H$. Thus orthogonality is left additive and so from Theorem 2.2.3, it follows that $\mathbb{X}$ is a Hilbert space. Next, consider the case $\mathbb{X}$ is not strictly convex. Then there exist $x, y \in S_{\mathbb{X}}$ such that $(1 - t)x + ty \in S_{\mathbb{X}}$ for all $t \in [0, 1]$. Let f be a linear functional with $f(x) = f(y) = \|f\|$. Let $z \in S_{\mathbb{X}}$ be such that $z \notin span\{x, y\}$. Consider $\mathbb{Y} = span\{x, y, z\}$. Let $P = \{u \in S_{\mathbb{Y}} : f(u) = \|f\|\}$. Then P contains the line segment $L[x, y] = \{(1 - t)x + ty : t \in [0, 1]\}$. Let $H_{\mathbb{Y}}$ be the hyperspace of $\mathbb{Y}$ such that $P \perp_B H_{\mathbb{Y}}$, where $H_{\mathbb{Y}}$ contains all elements at which f is zero. Then for any v and any scalar a there is a hyperspace H_a of $\mathbb{Y}$ with $H_a \perp_B (v + ax)$. As $a \to 0+$ (or as $a \to 0-$), the hyperspaces H_a will have at least one limit H_+ (or H_-)

in the sense that there exist sequences $\{a_n\}$ (or $\{b_n\}$) with $a_n \to 0+$ (or $b_n \to 0-$), $\lim_{a_n \to 0+} d(w, H_{a_n}) = 0$ (or $\lim_{b_n \to 0-} d(w, H_{b_n}) = 0$), if w is any fixed element of H_+ (or H_-). Since, at each point of unit norm in H_a, there is a supporting plane of $S_{\mathbb{Y}}$ parallel to $v + ax$, it follows that, if $v \in H_{\mathbb{Y}}$, then neither H_+ nor H_- crosses P and P consists of those and only those points of $S_{\mathbb{Y}}$ in a region containing x and bounded by H_+, H_-, and the two supporting lines of P parallel to v. But this is possible for arbitrary $v \in H_{\mathbb{Y}}$ only if P is singleton. This contradicts the fact that $\mathbb{X}$ is not strictly convex. Thus, the space has to be strictly convex if it satisfies the hypothesis and, so by the first case, it is a Hilbert space. □

For more on the basic theory of B–J orthogonality and related concepts readers can go through the articles [4, 5, 42].

Chapter 3
Operator Norm Attainment

3.1 Introduction

Linear operators lie at the very heart of functional analysis and operator theory. As mentioned in the Preface, this monograph aims at exploring the beautiful interrelation between analysis, algebra, and geometry in the space of bounded linear operators between Banach spaces. Geometry of operator spaces is intimately related with the norm attainment sets of operators. It is well known that the norm attainment set of a compact operator on a reflexive Banach space is always non-empty. On the other hand, it is rather trivial to observe that the norm attainment set of a bounded linear operator may very well be empty. In this context, James [58] proved that a Banach space is reflexive if and only if every bounded linear functional on the space attains its norm. This classical theorem, apart from being interesting in its own right, indicates the historical significance of the norm attainment set of bounded linear functionals. The importance of operator norm attainment was further highlighted in more recent times, in view of the pivotal result obtained by Bhatia and Semrl [17] regarding Birkhoff–James orthogonality of matrices and some related distance formulae. It was shown in [17] that if T, A are two operators on a finite-dimensional Hilbert space $\mathbb{H}$, then $T \perp_B A$ if and only if there exists an element x with $\|x\| = 1$ such that $\|Tx\| = \|T\|$ and $\langle Tx, Ax\rangle = 0$. We also refer the readers to a similar but independent work by Paul [99], broadly with the same objectives. We would like to consider such results on operator orthogonality as one of the major motivations behind exploring the norm attainment set of an operator. To proceed in our intended path, we begin with the easy yet useful observation that if T, A are bounded linear operators between Banach spaces and if there exists an element x with $\|x\| = 1$ and $\|Tx\| = \|T\|$ such that $Tx \perp_B Ax$ holds, then, for each scalar λ,

$$\|T + \lambda A\| \geq \|Tx + \lambda Ax\| \geq \|Tx\| = \|T\|$$

A. Mal et al., *Birkhoff–James Orthogonality and Geometry of Operator Spaces*, Infosys Science Foundation Series, https://doi.org/10.1007/978-981-99-7111-4_3

holds, which yields that $T \perp_B A$. This prompts us to study the structure and properties of the norm attainment set of a bounded linear operator between Banach spaces, in order to gain insights into operator orthogonality and its several implications. Throughout this chapter, we assume that the scalars are either real or complex unless mentioned otherwise explicitly. We illustrate in this chapter that B–J orthogonality plays a pivotal role in understanding the norm attainment set of a bounded linear operator between Banach spaces. Indeed, the key idea in this context is the preservation of B–J orthogonality by an operator at certain special points of the unit sphere of the domain space, which include the points at which the concerned operator attains norm. This is established by taking into account the geometry of the domain space and the codomain space, in terms of the positive part and the negative part of an element in a Banach space. It is also observed that the minimum norm attainment set of an operator can also be described by using the same principle of local preservation of B–J orthogonality. In fact, as far as the content of this chapter is concerned, such preservation properties form the backbone of the mathematical developments presented here. By following this very useful and interesting principle, we arrive at the conclusion that perhaps the best way to understand and describe the complete set of solutions to the above optimization problems is by using the concept of semi-inner-products which share a close relationship with B–J orthogonality. Moreover, we also illustrate that the theory developed in this chapter can be successfully applied to obtain concrete information on operator norm attainment in specific Banach spaces such as ℓ_p^2.

Let us now introduce a few notations and terminologies which will be frequently used.

For a bounded linear operator $T \in \mathbb{L}(\mathbb{X}, \mathbb{Y})$, we denote the set of all points of $S_{\mathbb{X}}$ at which T attains its norm by M_T, i.e.,

$$M_T = \{x \in S_{\mathbb{X}} : \|Tx\| = \|T\|\}.$$

Let $m(T)$ denote the minimum norm of T, i.e., $m(T) = \inf\{\|Tx\| : x \in S_{\mathbb{X}}\}$. We denote by m_T, the set of all points of $S_{\mathbb{X}}$ at which T attains its minimum norm, i.e.,

$$m_T = \{x \in S_{\mathbb{X}} : \|Tx\| = m(T)\}.$$

M_T and m_T are, respectively, known as the norm attainment set and the minimum norm attainment set of the operator T.

A bounded linear functional f on $\mathbb{X}$ is said to be a norming linear functional of $x \in \mathbb{X} \setminus \{0\}$, if $f(x) = \|x\|$ and $\|f\| = 1$. The Hahn–Banach Theorem ensures the existence of at least one norming linear functional for each non-zero x. A non-zero element $x \in \mathbb{X}$ is said to be a smooth point if x has exactly one norming linear functional. Observe that in this case, $\{y \in \mathbb{X} : x \perp_B y\}$ is a hyperspace of codimension one in $\mathbb{X}$. As a matter of fact, this geometrically suggestive property characterizes the smooth points of a Banach space. A Banach space $\mathbb{X}$ is said to be smooth if every non-zero vector in $\mathbb{X}$ is smooth. Another related geometric notion is that of strict convexity. $\mathbb{X}$ is said to be strictly convex, if $S_{\mathbb{X}}$ does not contain any non-trivial straight line segment. Note that smoothness of a Banach space is characterized by right uniqueness of B–J orthogonality on that space. Right additivity of B–J orthogonality also characterizes smoothness of a Banach space. On the other hand, strict convexity of a Banach space is characterized by left uniqueness of B–J orthogonality on that space. If a Banach space is two-dimensional, then strict convexity of the space is characterized by left additivity of B–J orthogonality. However, if the dimension of the Banach space is greater than two, then left additivity of B–J orthogonality characterizes Hilbert spaces. All these basic facts concerning B–J orthogonality, and more, can be found in the pioneering article [59].

In Sect. 3.2, we explore properties of the operator norm attainment set in light of B–J orthogonality preservation. We also study analogous properties for the minimum norm attainment set of an operator. In Sect. 3.3, we provide a complete characterization of the operator norm attainment set (minimum norm attainment set). In Sect. 3.4, we focus on studying the relation between M_T and m_T. We emphasize the use of this relation in characterizing reflexive Banach spaces and finite-dimensional Hilbert spaces. Finally, in Sect. 3.5, we obtain an optimal upper bound for the cardinality of the norm attainment set of operators on $\ell_p^2(\mathbb{R})$ ($p \in \mathbb{N} \setminus \{1, 2\}$) spaces, which are not scalar multiple of isometry.

3.2 Properties of the Norm Attainment Set (Minimum Norm Attainment Set)

In this section, we study the properties of the operator norm attainment set (minimum norm attainment set) from the perspective of B–J orthogonality. It turns out that the following geometrical notion, introduced in [135], plays a very crucial role in our exploration.

For any two points x, y in a real Banach space $\mathbb{X}$, we say that

$$y \in x^+ \text{ if } \|x + \lambda y\| \geq \|x\| \text{ for all } \lambda \geq 0.$$

Accordingly, we say that

$$y \in x^- \text{ if } \|x + \lambda y\| \geq \|x\| \text{ for all } \lambda \leq 0.$$

By $x^\perp$, we denote the set $\{y \in \mathbb{X} : x \perp_B y\}$. These notions were extended to the setting of complex Banach spaces in [107] in the following way: Let $x \in \mathbb{X}$ and $U = \{\alpha \in \mathbb{C} : |\alpha| = 1, \ \arg \alpha \in [0, \pi)\}$. For $\alpha \in U$, we define

$$x_\alpha^+ = \{y \in \mathbb{X} : \|x + \lambda y\| \geq \|x\| \ \forall \ \lambda = t\alpha, t \geq 0\},$$

$$x_\alpha^- = \{y \in \mathbb{X} : \|x + \lambda y\| \geq \|x\| \ \forall \ \lambda = t\alpha, t \leq 0\},$$

$$x_\alpha^\perp = \{y \in \mathbb{X} : \|x + \lambda y\| \geq \|x\| \ \forall\ \lambda = t\alpha, t \in \mathbb{R}\}.$$

If $\beta = e^{i\pi}\alpha$ then we define $x_\beta^+ = x_\alpha^-,\ x_\beta^- = x_\alpha^+, x_\beta^\perp = x_\alpha^\perp$. If $y \in x_\alpha^\perp$, then we write $x\perp_\alpha y$. In complex Banach spaces, the notions of x^+, x^- and $x^\perp$ are defied in the following way:

$$x^+ = \bigcap\{x_\alpha^+ : \alpha \in U\}, x^- = \bigcap\{x_\alpha^- : \alpha \in U\}, x^\perp = \bigcap\{x_\alpha^\perp : \alpha \in U\}.$$

Let us begin with some basic properties of these notions. The proofs of these results are quite easy to verify and are therefore omitted.

Proposition 3.2.1 ([107, Proposition 2.1]) *Let $\mathbb{X}$ be a complex Banach space, let $x, y \in \mathbb{X}$ and let $\alpha \in U$. Then the following are true:*
(i) Either $y \in x_\alpha^+$ or $y \in x_\alpha^-$.
(ii) $x \perp_\alpha y$ if and only if $y \in x_\alpha^+$ and $y \in x_\alpha^-$.
(iii) $y \in x_\alpha^+$ implies that $\eta y \in (\mu x)_\alpha^+$ for all $\eta, \mu > 0$.
(iv) $y \in x_\alpha^+$ implies that $-y \in x_\alpha^-$ and $y \in (-x)_\alpha^-$.
(v) $y \in x_\alpha^-$ implies that $\eta y \in (\mu x)_\alpha^-$ for all $\eta, \mu > 0$.
(vi) $y \in x_\alpha^-$ implies that $-y \in x_\alpha^+$ and $y \in (-x)_\alpha^+$.
(vii) $y \in x_\alpha^+$ implies that $\beta y \in (\beta x)_\alpha^+$ for all $\beta \in \mathbb{C}$.
(viii) $y \in x_\alpha^-$ implies that $\beta y \in (\beta x)_\alpha^-$ for all $\beta \in \mathbb{C}$.

Proposition 3.2.2 ([134, Proposition 2.2]) *Let $\mathbb{X}$ be a Banach space and let $x, y \in \mathbb{X}$. Then the following are true:*
(i) $x \perp_B y$ if and only if $y \in x^+$ and $y \in x^-$.
(ii) $y \in x^+$ implies that $\eta y \in (\mu x)^+$ for all $\eta, \mu > 0$.
(iii) $y \in x^+$ implies that $-y \in x^-$ and $y \in (-x)^-$.
(iv) $y \in x^-$ implies that $\eta y \in (\mu x)^-$ for all $\eta, \mu > 0$.
(v) $y \in x^-$ implies that $-y \in x^+$ and $y \in (-x)^+$.

Moreover, if $\mathbb{X}$ is a real Banach space, then, for each $x, y \in \mathbb{X}$, either $y \in x^+$ or $y \in x^-$.

Next, we state an easy observation that relates the set $(Tx)^\perp$ with $x^\perp$ for $x \in M_T$.

Proposition 3.2.3 *Let $\mathbb{X}, \mathbb{Y}$ be Banach spaces. Let $0 \neq T \in \mathbb{L}(\mathbb{X}, \mathbb{Y})$ and $x \in M_T$. Then, for any $y \in \mathbb{X}$, $Tx \perp_B Ty \Rightarrow x \perp_B y$.*

Proof Since $x \in M_T$ and $Tx\perp_B Ty$, we get for all scalars λ,

$$\|T\|\|x\| = \|Tx\| \leq \|Tx + \lambda Ty\| \leq \|T\|\|x + \lambda y\|.$$

This implies that $\|x + \lambda y\| \geq \|x\|$ for all scalars λ and so $x\perp_B y$. □

An easy application of Proposition 3.2.3 gives a description of the kernel of an operator in terms of its norm attainment set.

Corollary 3.2.4 ([134, Corollary 2.3.1]) *Let $\mathbb{X}$, $\mathbb{Y}$ be Banach spaces and let $T \in \mathbb{L}(\mathbb{X}, \mathbb{Y})$. Then*

$$\ker(T) \subseteq \bigcap_{x \in M_T} x^{\perp}.$$

In particular, if $\mathbb{X}$ is a two-dimensional strictly convex Banach space and $\mathbb{Y}$ is a two-dimensional Banach space, then any linear operator in $\mathbb{L}(\mathbb{X}, \mathbb{Y})$ attaining norm at two linearly independent points must be invertible.

Proof If $M_T = \emptyset$, then $\ker(T) \subseteq \bigcap_{x \in M_T} x^{\perp}$ holds trivially. Consider $M_T \neq \emptyset$. Let $z \in ker(T)$. Then we have $Tz = 0$ and so $Tx \perp_B Tz$ for all $x \in M_T$. From Proposition 3.2.3, it follows that $x \perp_B z$, for each $x \in M_T$. Thus, we have

$$\ker(T) \subseteq \bigcap_{x \in M_T} x^{\perp}.$$

Next, assume that $\mathbb{X}$ is a two-dimensional strictly convex Banach space and $\mathbb{Y}$ is a two-dimensional Banach space. Let $T \in \mathbb{L}(\mathbb{X}, \mathbb{Y})$. From Theorem 2.2.3, it follows that B–J orthogonality is left additive. Let $x, y \in M_T$ be such that $\{x, y\}$ is linearly independent. We first claim that $x^{\perp} \cap y^{\perp} = \{0\}$. Let $z \in x^{\perp} \cap y^{\perp}$. Then $x \perp_B z$, $y \perp_B z$ and so by left additivity of B–J orthogonality, we get $\alpha x + \beta y \perp_B z$ for all scalars α, β. Therefore, $z \perp_B z$ and so we must have $z = 0$. Then, from the first part we get, $\ker(T) \subset (x^{\perp} \cap y^{\perp}) = \{0\} \subset \ker(T)$. □

It is easy to observe that the reverse implication in Proposition 3.2.3 does not hold in general. The following example proves this claim.

Example 3.2.5 Consider the linear operator $T \in \mathbb{L}(\ell_{\infty}^2)$ given by $T(1, 1) = (1, 0)$ and $T(1, -1) = (\frac{1}{2}, \frac{1}{2})$. Then it is easy to check that $(1, 1) \in M_T$. Moreover, note that $(1, 1) \perp_B (-\frac{1}{2}, 1)$, whereas $T(1, 1) \not\perp_B T(-\frac{1}{2}, 1)$.

However, the reverse implication in Proposition 3.2.3 holds if we assume that $x \in M_T$ is a smooth point of $\mathbb{X}$.

Proposition 3.2.6 *Let $\mathbb{X}$, $\mathbb{Y}$ be Banach spaces, let $T \in \mathbb{L}(\mathbb{X}, \mathbb{Y})$ and let $x \in M_T$. Suppose that x is smooth. Then, for any $y \in \mathbb{X}$, $x \perp_B y \Rightarrow Tx \perp_B Ty$. In other words, $T(x^{\perp}) \subseteq (Tx)^{\perp}$.*

Proof Without loss of generality, assume that $\|T\| = 1$. The smoothness of x gives that x has a unique norming linear functional, i.e., there exists a unique functional $f \in S_{\mathbb{X}^*}$ such that $f(x) = \|x\| = 1$. Suppose that $x \perp_B y$. Then, from Theorem 2.1.14, it follows that $f(y) = 0$. Suppose that $g \in S_{\mathbb{Y}^*}$ is a norming linear functional of Tx, i.e., $g(Tx) = \|Tx\| = \|T\| = 1$. Then clearly $g \circ T \in S_{\mathbb{Y}^*}$ and $g \circ T(x) = 1$. Now, the uniqueness of f yields $f = g \circ T$. Thus, $f(y) = 0$ implies that $g(Ty) = 0$. Again using Theorem 2.1.14, we get $Tx \perp_B Ty$. This completes the proof. □

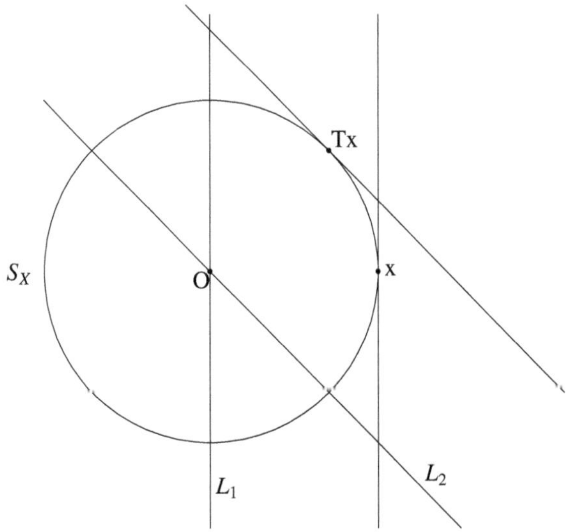

Fig. 3.1 Necessary condition for operator norm attainment

Remark 3.2.7 Let us discuss a geometrical interpretation of Proposition 3.2.6. Suppose $\mathbb{X}$, $\mathbb{Y}$ are Banach spaces and $T \in \mathbb{L}(\mathbb{X}, \mathbb{Y})$. If $x \in M_T$ is smooth, then the image of the hyperspace $x^\perp$ under T is a subset of the set $(Tx)^\perp$. In particular, this provides an insight on M_T without knowing the action of T on the whole space $\mathbb{X}$. For instance, if $x, y \in S_{\mathbb{X}}$ are such that $x \perp_B y$ and $Tx \not\perp_B Ty$ then Proposition 3.2.6 ensures that $x \notin M_T$, irrespective of the action of T on other points.

Observe that Proposition 3.2.6 gives a necessary condition for $x \in M_T$, whenever x is smooth. With the following diagram, we illustrate this condition, assuming that T is defined on a two-dimensional, smooth real Banach space. Without loss of generality, we assume that $\|T\| = 1$.

In Fig. 3.1, $x^\perp \equiv L_1$ and $(Tx)^\perp \equiv L_2$. Proposition 3.2.6 states that if $x \in M_T$ then $T(L_1) \subseteq L_2$.

Recall that a bounded linear operator T on a Banach space is said to satisfy the Daugavet equation if $\|I + T\| = 1 + \|T\|$ holds, where I is the identity operator. The next corollary ensures the existence of an invariant subspace of codimension one of a bounded linear operator satisfying the Daugavet equation, whenever the underlying space is finite-dimensional, smooth and strictly convex.

Corollary 3.2.8 ([134, Corollary 2.3.2]) *Let $\mathbb{X}$ be a Banach space and let $T \in \mathbb{L}(\mathbb{X})$ satisfies the Daugavet equation $\|I + T\| = 1 + \|T\|$. If $\mathbb{X}$ is smooth and strictly convex then $\mathbb{X}$ has a $T-$invariant subspace of codimension one.*

Proof The result holds trivially if $T = 0$. Assume that T is non-zero. We note from [2, Lemma 2.1] that T satisfies the Daugavet equation if and only if αT satisfies the Daugavet equation, where $\alpha > 0$. Without loss of generality we assume that $\|T\| = 1$. Since $\mathbb{X}$ is finite-dimensional, there exists $x \in S_{\mathbb{X}}$ such that $\|I + T\| = \|(I + T)x\|$. We claim that $Tx = x$. Observe that, $2 = \|I + T\| = \|(I + T)x\| = \|x + Tx\| \leq \|x\| + \|Tx\| \leq 1 + \|T\| = 2$. Since $\mathbb{X}$ is strictly convex, we get, $Tx = kx$, for some $k \geq 0$. Then, from $2 = \|x + Tx\| \leq \|x\| + \|Tx\| = 1 + \|Tx\| \leq 2$ we infer that $k = \|Tx\| = 1$. This establishes our claim.

Thus, we have $1 = \|T\| = \|Tx\|$. This proves that $x \in M_T$. Since $\mathbb{X}$ is smooth and $x \in M_T$, from Proposition 3.2.6, it follows that, $T(x^{\perp}) \subseteq (Tx)^{\perp} = x^{\perp}$. Since $\mathbb{X}$ is smooth, $x^{\perp}$ is a subspace of $\mathbb{X}$, having codimension one. Thus, $x^{\perp}$ is a $T-$invariant subspace of $\mathbb{X}$, having codimension one. □

Remark 3.2.9 Observe that the proof of Corollary 3.2.8 guarantees the existence of a non-trivial fixed point of a norm one operator T, if T satisfies the Daugavet equation and T is defined on a strictly convex, finite-dimensional Banach space.

If we omit the smoothness condition of x in Proposition 3.2.6, then we get the following weaker conclusion.

Proposition 3.2.10 ([127, Lemma 2.1]) *Let $\mathbb{X}, \mathbb{Y}$ be Banach spaces and $T \in \mathbb{L}(\mathbb{X}, \mathbb{Y})$. Let $x \in M_T$ and let $y = Tx$. Then there exist hyperspaces H_x, H_y in $\mathbb{X}$ and $\mathbb{Y}$, respectively, such that $x \perp_B H_x$ and $y \perp_B H_y$ with $T(H_x) \subseteq H_y$.*

Proof The result holds trivially if $T = 0$. Suppose T is non-zero. Since B–J orthogonality is homogeneous, without loss of generality, we assume that $\|T\| = 1$. Let $x \in M_T$. For $y(= Tx)$, there exists a linear functional $g \in S_{\mathbb{Y}^*}$ such that $g(Tx) = \|Tx\| = 1$. Let $\ker(g) = H_y$. Then by Theorem 2.1.14, we have $Tx \perp_B H_y$. Now, $g \circ T : \mathbb{X} \to \mathbb{K}$ is a linear functional with $g \circ T(x) = \|Tx\| = 1 = \|x\|$ and $\|g \circ T\| \leq \|g\|\|T\| = \|T\| = 1$. Let $H_x = \ker(g \circ T)$. Again using Theorem 2.1.14, we get, $x \perp_B H_x$. Let $h \in H_x$. Then $g \circ T(h) = 0 \Rightarrow g(Th) = 0 \Rightarrow Th \in H_y$. Since this is true for all $h \in H_x$, we have $T(H_x) \subseteq H_y$. This completes the proof. □

Remark 3.2.11 Note that, Proposition 3.2.10 may not be true for all hyperspaces. To be precise, given $T \in \mathbb{L}(\mathbb{X}, \mathbb{Y})$ and $x \in M_T$, there may exist a hyperspace H in $\mathbb{X}$ such that $x \perp_B H$ but $Tx \not\perp_B T(H)$. Let us illustrate the scenario by furnishing the following example.

Consider $T : \ell^2_\infty(\mathbb{R}) \to \ell^2_\infty(\mathbb{R})$ defined by

$$T(1, 1) = (0, 1), \quad T(-1, 1) = (-1, 0).$$

Then $M_T = \{\pm(1, 1), \pm(-1, 1)\}$. Here, we have $(1, 1) \in M_T$ and $(1, 1) \perp_B (0, 1)$ but $T(1, 1) = (0, 1) \not\perp_B T(0, 1) = (-\frac{1}{2}, \frac{1}{2})$. Therefore, taking $x = (1, 1)$ and H to be the one-dimensional subspace spanned by $(0, 1)$, we see that $x \perp_B H$ but $Tx \not\perp_B T(H)$.

If we replace the underlying field $\mathbb{R}$ by $\mathbb{C}$ in this example, then

$$M_T = \bigcup_{\theta \in [0, 2\pi)} e^{i\theta}\{(1, 1), (-1, 1)\}.$$

Once again, we have $(1, 1) \in M_T$ and $(1, 1) \perp_B (0, 1)$ but $T(1, 1) = (0, 1) \not\perp_B T(0, 1) = (-\frac{1}{2}, \frac{1}{2})$. Therefore, with the same choice of x and H, it follows that $x \perp_B H$ but $Tx \not\perp_B T(H)$.

Combining Propositions 3.2.3 and 3.2.6, we immediately get a nice and useful property of a smooth point in the norm attainment set of a bounded linear operator.

Theorem 3.2.12 ([134, Theorem 2.4]) *Let $\mathbb{X}$, $\mathbb{Y}$ be Banach spaces, let $T \in \mathbb{L}(\mathbb{X}, \mathbb{Y})$ and let $x \in M_T$. Suppose that x is smooth. Then, for any $y \in \mathbb{X}$, $x \perp_B y \Leftrightarrow Tx \perp_B Ty$.*

Theorem 3.2.12 also ensures that T preserves B–J orthogonality at x, whenever $x \in M_T$ and x is smooth. As a consequence of Theorem 3.2.12, together with the well-known fact that a compact linear operator on a reflexive Banach space must attain its norm, we derive the following theorem.

Theorem 3.2.13 ([134, Theorem 2.5]) *Let $\mathbb{X}$ be a reflexive, smooth Banach space and let $\mathbb{Y}$ be an arbitrary Banach space. Let $T \in \mathbb{K}(\mathbb{X}, \mathbb{Y})$. Then there exists $x \in S_{\mathbb{X}}$ such that T preserves B–J orthogonality at x, in both directions, i.e., for any $y \in \mathbb{X}$, $x \perp_B y \Leftrightarrow Tx \perp_B Ty$.*

Theorem 3.2.13 may be compared to a result by Koldobsky [72], which states that an isometry on $\mathbb{X}$ preserves B–J orthogonality at every point of $\mathbb{X}$, and conversely.

We now proceed towards a geometric characterization of smooth Banach spaces, in terms of operator norm attainment and B–J orthogonality. To achieve our goal, we first prove the following result:

Theorem 3.2.14 ([134, Theorem 2.6]) *Let $\mathbb{X}$ be a Banach space. If for every $T \in \mathbb{L}(\mathbb{X})$ and for every $x \in M_T$, it follows that $T(x^{\perp}) \subseteq (Tx)^{\perp}$, then $\mathbb{X}$ is smooth.*

Proof Assume that, for every $T \in \mathbb{L}(\mathbb{X})$ and for every $x \in M_T$, $T(x^{\perp}) \subseteq (Tx)^{\perp}$. If possible let $\mathbb{X}$ be not smooth. Then there exists $x \in S_{\mathbb{X}}$ such that $x \perp_B H_1$ and $x \perp_B H_2$, where H_1, H_2 are two different hyperspaces of codimension one. Clearly, any point $z \in \mathbb{X}$ can be written as $z = \alpha x + h$, where α is a scalar and $h \in H_1$. Define $T : \mathbb{X} \to \mathbb{X}$ as

$$T(\alpha x + h) = \alpha x, \text{ for each scalar } \alpha \text{ and for each } h \in H_1.$$

Clearly, T is well-defined and linear. Since $x \perp_B H_1$, it is easy to check that T is bounded and $x \in M_T$. Choose $y \in H_2 \setminus H_1$. Let $y = \alpha_0 x + h_0$, where α_0 is a scalar and $h_0 \in H_1$. Since $y \in H_2 \setminus H_1$, we have $y \in x^{\perp}$ and $\alpha_0 \neq 0$. By the hypothesis, we have

$$Tx = x \perp_B Ty = \alpha_0 x.$$

This is a contradiction as $x \in S_{\mathbb{X}}$ and $\alpha_0 \neq 0$. This completes the proof. □

Combining Proposition 3.2.6 and Theorem 3.2.14, we immediately get the following characterization of smooth Banach spaces.

Theorem 3.2.15 ([134, Theorem 2.7]) *A Banach space $\mathbb{X}$ is smooth if and only if for every $T \in \mathbb{L}(\mathbb{X})$ and for every $x \in M_T$, we have $T(x^{\perp}) \subseteq (Tx)^{\perp}$.*

The following theorem gives another important property of norm attainment set of a bounded linear operator on a complex Banach space.

Theorem 3.2.16 *Let $\mathbb{X}, \mathbb{Y}$ be complex Banach spaces, $T \in \mathbb{L}(\mathbb{X}, \mathbb{Y})$ be a non-zero operator and $x \in M_T$. Then, for each $\alpha \in U$,*
(i) $T(x_\alpha^+ \setminus x_\alpha^{\perp}) \subseteq (Tx)_\alpha^+ \setminus (Tx)_\alpha^{\perp}$,
(ii) $T(x_\alpha^- \setminus x_\alpha^{\perp}) \subseteq (Tx)_\alpha^- \setminus (Tx)_\alpha^{\perp}$.

Proof Let us first prove (i). Let $\alpha \in U$ and $z \in x_\alpha^+ \setminus x_\alpha^{\perp}$ be chosen arbitrarily. If possible, suppose that $Tz \in (Tx)_\alpha^-$. Then, for each $t \leq 0$,

$$\|T\|\|x + t\alpha z\| \geq \|Tx + t\alpha Tz\| \geq \|Tx\| = \|T\|\|x\|,$$

which implies that $\|x + t\alpha z\| \geq \|x\|$. Thus, $z \in x_\alpha^-$, contradicting that $z \notin x_\alpha^{\perp}$. Therefore, $Tz \notin (Tx)_\alpha^-$, i.e., $Tz \in (Tx)_\alpha^+ \setminus (Tx)_\alpha^{\perp}$. This completes the proof of (i). The proof of (ii) follows using similar arguments. □

Now we turn our attention to the properties of the minimum norm attainment set of a bounded linear operator between Banach space. In this context, we first obtain a necessary condition for a point on the unit sphere to be in the minimum norm attainment set of a bounded linear operator.

Theorem 3.2.17 ([127, Theorem 2.2]) *Let $\mathbb{X}$ and $\mathbb{Y}$ be Banach spaces. Let $T \in \mathbb{L}(\mathbb{X}, \mathbb{Y})$ be non-zero and let $x \in m_T$. Then*
(i) $T(x^+) \subseteq (Tx)^+$.
(ii) $T(x^-) \subseteq (Tx)^-$.
(iii) $T(x^{\perp}) \subseteq (Tx)^{\perp}$.

Proof (i) We first consider both $\mathbb{X}$ and $\mathbb{Y}$ to be complex Banach spaces. Let $y \in x^+$. Then $y \in x_\alpha^+$ for each $\alpha \in U$. Then $\|x + \lambda y\| \geq \|x\| = 1$ for all $\lambda = t\alpha, t \geq 0$. Since $x \in m_T$, it follows that, for any $t \geq 0$, we must have

$$\|Tx\| \leq \|T\left(\frac{x + t\alpha y}{\|x + t\alpha y\|}\right)\| = \frac{\|Tx + t\alpha Ty\|}{\|x + t\alpha y\|} \leq \|Tx + t\alpha Ty\|.$$

This implies that $Ty \in (Tx)_\alpha^+$ and so $T(x_\alpha^+) \subseteq (Tx)_\alpha^+$. This holds for each $\alpha \in U$ and so $T(x^+) \subseteq (Tx)^+$.

For real Banach spaces, the result follows by noting that $x^+ = x_\alpha^+$ with $\alpha = 1$.

The proofs of (ii) and (iii) follow from similar arguments. □

Remark 3.2.18 We would like to remark here that in Theorem 3.2.17, if we assume $x \in M_T$ instead of $x \in m_T$, then, from Example 3.2.5 it is clear that Theorem 3.2.17 (iii) may not be true in general. Also observe that Theorem 3.2.17 (iii) yields for $x \in m_T$, $x \perp_B y \Rightarrow Tx \perp_B Ty$. However, it can be easily observed that in contrast to Proposition 3.2.3, for $x \in m_T$, the reverse implication may not be true, i.e., $Tx \perp_B Ty$ may not imply that $x \perp_B y$. For example, consider the operator $T \in \mathbb{L}(\ell_\infty^2(\mathbb{R}))$ defined by $T(1,0) = (0,0)$ and $T(1, \frac{1}{2}) = (1, \frac{1}{2})$. Then clearly $(1,0) \in m_T$, $T(1,0) \perp_B T(1, \frac{1}{2})$. However, $(1,0) \not\perp_B (1, \frac{1}{2})$.

By virtue of Theorem 3.2.13, we conclude that every compact operator on a reflexive, smooth Banach space preserves B–J orthogonality at some point on the unit sphere. In the next corollary, we show that if we consider finite-dimensional Banach spaces, then the smoothness of the space is not necessary to obtain a point on the unit sphere where a bounded linear operator preserves B–J orthogonality.

Corollary 3.2.19 ([127, Corollary 2.2.1]) *Let $\mathbb{X}$ be a finite-dimensional Banach space and $\mathbb{Y}$ be an arbitrary Banach space. Suppose $T \in \mathbb{L}(\mathbb{X}, \mathbb{Y})$. Then there exists $x \in S_{\mathbb{X}}$ such that T preserves B–J orthogonality at x, i.e., $x \perp_B y \Rightarrow Tx \perp_B Ty$.*

Proof Let $m(T) = inf\{\|Tx\| : \|x\| = 1\}$. As the space $\mathbb{X}$ is finite-dimensional there exists $x_0 \in S_{\mathbb{X}}$ such that $\|Tx_0\| = m(T)$. Thus, $x_0 \in m_T$. Now, by Theorem 3.2.17, for any $y \in \mathbb{X}$, $x_0 \perp_B y \Rightarrow Tx_0 \perp_B Ty$, i.e., T preserves B–J orthogonality at x_0. □

As an immediate consequence of Theorem 3.2.17, we obtain the next corollary, which should be compared to Proposition 3.2.10.

Corollary 3.2.20 ([127, Corollary 2.2.2]) *Let $\mathbb{X}$, $\mathbb{Y}$ be normed linear spaces and let $T \in \mathbb{L}(\mathbb{X}, \mathbb{Y})$. If $x \in m_T$, then, for any hyperspace H_x in $\mathbb{X}$, with $x \perp_B H_x$, there exists a hyperspace H_y in $\mathbb{Y}$ such that $y(= Tx) \perp_B H_y$ with $T(H_x) \subseteq H_y$.*

3.3 Norm Attainment Set (Minimum Norm Attainment Set) in Terms of s.i.p

In this section, we will see that the notion of semi-inner-product (s.i.p.) plays a central role in the study of the norm attainment sets (minimum norm attainment sets) of operators on Banach spaces.

Definition 3.3.1 ([40, 81]) Let $\mathbb{X}$ be a Banach space. A function $[\ ,\ ] : \mathbb{X} \times \mathbb{X} \to \mathbb{K}(= \mathbb{C}, \mathbb{R})$ is a semi-inner-product if for any $\alpha, \beta \in \mathbb{K}$ and for any $x, y, z \in \mathbb{X}$, it satisfies the following properties:
(i) $[\alpha x + \beta y, z] = \alpha[x, z] + \beta[y, z]$,
(ii) $[x, x] > 0$, whenever $x \neq 0$,
(iii) $|[x, y]|^2 \leq [x, x][y, y]$,
(iv) $[x, \alpha y] = \overline{\alpha}[x, y]$.

Giles proved in [40] that every normed space $(\mathbb{X}, \|\|)$ can be represented as an s.i.p. space $(\mathbb{X}, [\ ,\ ])$ such that, for all $x \in \mathbb{X}$, we have $[x, x] = \|x\|^2$. In general, there can be many compatible s.i.p. corresponding to a given norm. Whenever we speak of a s.i.p. $[\ ,\ ]$ in context of a normed space $\mathbb{X}$, we implicitly assume that $[\ ,\ ]$ is compatible with the norm, i.e., for all $x \in \mathbb{X}$, we have $[x, x] = \|x\|^2$. Lumer stated in [81] that there exists a unique s.i.p. on a normed space if and only if the space is smooth. We now obtain a characterization of the norm attainment set of a bounded linear operator defined between Banach spaces.

Theorem 3.3.2 ([127, Theorem 2.1]) *Let $\mathbb{X}, \mathbb{Y}$ be Banach spaces and let $T \in \mathbb{L}(\mathbb{X}, \mathbb{Y})$. Let $x \in S_{\mathbb{X}}$. Then $x \in M_T$ if and only if there exist two semi-inner-products $[\ ,\ ]_{\mathbb{X}}$ and $[\ ,\ ]_{\mathbb{Y}}$ on $\mathbb{X}$ and $\mathbb{Y}$, respectively, such that, for any $z \in \mathbb{X}$,*

$$[Tz, Tx]_{\mathbb{Y}} = \|T\|^2[z, x]_{\mathbb{X}}.$$

Proof The result holds trivially if $T = 0$. It is easy to see that if the result holds for T with $\|T\| = 1$ then the same holds for any non-zero T. We first prove the sufficient part. Let $x \in S_{\mathbb{X}}$ be such that, for any $z \in \mathbb{X}$, $[Tz, Tx]_{\mathbb{Y}} = \|T\|^2[z, x]_{\mathbb{X}}$. Taking $z = x$, we obtain, $[Tx, Tx]_{\mathbb{Y}} = \|T\|^2$, i.e., $\|Tx\| = \|T\|$. This proves that $x \in M_T$.

Next, we prove the necessary part. Let $x \in M_T$ and $y = Tx$. Then, from Proposition 3.2.10, it follows that there exist hyperspaces H_x, H_y in $\mathbb{X}$ and $\mathbb{Y}$, respectively, such that $x \perp_B H_x$ and $Tx \perp_B H_y$ with $T(H_x) \subseteq H_y$. Since $x \perp_B H_x$, there exists a linear functional $\psi_x : \mathbb{X} \to \mathbb{K}$ such that $\psi_x(x) = \|x\|$ and $\ker(\psi_x) = H_x$. Since T is non-zero and $x \in M_T$, we have $Tx \neq 0$. So we can find a linear functional ψ_y on $\mathbb{Y}$ such that $\psi_y(y) = \|y\| = \|Tx\| = 1$ and $\ker(\psi_y) = H_y$. By the Hahn–Banach Theorem, for each $u \in S_{\mathbb{X}}$ and $v \in S_{\mathbb{Y}}$ there exists at least one $f_u \in S_{\mathbb{X}^*}$ such that $f_u(u) = 1$ and at least one $g_v \in S_{\mathbb{Y}^*}$ such that $g_v(v) = 1$. Define two *semi-inner-products* on $\mathbb{X}$ and $\mathbb{Y}$ as :

For each $z, u \in \mathbb{X}$, $[z, u]_{\mathbb{X}} = f_u(z)$, with the additional restriction that if $u = x$ then we take $f_u = \psi_x$. Moreover, for any $\lambda \in \mathbb{K}$, we choose $f_{\lambda u} = \bar{\lambda} f_u$.

For each $w, v \in \mathbb{Y}$, we define $[w, v]_{\mathbb{Y}} = g_v(w)$, with the additional restriction that if $v = y$ then we take $g_v = \psi_y$. Moreover, for any $\lambda \in \mathbb{K}$, we choose $g_{\lambda v} = \bar{\lambda} g_v$.

Observe that $[\ ,\ ]_{\mathbb{X}}$ and $[\ ,\ ]_{\mathbb{Y}}$ are indeed *semi-inner-products* on $\mathbb{X}$ and $\mathbb{Y}$, respectively. Let $z \in \mathbb{X}$ be arbitrary. Then $z = \alpha x + h$, for some $\alpha \in \mathbb{K}$ and $h \in H_x$. Moreover, we have

$$[z, x]_{\mathbb{X}} = [\alpha x + h, x]_{\mathbb{X}} = \alpha[x, x]_{\mathbb{X}} + [h, x]_{\mathbb{X}} = \alpha\|x\|^2 = \alpha.$$

Also,

$$[Tz, Tx]_{\mathbb{Y}} = [\alpha Tx + Th, Tx]_{\mathbb{Y}} = \alpha[Tx, Tx]_{\mathbb{Y}} + [Th, Tx]_{\mathbb{Y}} = \alpha\|Tx\|^2 = \|T\|^2[z, x]_{\mathbb{X}}.$$

This holds for all $z \in \mathbb{X}$. This completes the proof. □

Note that if T is a scalar multiple of isometry, also called a similarity, between Banach spaces, then T attains norm at each point on the unit sphere. Theorem 3.3.2 immediately gives a characterization of isometric operators between Banach spaces.

Corollary 3.3.3 ([132, Corollary 2.2.3]) *Let* $\mathbb{X}$, $\mathbb{Y}$ *be Banach spaces and let* $T \in \mathbb{L}(\mathbb{X}, \mathbb{Y})$ *be non-zero. Then* T *is a similarity if and only if given any* $x \in \mathbb{X}$ *and any* $z \in S_{\mathbb{X}}$, *there exist two s.i.p.* $[\ ,\ ]_{\mathbb{X}}$ *and* $[\ ,\ ]_{\mathbb{Y}}$ *on* $\mathbb{X}$ *and* $\mathbb{Y}$, *respectively, such that*

$$[Tx, Tz]_{\mathbb{Y}} = \|T\|^2[x, z]_{\mathbb{X}}.$$

Keeping in mind that the only s.i.p. on a Hilbert space is the inner product of the space, we get the following characterization of the norm attainment set of a bounded linear operator defined between Hilbert spaces.

Theorem 3.3.4 ([133, Theorem 2.1]) *Let* $\mathbb{H}_1, \mathbb{H}_2$ *be Hilbert spaces and let* $T \in \mathbb{L}(\mathbb{H}_1, \mathbb{H}_2)$. *Suppose* $x \in S_{\mathbb{H}_1}$. *Then the following are equivalent:*
(i) $x \in M_T$.
(ii) (a) For any $y \in \mathbb{H}_1$, $\langle x, y\rangle = 0$ *implies that* $\langle Tx, Ty\rangle = 0$,
(b) $\sup\{\|Ty\| : \ \|y\| = 1,\ \langle x, y\rangle = 0\} \leq \|Tx\|$.
(iii) $\langle Tx, Ty\rangle = \|T\|^2\langle x, y\rangle$ *for all* $y \in \mathbb{H}_1$.

Proof First, observe that the equivalence of (i) and (iii) follows from Theorem 3.3.2. We only prove the equivalence of (i) and (ii). Suppose that (i) holds, i.e., $x \in M_T$. It follows from the definition of $\|T\|$ that (b) of (ii) holds true. We also note that every Hilbert space is trivially smooth. Since B–J orthogonality coincides with the usual inner product orthogonality in a Hilbert space, it follows from Proposition 3.2.6 that (a) of (ii) holds true.

Now, suppose $x \in S_{\mathbb{H}_1}$ be such that (a) and (b) of (ii) are satisfied. Let $z \in S_{\mathbb{H}_1}$ be chosen arbitrarily. It is easy to see that z can be written as $z = \alpha x + \beta h$, where $\langle x, h\rangle = 0$, $h \in S_{\mathbb{H}_1}$ and α, β are scalars. Clearly, $1 = \|z\|^2 = |\alpha|^2 + |\beta|^2$. By (a), we get $\langle Tx, Th\rangle = 0$. Now, using (b), we get

$$\|Tz\|^2 = \|\alpha Tx + \beta Th\|^2 = |\alpha|^2\|Tx\|^2 + |\beta|^2\|Th\|^2 \leq (|\alpha|^2 + |\beta|^2)\|Tx\|^2 = \|Tx\|^2.$$

This clearly gives that $x \in M_T$, proving (i). □

In the next theorem we observe that the norm attainment set of a bounded linear operator between Hilbert spaces assumes a very nice form.

Theorem 3.3.5 ([133, Theorem 2.2]) *Let* $\mathbb{H}_1, \mathbb{H}_2$ *be Hilbert spaces and* $T \in \mathbb{L}(\mathbb{H}_1, \mathbb{H}_2)$. *If* $M_T \neq \emptyset$, *then* M_T *is the unit sphere of a subspace. In particular,* $M_T = S_H$, *where* $H = \ker(T^*T - \|T\|^2 I)$. *Moreover,* M_T *is the unit sphere of eigen space of the largest eigenvalue (which is equal to* $\|T\|^2$*) of* T^*T.

Proof First, suppose that $x \in M_T$. Then

$$\|T\|^2 = \|Tx\|^2 = \langle Tx, Tx\rangle = \langle T^*Tx, x\rangle \leq \|T^*Tx\|\|x\| \leq \|T\|^2.$$

Thus, the equality condition of Cauchy-Schwarz inequality yields that $T^*Tx = \|T\|^2 x$, i.e., $x \in \ker(T^*T - \|T\|^2 I)$.
Conversely, suppose that $x \in \ker(T^*T - \|T\|^2 I)$ and $\|x\| = 1$. Then $\|Tx\|^2 = \langle T^*Tx, x\rangle = \|T\|^2$, i.e., $x \in M_T$. This completes the proof of the first part.
From the above arguments, it is clear that $\|T\|^2$ is an eigenvalue of T^*T. Now, if $x \in M_T$ and λ is an eigenvalue of T^*T corresponding to the eigenvector z with $\|z\| = 1$, then

$$\|T\|^2 = \|Tx\|^2 \geq \|Tz\|^2 = \langle T^*Tz, z\rangle = \langle \lambda z, z\rangle = \lambda.$$

This completes the proof. □

Now, we obtain a characterization of the minimum norm attainment set of a bounded linear operator between Banach spaces. It is apparent that the following result is analogous to Theorem 3.3.2.

Theorem 3.3.6 ([127, Theorem 2.6]) *Let $\mathbb{X}, \mathbb{Y}$ be Banach spaces and let $T \in \mathbb{L}(\mathbb{X}, \mathbb{Y})$. Let $x \in S_{\mathbb{X}}$. Then $x \in m_T$ if and only if there exist two semi-inner-products $[\ ,\ ]_{\mathbb{X}}$ and $[\ ,\ ]_{\mathbb{Y}}$ on $\mathbb{X}$ and $\mathbb{Y}$, respectively, such that, for any $y \in \mathbb{X}$,*

$$[Ty, Tx]_{\mathbb{Y}} = m^2(T)[y, x]_{\mathbb{X}}.$$

Proof We first prove the sufficient part. Let $x \in S_{\mathbb{X}}$ satisfies $[Ty, Tx]_{\mathbb{Y}} = m^2(T)[y, x]_{\mathbb{X}}$ for any $y \in S_{\mathbb{X}}$. Letting $y = x$ we obtain, $[Tx, Tx]_{\mathbb{Y}} = m^2(T)$, i.e., $\|Tx\| = m(T)$. This implies that $x \in m_T$.
We next prove the necessary part. The result holds trivially if $T = 0$. Assume that T is non-zero. Let $x \in m_T$. Let $y \in \mathbb{X}$ be arbitrary. If $m(T) = 0$ then $\|Tx\| = 0 \Rightarrow Tx = 0$, and therefore, the theorem holds true. Suppose $m(T) > 0$. Then applying Corollary 3.2.20 and using similar arguments, as done in the proof of Theorem 3.3.2, we can complete the proof. □

If a bounded linear operator is defined between Hilbert spaces, then the minimum norm attainment set of the operator can be characterized in the following way. The proof follows using similar arguments as Theorem 3.3.4, and it is therefore omitted.

Theorem 3.3.7 ([127, Theorem 2.4]) *Let $\mathbb{H}_1, \mathbb{H}_2$ be Hilbert spaces and let $T \in \mathbb{L}(\mathbb{H}_1, \mathbb{H}_2)$. Given any $x \in S_{\mathbb{H}_1}$, the following are equivalent:*
(i) $x \in m_T$.
(ii) (a) Given any $y \in \mathbb{H}_1$, $\langle x, y\rangle = 0$ implies that $\langle Tx, Ty\rangle = 0$,
(b) $\inf\{\|Ty\| : \|y\| = 1, \langle x, y\rangle = 0\} \geq \|Tx\|$.
(iii) $\langle Tx, Ty\rangle = m^2(T)\langle x, y\rangle$, for every $y \in \mathbb{H}_1$.

Analogous to Theorem 3.3.5, we get the following theorem related to minimum norm attainment set of an operator defined between Hilbert spaces.

Theorem 3.3.8 ([127, Theorem 2.5]) *Let $\mathbb{H}_1, \mathbb{H}_2$ be Hilbert spaces and let $T \in \mathbb{L}(\mathbb{H}_1, \mathbb{H}_2)$. If $m_T \neq \emptyset$, then m_T is the unit sphere of a subspace. In particular, $m_T = S_H$, where $H = \ker(T^*T - m^2(T)I)$. Moreover, m_T is the unit sphere of eigen space of the least eigenvalue (which is equal to $m^2(T)$) of T^*T.*

Proof First, suppose that $x \in m_T$. Then by Theorem 3.3.7 (iii), $\langle Tx, Ty\rangle = m^2(T)\langle x, y\rangle$, for every $y \in \mathbb{H}_1$. In other words $\langle T^*Tx, y\rangle = m^2(T)\langle x, y\rangle$. In particular this is true for $y = T^*Tx$, i.e.,

$$\|T^*Tx\|^2 = \langle T^*Tx, T^*Tx\rangle = m^2(T)\langle x, T^*Tx\rangle = m^2(T)\langle Tx, Tx\rangle = m^4(T).$$

Thus,

$$\|T^*Tx\| = m^2(T) = \|Tx\|^2 = \langle Tx, Tx\rangle = \langle T^*Tx, x\rangle \leq \|T^*Tx\|\|x\| = m^2(T).$$

Now, the equality condition of Cauchy-Schwarz inequality yields that $T^*Tx = m^2(T)x$, i.e., $x \in \ker(T^*T - m^2(T)I)$.
Conversely, suppose that $x \in \ker(T^*T - m^2(T)I)$ and $\|x\| = 1$. Then $T^*Tx = m^2(T)x$, which implies that $\|Tx\|^2 = \langle T^*Tx, x\rangle = m^2(T)$, i.e., $x \in m_T$. This completes the proof of the first part.
From the proof of the first part, it is clear that $\|T\|^2$ is an eigenvalue of T^*T. Now, if $x \in m_T$ and λ is an eigenvalue of T^*T corresponding to the eigenvector z with $\|z\| = 1$, then

$$m^2(T) = \|Tx\|^2 \leq \|Tz\|^2 = \langle T^*Tz, z\rangle = \langle \lambda z, z\rangle = \lambda.$$

This completes the proof. □

3.4 Relation Between the Norm Attainment Set and the Minimum Norm Attainment Set of an Operator

In this section, we study the relation between the norm attainment set and the minimum norm attainment set of a bounded linear operator between Banach spaces (Hilbert spaces). First, we consider the operators between Hilbert spaces. Recall from the previous section that, for operators between Hilbert spaces, both the norm attainment set and minimum norm attainment set are unit sphere of subspaces, if they are non-empty. The next theorem ensures that these subspaces are either equal or orthogonal.

Theorem 3.4.1 ([127, Theorem 3.1]) *Let $\mathbb{H}_1, \mathbb{H}_2$ be Hilbert spaces and let $T \in \mathbb{L}(\mathbb{H}_1, \mathbb{H}_2)$ be such that T is not a scalar multiple of an isometry. Then $m_T \subseteq (M_T)^\perp$, provided both M_T and m_T are non-empty.*

Proof Observe that T is not a scalar multiple of an isometry and so we must have $\|T\| > m(T)$. Let $y \in m_T$ and $x \in M_T$. Then there exists a unique hyperspace H_x such that $x \perp_B H_x$. We can write $y = \alpha x + h$, where $h \in H_x$ and α is a scalar. If $\alpha = 0$ then clearly $y = h \in (M_T)^{\perp}$. If possible, suppose that $\alpha \neq 0$. Now, $1 = \|y\|^2 = \langle \alpha x + h, \alpha x + h \rangle = |\alpha|^2 + \|h\|^2$, since $\langle x, h \rangle = 0$. Moreover, from Proposition 3.2.10, it follows that $\langle Tx, Th \rangle = 0$. Now,

$$\begin{aligned}\|Ty\|^2 &= \langle \alpha Tx + Th, \alpha Tx + Th \rangle \\ &= |\alpha|^2 \|Tx\|^2 + \|Th\|^2 \\ &= |\alpha|^2 \|T\|^2 + \|h\|^2 \|T\left(\frac{h}{\|h\|}\right)\|^2 \\ &> |\alpha|^2 m^2(T) + \|h\|^2 m^2(T) \\ &= (|\alpha|^2 + \|h\|^2) m^2(T) = m^2(T).\end{aligned}$$

This is a contradiction as $y \in m_T$. Therefore, $\alpha = 0$. Thus, for each $x \in M_T$, we get $\langle y, x \rangle = 0$ and so $y \in (M_T)^{\perp}$. As $y \in m_T$ was chosen arbitrarily, this completes the proof of the theorem. □

Note that if a linear operator is a scalar multiple of an isometry, then it attains its norm at each point on the unit sphere. Therefore, from Theorem 3.4.1, we can conclude that if $T \in \mathbb{L}(\mathbb{H}_1, \mathbb{H}_2)$, where $\mathbb{H}_1, \mathbb{H}_2$ are finite-dimensional Hilbert spaces, then either $M_T = m_T = S_{\mathbb{H}_1}$ or $M_T \perp_B m_T$ and $m_T \perp_B M_T$. However, the following example indicates that this is not true in general for operators between Banach spaces.

Example 3.4.2 Consider $(\mathbb{R}^2, \|.\|)$, whose unit sphere is given by the regular hexagon with vertices at $\pm(1, 0), \pm(\frac{1}{2}, \frac{\sqrt{3}}{2}), \pm(-\frac{1}{2}, \frac{\sqrt{3}}{2})$. It is easy to observe that B–J orthogonality is symmetric for this Banach space, though it is not an inner product space. Consider the linear operator $T = \begin{pmatrix} \frac{3}{4} & -\frac{\sqrt{3}}{4} \\ \frac{\sqrt{3}}{4} & \frac{3}{4} \end{pmatrix}$. It follows immediately that $\|T\| = 1$ and $m(T) = \frac{3}{4}$. It is also easy to check that $\pm(1, 0), \pm(\frac{1}{2}, \frac{\sqrt{3}}{2}), \pm(-\frac{1}{2}, \frac{\sqrt{3}}{2}) \in M_T$ and $\pm(\frac{3}{4}, \frac{\sqrt{3}}{4}), \pm(0, \frac{\sqrt{3}}{2}), \pm(-\frac{3}{4}, \frac{\sqrt{3}}{4}) \in m_T$. Therefore, in this case, $M_T \not\perp_B m_T$.

Our next goal is to generalize Theorem 3.4.1 in the setting of Banach spaces. Towards this, we first consider rank one linear operators on a reflexive Banach space.

Theorem 3.4.3 ([127, Theorem 3.2]) *Let $\mathbb{X}$ be a reflexive Banach space and let $\mathbb{Y}$ be any Banach space. Let $T \in \mathbb{L}(\mathbb{X}, \mathbb{Y})$ be a rank one linear operator. Then $x \in M_T$ for some $x \in S_{\mathbb{X}}$ and $m_T = H_x \bigcap S_{\mathbb{X}}$, where H_x is a hyperspace of $\mathbb{X}$ such that $x \perp_B H_x$. In addition, if $\mathbb{X}$ is strictly convex, then $M_T = \{\alpha x : |\alpha| = 1\}$.*

Proof Without loss of generality, we assume that $\|T\| = 1$. Since T is a rank one operator on a reflexive space, T attains its norm at some element $x \in S_{\mathbb{X}}$. Let $y = Tx$. Then by Proposition 3.2.10, there exist hyperspaces H_x and H_y in $\mathbb{X}$ and

$\mathbb{Y}$, respectively, such that $x \perp_B H_x$, $y \perp_B H_y$ and $T(H_x) \subset H_y$. We further note that $Tx \neq 0$. We claim that $Tz = 0$ for all $z \in H_x$. If possible let $z \in H_x$ be such that $Tz \neq 0$. Now, $x \in M_T$ and $Tx \perp_B Tz$ imply that $\{Tx, Tz\}$ is linearly independent, contradicting the fact that T is of rank one. Thus, $Tz = 0$ for all $z \in H_x$ and so $H_x \cap S_{\mathbb{X}} \subset m_T$. Next, let $z \in m_T$. Then z can be expressed as $z = \alpha x + h$ for some scalar α and $h \in H_x$. Clearly, $0 = Tz = \alpha Tx + Th = \alpha Tx$, so that $\alpha = 0$ and hence $z = h \in H_x \cap S_{\mathbb{X}}$. Thus, $m_T \subset H_x \cap S_{\mathbb{X}}$. This proves that $m_T = H_x \bigcap S_{\mathbb{X}}$, which completes the proof of the first part of the theorem.

Next, assume that $\mathbb{X}$ is strictly convex. We show that $M_T = \{\alpha x : |\alpha| = 1\}$. Let $w \in S_{\mathbb{X}}$. Clearly, w can be written as $w = \alpha x + h$, for some scalar α and some $h \in H_x$. Since $\mathbb{X}$ is strictly convex and $x \perp_B h$, we have $1 = \|w\| = \|\alpha x + h\| \geq |\alpha|$ and $|\alpha| = 1$ if and only if $h = 0$. Now, $\|Tw\| = \|T(\alpha x + h)\| = |\alpha|\|Tx\| = |\alpha| \leq 1$ and equality holds if and only if $h = 0$. Therefore, $M_T = \{\alpha x : |\alpha| = 1\}$. This completes the proof of the theorem. □

The following theorem gives a characterization of reflexive Banach space in terms of the norm attainment set and the minimum norm attainment set of a bounded linear operator.

Theorem 3.4.4 ([127, Theorem 3.3]) *Let $\mathbb{X}$ be a Banach space. Then $\mathbb{X}$ is reflexive if and only if for any closed hyperspace H of $\mathbb{X}$, there exists a rank one linear operator $T \in \mathbb{L}(\mathbb{X})$ such that*
(i) $x \in M_T$, for some $x \in S_{\mathbb{X}}$.
(ii) $m_T = H \bigcap S_{\mathbb{X}}$.

Proof We first prove the necessary part. Let H be a closed hyperspace of $\mathbb{X}$. Then, from Theorem 2.4.1, it follows that there exists a unit vector x such that $x \perp_B H$. Let $z \in \mathbb{X}$. Then $z = \alpha x + h$, where α is a scalar and $h \in H$. Let us define $T : \mathbb{X} \to \mathbb{X}$ as $T(\alpha x + h) = \alpha y$, where $y \in S_{\mathbb{X}}$ is fixed. Clearly, T is a rank one linear operator. Since $x \perp_B H$, it is easy to see that T is bounded and $x \in M_T$. So from Theorem 3.4.3, it follows that $m_T = H \bigcap S_{\mathbb{X}}$. This completes the proof of the necessary part. We next prove the sufficient part. Let H be a closed hyperspace of $\mathbb{X}$. By the hypothesis, there exists a rank one linear operator $T \in \mathbb{L}(\mathbb{X})$ such that
(i) $x \in M_T$, for some $x \in S_{\mathbb{X}}$.
(ii) $m_T = H \bigcap S_{\mathbb{X}}$.
Since T is a rank one linear operator $x \in M_T$ we get $m(T) = 0$. Since $m_T = H \bigcap S_{\mathbb{X}}$, it follows that $Th = 0$ for all $h \in H$. Thus, we get $x \in M_T$ and $Tx \perp_B Th$ for all $h \in H$. Applying Proposition 3.2.3, we get $x \perp_B h$ for all $h \in H$, i.e., $x \perp_B H$. Thus, for each closed hyperspace H of $\mathbb{X}$, there exists an element $x \in S_{\mathbb{X}}$ such that $x \perp_B H$. Therefore, it follows from Theorem 2.4.1 that $\mathbb{X}$ is reflexive. This completes the proof of the sufficient part. □

In addition to the above Theorem 3.4.4, if we further assume that the space is strictly convex, then the characterization of reflexivity of the space assumes the following form.

Theorem 3.4.5 ([127, Theorem 3.4]) *Let $\mathbb{X}$ be a strictly convex Banach space. Then $\mathbb{X}$ is reflexive if and only if for any closed hyperspace H of $\mathbb{X}$, there exists a rank one linear operator $T \in \mathbb{L}(\mathbb{X})$ such that*
(i) $M_T = \{\alpha x : |\alpha| = 1\}$, for some $x \in S_{\mathbb{X}}$.
(ii) $m_T = H \bigcap S_{\mathbb{X}}$.

Next, we characterize two-dimensional real Hilbert spaces among strictly convex Banach spaces.

Theorem 3.4.6 ([127, Theorem 3.5]) *A two-dimensional strictly convex real Banach space $\mathbb{X}$ is a Hilbert space if and only if for any $T \in \mathbb{L}(\mathbb{X})$, either (i) or (ii) holds:*
(i) $M_T = m_T = S_{\mathbb{X}}$.
(ii) $M_T \perp_B m_T$ and $m_T \perp_B M_T$.

Proof We first prove the necessary part. Let $\mathbb{X}$ be a two-dimensional Hilbert space. Let $T \in \mathbb{L}(\mathbb{X})$. If T is a scalar multiple of an isometry then clearly $M_T = m_T = S_X$ and if T is not a scalar multiple of an isometry then from Theorem 3.4.1, it follows that $M_T \perp_B m_T$ and $m_T \perp_B M_T$. This completes the proof of the necessary part. We next prove the sufficient part. We first claim that, for any $T \in L(\mathbb{X})$, $M_T = \pm D$, where D is a connected subset of $S_{\mathbb{X}}$. If (i) holds, i.e., $M_T = S_{\mathbb{X}}$, then the claim is trivially true. Next, suppose (ii) holds. We show that $M_T = \{\pm x\}$ for some $x \in S_{\mathbb{X}}$. If possible, suppose that $x, y \in M_T$, where $y \neq \pm x$. Let $z \in m_T$. Clearly, $z \neq \pm x, \ \pm y$, as T is not a scalar multiple of an isometry. Therefore, z can be written as $z = \alpha x + \beta y$, where α, β are non-zero scalars. By (ii), we have $x \perp_B z$ and $y \perp_B z$. Now, $\mathbb{X}$ being a two-dimensional strictly convex Banach space, by applying Theorem 2.2.3 and using the homogeneity property of B–J orthogonality, we get $\alpha x + \beta y \perp_B z$. Thus, $z \perp_B z$, which is possible only if $z = 0$. However, this clearly contradicts that $z \in m_T \subset S_{\mathbb{X}}$. This completes the proof of the fact that if (ii) holds then T must attain norm only at one pair of points. Therefore, in any case, $M_T = \pm D$, where D is a connected subset of $S_{\mathbb{X}}$. It now follows from [120, Theorem 2.2] that $\mathbb{X}$ is a Hilbert space. This completes the proof of the sufficient part. □

If we assume that the dimension of the space is strictly greater than two, then an analogous characterization of Hilbert space can be obtained without assuming the strict convexity of the space. Moreover, the result is true for real as well as complex space.

Theorem 3.4.7 ([127, Theorem 3.6]) *Let $\mathbb{X}$ be a finite-dimensional Banach space having dimension strictly greater than 2. Then $\mathbb{X}$ is a Hilbert space if and only if for any $T \in \mathbb{L}(\mathbb{X})$, either (i) or (ii) holds:*
(i) $M_T = m_T = S_{\mathbb{X}}$.
(ii) $M_T \perp_B m_T$ and $m_T \perp_B M_T$.

Proof The proof of the necessary part follows similarly as in Theorem 3.4.6. We only prove the sufficient part. We claim that B–J orthogonality is symmetric in $\mathbb{X}$. Let $x, y \in S_{\mathbb{X}}$ be such that $x \perp_B y$. Then there exists a hyperplane H containing y such that $x \perp_B H$. Clearly, any element $z \in \mathbb{X}$ can be written as $z = \alpha x + h$, where α is a scalar $h \in H$. Define a linear operator $T : \mathbb{X} \to \mathbb{X}$ as follows:

$$T(\alpha x + h) = \alpha x, \text{ for each scalar } \alpha \text{ and for each } h \in H.$$

Then T is a bounded linear operator on $\mathbb{X}$. Since $x \perp_B H$, it is easy to see that $x \in M_T$ and $y \in m_T$. Clearly, $M_T \neq S_{\mathbb{X}}$. Therefore, (ii) holds. So, we have $y \perp_B x$. Thus, $x, y \in S_{\mathbb{X}}$ and $x \perp_B y$ imply that $y \perp_B x$. Thus, using the homogeneity property of B–J orthogonality we infer that B–J orthogonality is symmetric in $\mathbb{X}$. Since $\dim(\mathbb{X}) > 2$, it follows from [60] that $\mathbb{X}$ is a Hilbert space. This completes the proof. □

3.5 Norm Attainment Sets (Minimal Norm Attainment Sets) of Operators on ℓ_p Spaces

From Sect. 3.3, it follows that the norm attainment set of an operator on a Hilbert space assumes a nice form. In fact, in this case, the norm attainment set is the unit sphere of a subspace of the domain space. However, the norm attainment set of an operator on arbitrary Banach space behaves differently. Note that, if the unit sphere of a Banach space contains a straight line segment, then it is easy to construct an operator on that space for which the straight line segment is a part of the norm attainment set. For example, suppose that in a Banach space $\mathbb{X}$, there exist $x, y \in S_{\mathbb{X}}$ such that $x \neq y$ and $tx + (1-t)y \in S_{\mathbb{X}}$ for all $t \in [0, 1]$. Suppose $z = \frac{1}{2}(x + y)$ and H is a hyperspace in $\mathbb{X}$ such that $z \perp_B H$. Define $T : \mathbb{X} \to \mathbb{X}$ by $T(az + h) = az$ for all scalars a and $h \in H$. Now, it is straightforward to check that T is a bounded linear operator and $\{tx + (1-t)y : t \in [0, 1]\} \subseteq M_T$. On the other hand, from the following theorem proved in [127], we see that if an operator is defined on ℓ_p^2 $(p \in \mathbb{N} \setminus \{1\})$, then the norm attainment set is either $S_{\ell_p^2}$ or it contains countably many points.

Theorem 3.5.1 ([127, Theorem 2.3]) *Let* $\mathbb{X} = \ell_p^2(\mathbb{R})$ $(p \in \mathbb{N} \setminus \{1\})$ *and let* $T \in \mathbb{L}(\mathbb{X})$ *be such that* T *is not a scalar multiple of an isometry. Then* $|M_T| \leq 4(4p - 3)$.

Very recently, in [89], Theorem 3.5.1 is improved and the optimal upper bound for $|M_T|$ is obtained, where $T \in \mathbb{L}(\ell_p^2(\mathbb{R}))$ $(1 < p < \infty)$. In the next few theorems, we obtain the optimal upper bound for $|M_T|$ and $|m_T|$, where $T \in \mathbb{L}(\ell_p^2(\mathbb{R}))$ $(1 < p < \infty, p \neq 2)$. First, we consider the case $T = \begin{pmatrix} a & b \\ c & d \end{pmatrix}$, where at least one of a, b, c, d is zero.

Theorem 3.5.2 ([89, Theorem 2.3]) *Let* $\mathbb{X} = \ell_p^2(\mathbb{R})$ $(1 < p < \infty)$. *Let* $T \in \mathbb{L}(\mathbb{X})$ *be such that* T *is not a scalar multiple of an isometry. Suppose at least one of the four elements of the corresponding matrix of* T *with respect to standard ordered basis is zero. Then* $|M_T| = |m_T| = 2$.

Proof We provide the details of the proof that $|m_T| = 2$. The proof of $|M_T| = 2$ is similar. Suppose with respect to the standard ordered basis of ℓ_p^2, the matrix representation of T is $\begin{pmatrix} a & b \\ c & d \end{pmatrix}$. We prove the theorem considering $b = 0$. Other cases can be proved similarly. Note that if $d = 0$, then $m_T = \{\pm(0, 1)\}$, i.e., $|m_T| = 2$. If either $a = 0$ or $c = 0$, then it can be easily checked that $|m_T| = 2$. Therefore, we assume that $acd \neq 0$. Thus, T is invertible, whence $m(T) > 0$. Without any loss of generality, we may assume that $m(T) = 1$. The unique semi-inner-product $[\ ,\ ]_p$ on ℓ_p^2 is defined as follows :

$$[(x_1, x_2), (y_1, y_2)]_p = \frac{x_1 y_1 |y_1|^{p-2} + x_2 y_2 |y_2|^{p-2}}{\|(y_1, y_2)\|_p^{p-2}}.$$

Suppose $(x, y) \in m_T$. Since $m(T) = 1$, from Theorem 3.3.6 we get for any $(u, v) \in \mathbb{X}$, $[T(u, v), T(x, y)]_p = [(u, v), (x, y)]_p$. Particularly, this holds for $(u, v) = (d, -c)$ and hence we have

$$a^2 dx|ax|^{p-2} + cy|y|^{p-2} = dx|x|^{p-2}. \tag{3.5.1}$$

Now, using the fact that if $(x, y) \in m_T$ and $(x, y) \perp_B (u, v)$ then $T(x, y) \perp_B T(u, v)$, we conclude that neither $\pm(1, 0)$ nor $\pm(0, 1)$ can be in m_T. Since $(x, y) \notin \{\pm(1, 0), \pm(0, 1)\}$, we get $y = kx$ for some $k \in \mathbb{R} \setminus \{0\}$. Therefore, from Eq. (3.5.1), we have

$$k|k|^{p-2} = \frac{d}{c}(1 - |a|^p). \tag{3.5.2}$$

Note that if $|a| = 1$, then $\|T(-d, c)\| < \|(-d, c)\|$, which contradicts that $m(T) = 1$. Hence, $|a| \neq 1$, and so $\frac{d}{c}(1 - |a|^p) \neq 0$. For $\frac{d}{c}(1 - |a|^p) > 0$, we have exactly one real positive solution for k and for $\frac{d}{c}(1 - |a|^p) < 0$, we have exactly one real negative solution for k. Therefore, we have 2 possibilities for (x, y). Hence, $|m_T| = 2$, completing the proof. □

Next, we consider the case $T = \begin{pmatrix} a & b \\ c & d \end{pmatrix}$, where either $a, b, c, d > 0$ or exactly two of a, b, c, d are negative and the other two are positive. The following easy lemma is used in the next theorem.

Lemma 3.5.3 ([89, Lemma 2.4]) *Let $\mathbb{X}$ be a Banach space, $T \in \mathbb{L}(\mathbb{X})$ and J an onto isometry. Then $M_{JT} = M_T$, $M_{TJ} = J^{-1}(M_T)$, $m_{JT} = m_T$ and $m_{TJ} = J^{-1}(m_T)$. Therefore, $|M_{TJ}| = |M_T| = |M_{JT}|$ and $|m_{TJ}| = |m_T| = |m_{JT}|$.*

Theorem 3.5.4 ([89, Theorem 2.5]) *Let $\mathbb{X} = \ell_p^2(\mathbb{R})$ $(1 < p < \infty)$. Let $T \in \mathbb{L}(\mathbb{X})$ be such that $T = \begin{pmatrix} a & b \\ c & d \end{pmatrix}$, where either $a, b, c, d > 0$ or exactly two of a, b, c, d are negative and the other two are positive. Then $|M_T| = |m_T| = 2$.*

Proof We only prove that $|M_T| = 2$. Using similar arguments it can be proved that $|m_T| = 2$. Without any loss of generality, we may assume that $\|T\| = 1$. We only consider the case $a, b, c, d > 0$. The other cases follow as immediate consequences of this case and Lemma 3.5.3. Therefore, consider that $a, b, c, d > 0$. First, we prove that neither $\pm(1, 0)$ nor $\pm(0, 1)$ can be in M_T. If possible, assume that $(1, 0) \in M_T$. Since $(1, 0)\perp_B(0, 1)$, we get $(a, c)\perp_B(b, d)$, a contradiction, since a, b, c, d are positive. Hence, $\pm(1, 0) \notin M_T$. Using similar arguments, we can prove that $\pm(0, 1) \notin M_T$. On the other hand, since a, b, c, d are all positive, T can attain its norm only at elements of the form (x, y) for which $xy > 0$. Recall that, the unique semi-inner-product $[\ ,\ ]_p$ on ℓ_p^2 is defined as follows :

$$[(x_1, x_2), (y_1, y_2)]_p = \frac{x_1 y_1 |y_1|^{p-2} + x_2 y_2 |y_2|^{p-2}}{\|(y_1, y_2)\|_p^{p-2}}.$$

Since $(x, y) \in M_T$, from Theorem 3.3.2, we get $[T(u, v), T(x, y)]_p = \|T\|^2[(u, v), (x, y)]_p$ for each $(u, v) \in \mathbb{X}$. Particularly, this holds for $(u, v) = (d, -c)$ and hence we have

$$(ad - bc)(ax + by)|ax + by|^{p-2} = dx|x|^{p-2} - cy|y|^{p-2}. \tag{3.5.3}$$

Suppose $x, y > 0$. Now from Eq. (3.5.3), we get

$$(ad - bc)(ax + by)^{p-1} = dx^{p-1} - cy^{p-1}. \tag{3.5.4}$$

First, consider $det(T) > 0$, i.e., $(ad - bc) > 0$. Since $x, y > 0$, we have $y = kx$, for some $k > 0$. Therefore, from Eq. (3.5.4), it follows that

$$(ad - bc)(a + bk)^{p-1} + ck^{p-1} - d = 0. \tag{3.5.5}$$

Consider $f : [0, \infty) \longrightarrow \mathbb{R}$ defined as

$$f(k) = (ad - bc)(a + bk)^{p-1} + ck^{p-1} - d.$$

Since $a, b, c, d > 0$ and $1 < p < \infty$, we get that f is a strictly increasing function. Since $\|T\| = 1$, we have $0 < a, b, c, d < 1$. Hence, $(ad - bc)a^{p-1} < ad - bc < ad < d$ and so $(ad - bc)a^{p-1} - d < 0$. Therefore, $f(0) < 0$ and for large values of k, f takes on positive values. Thus, f is strictly increasing, continuous and takes on both negative and positive values. Hence, $f(k) = 0$ has a unique real solution.

When $det(T) < 0$, the operator $S = \begin{pmatrix} a & b \\ c & d \end{pmatrix}\begin{pmatrix} 0 & 1 \\ 1 & 0 \end{pmatrix} = \begin{pmatrix} b & a \\ d & c \end{pmatrix}$ satisfies $det(S) = -det(T) > 0$. Now, $\begin{pmatrix} 0 & 1 \\ 1 & 0 \end{pmatrix}$ is an isometry on $\mathbb{X}$. Therefore, by Lemma 3.5.3, we have $|M_S| = |M_T| = 2$.

Finally, consider $det(T) = 0$, i.e., $ad - bc = 0$. Since $x, y > 0$, we get $y = kx$ for some $k > 0$. Accordingly Eq. (3.5.4) implies that $d - ck^{p-1} = 0$. Hence, k has a unique positive real solution. Thus, in each possible case, we have $|M_T| = 2$. $\square$

In the following theorem, we obtain a bound of $|M_T|$ and $|m_T|$ for $T = \begin{pmatrix} a & b \\ c & d \end{pmatrix}$, where $abcd \neq 0$ and exactly three of a, b, c, d are of same sign. The next two lemmas are used to achieve the goal.

Lemma 3.5.5 ([89, Lemma 2.6]) *Suppose $\mathbb{X}$, $\mathbb{Y}$ are smooth, strictly convex, reflexive Banach spaces. Let $T \in \mathbb{L}(\mathbb{X}, \mathbb{Y})$ be invertible. Then $|m_T| = |m_{T^*}|$.*

Proof We first prove that $m(T) = m(T^*)$. Let $y_n^* \in S_{\mathbb{Y}^*}$ be such that $\|T^* y_n^*\| \to m_{T^*}$. Since $\mathbb{Y}$ is strictly convex, there exists unique $y_n \in S_{\mathbb{Y}}$ such that $y_n^*(y_n) = 1$. Now,

$$\begin{aligned}
\|T(T^{-1}y_n)\| = \|y_n\| = 1 &= y_n^*(y_n) \\
&= y_n^*(T(T^{-1}y_n)) \\
&= T^* y_n^*(T^{-1}y_n) \\
&\leq \|T^* y_n^*\| \|T^{-1}y_n\| \\
\Rightarrow m(T) \leq \|T\Big(\frac{T^{-1}y_n}{\|T^{-1}y_n\|}\Big)\| &\leq \|T^* y_n^*\| \\
&\Rightarrow m(T) \leq m(T^*) \text{ (taking limit } n \to \infty).
\end{aligned}$$

Similarly, we have $m(T^*) \leq m(T)$. Thus, $m(T) = m(T^*)$. First, assume that $m_{T^*} \neq \emptyset$ and $y^* \in m_{T^*}$. Suppose y is the unique vector in $S_{\mathbb{Y}}$ such that $y^*(y) = 1$. Using similar arguments, we have

$$m(T) \leq \|T\Big(\frac{T^{-1}y}{\|T^{-1}y\|}\Big)\| \leq \|T^* y^*\| = m(T^*) = m(T),$$

i.e., $\frac{T^{-1}y}{\|T^{-1}y\|} \in m_T$. Define $g : m_{T^*} \to m_T$ by $g(y^*) = \frac{T^{-1}y}{\|T^{-1}y\|}$. We prove that g is one to one. Suppose $y_1^*, y_2^* \in m_{T^*}$ are such that $g(y_1^*) = g(y_2^*)$. Let for $i = 1, 2$, $y_i \in S_{\mathbb{Y}}$ be such that $y_i^*(y_i) = 1$. Then

$$\begin{aligned}
g(y_1^*) &= g(y_2^*) \\
\Rightarrow \frac{T^{-1}y_1}{\|T^{-1}y_1\|} &= \frac{T^{-1}y_2}{\|T^{-1}y_2\|} \\
\Rightarrow T\Big(\frac{T^{-1}y_1}{\|T^{-1}y_1\|}\Big) &= T\Big(\frac{T^{-1}y_2}{\|T^{-1}y_2\|}\Big) \\
\Rightarrow \frac{y_1}{\|T^{-1}y_1\|} &= \frac{y_2}{\|T^{-1}y_2\|} \\
\Rightarrow \|\frac{y_1}{\|T^{-1}y_1\|}\| &= \|\frac{y_2}{\|T^{-1}y_2\|}\| \\
\Rightarrow \|T^{-1}y_1\| = \|T^{-1}y_2\| \ \& \ y_1 &= y_2.
\end{aligned}$$

Hence, $y_1^*(y_1) = 1$ and $y_2^*(y_1) = 1$. Thus, y_1^*, y_2^* are norming linear functionals of y_1. Since $\mathbb{Y}$ is smooth, we get $y_1^* = y_2^*$, i.e., g is one to one. Therefore, $|m_{T^*}| \leq |m_T|$. Similarly, we have $|m_{T^*}| \leq |m_{T^{**}}| = |m_T|$. Hence, if $m_{T^*} \neq \emptyset$, then $|m_{T^*}| = |m_T|$. Now, assume that $m_{T^*} = \emptyset$. If possible, suppose that there exists $x \in m_T$. Suppose $x^* \in S_{\mathbb{X}^*}$ is the unique norming linear functional of x, i.e., $x^*(x) = 1$. Now, we can easily check that $\frac{(T^*)^{-1}x^*}{\|(T^*)^{-1}x^*\|} \in m_{T^*}$. It contradicts that $m_{T^*} = \emptyset$. Therefore, if $m_{T^*} = \emptyset$, then $m_T = \emptyset$. Clearly, in this case $|m_T| = |m_{T^*}|$ holds trivially, completing the proof. □

Lemma 3.5.6 ([89, Lemma 2.7]) *Suppose $\mathbb{X}$, $\mathbb{Y}$ are smooth, strictly convex, reflexive Banach spaces. Let $T \in \mathbb{L}(\mathbb{X}, \mathbb{Y})$ be non-zero. Then $|M_T| = |M_{T^*}|$.*

Proof Without any loss of generality, we may assume that $\|T\| = 1$. First, assume that $M_T \neq \emptyset$. Suppose $x \in M_T$ and y^* is the unique norming linear functional of Tx. Now

$$\|T^*\| \geq \|T^*y^*\| \geq T^*y^*(x) = y^*(Tx) = \|Tx\| = \|T\| = \|T^*\|.$$

Thus, $\|T^*y^*\| = \|T^*\|$, i.e., $y^* \in M_{T^*}$. Define a function $g : M_T \to M_{T^*}$ by $g(x) = y^*$. Let $g(x_1) = g(x_2)$ for some $x_1, x_2 \in M_T$. Then $y_1^* = y_2^*$. Therefore, $y_1^*(Tx_1) = \|Tx_1\| = 1$ and $y_1^*(Tx_2) = \|Tx_2\| = 1$, i.e., $(y_1^* \circ T)(tx_1 + (1-t)x_2) = 1$ for all $t \in [0, 1]$. Thus, $\|tx_1 + (1-t)x_2\| = 1$ for all $t \in [0, 1]$. Since $\mathbb{X}$ is strictly convex, we get $x_1 = x_2$. Hence, g is one to one, i.e., $|M_T| \leq |M_{T^*}|$. Similarly, $|M_{T^*}| \leq |M_{T^{**}}| = |M_T|$ gives that $|M_T| = |M_{T^*}|$.
Now, assume that $M_T = \emptyset$. If possible, let there exist $y^* \in M_{T^*}$. Since $\mathbb{X}$ is strictly convex and reflexive, we get unique $x \in S_{\mathbb{X}}$ such that $T^*y^*(x) = \|T^*y^*\|$. Note that

$$\|T\| \geq \|Tx\| \geq y^*(Tx) = T^*y^*(x) = \|T^*y^*\| = \|T^*\| = \|T\|.$$

Therefore, $\|Tx\| = \|T\|$, i.e., $x \in M_T$, a contradiction. Hence, if $M_T = \emptyset$, then $M_{T^*} = \emptyset$ and in this case we get $|M_T| = |M_{T^*}|$ trivially, completing the proof. □

Theorem 3.5.7 ([89, Theorem 2.8]) *Let $\mathbb{X} = \ell_p^2(\mathbb{R})$ $(1 < p < \infty, p \neq 2)$. Suppose $T \in \mathbb{L}(\mathbb{X})$ is such that $T = \begin{pmatrix} a & b \\ c & d \end{pmatrix}$, where exactly three of a, b, c, d are of same sign. Then*
(I) $|M_T| \leq 4$, (II) $|m_T| \leq 4$.

Proof We only prove the theorem for the case $T = \begin{pmatrix} -a & b \\ c & d \end{pmatrix}$, where $a, b, c, d > 0$. Lemma 3.5.3 can be used to prove the theorem for the remaining cases.

(I) We first show that $|M_T| \leq 4$ considering $1 < p < 2$. Let $\|T\| = 1$. We write $\alpha = ad + bc$, for our convenience. Recall that the unique semi-inner-product $[\ ,\ ]_p$ on ℓ_p^2 is as follows:

$$[(x_1, x_2), (y_1, y_2)]_p = \frac{x_1 y_1 |y_1|^{p-2} + x_2 y_2 |y_2|^{p-2}}{\|(y_1, y_2)\|_p^{p-2}}.$$

Since $(x, y) \in M_T$,

$$[T(u, v), T(x, y)]_p = [(u, v), (x, y)]_p \quad (3.5.6)$$

for each $(u, v) \in \mathbb{X}$. Particularly, for $(u, v) = (-d, c)$ and $(u, v) = (b, a)$, we have the following equations:

$$\alpha(-ax + by)| - ax + by|^{p-2} = -dx|x|^{p-2} + cy|y|^{p-2}, \quad (3.5.7)$$

$$\alpha(cx + dy)|cx + dy|^{p-2} = bx|x|^{p-2} + ay|y|^{p-2}. \quad (3.5.8)$$

First, consider that $x < 0$, $y \geq 0$. Then there exists $k \geq 0$ such that $y = -kx$. For this case, from Eq. (3.5.7), it follows that

$$\alpha(a + bk)^{p-1} = d + ck^{p-1}. \quad (3.5.9)$$

Consider $f : [0, \infty) \longrightarrow \mathbb{R}$ defined as follows

$$f(k) = \frac{\alpha(a + bk)^{p-1}}{d + ck^{p-1}} - 1.$$

Now from

$$f'(k) = \frac{(p-1)\alpha(a + bk)^{p-2}(bdk^{2-p} - ac)}{(d + ck^{p-1})^2 k^{2-p}},$$

we get f is strictly increasing on $\left((\frac{ac}{bd})^{\frac{1}{2-p}}, \infty\right)$ and strictly decreasing on $\left[0, (\frac{ac}{bd})^{\frac{1}{2-p}}\right)$. Thus, f can have at most two solutions and so there are at most two $(x, y) \in M_T$ such that $x < 0$, $y \geq 0$.

Next, consider that $x \geq 0$, $y > 0$. Then there exists $k \geq 0$ such that $x = ky$. For this case, from Eq. (3.5.8), it follows that

$$\alpha(ck + d)^{p-1} = bk^{p-1} + a. \quad (3.5.10)$$

Consider $g : [0, \infty) \longrightarrow \mathbb{R}$ defined as follows

$$g(k) = \frac{\alpha(ck + d)^{p-1}}{bk^{p-1} + a} - 1.$$

Now from

$$g'(k) = \frac{(p-1)\alpha(ck+d)^{p-2}(ack^{2-p} - bd)}{(bk^{p-1}+a)^2 k^{2-p}},$$

we get g is strictly increasing on $\left((\frac{bd}{ac})^{\frac{1}{2-p}}, \infty\right)$ and strictly decreasing on $\left[0, (\frac{bd}{ac})^{\frac{1}{2-p}}\right)$. Thus, g can have at most two solutions and so there are at most two $(x, y) \in M_T$ such that $x \geq 0, y > 0$.
Observe that, if either g or f has no solution, then $|M_T| \leq 4$. Note that

$$\lim_{k\to\infty} f(k) = \frac{\alpha b^{p-1} - c}{c}, \quad \lim_{k\to\infty} g(k) = \frac{\alpha c^{p-1} - b}{b}.$$

Since $M_T \neq \emptyset$, either f or g must have a solution. Thus, at least one of $f(0)$, $\lim_{k\to\infty} f(k)$, $g(0)$, $\lim_{k\to\infty} g(k)$ must be greater than or equal to zero. Without any loss of generality, we may assume that $f(0) \geq 0$.
We now take care of the following cases separately:

$$(A)\ f(0) > 0, \ (B)\ f(0) = 0.$$

(A) Suppose $f(0) > 0$, i.e., $\alpha a^{p-1} - d > 0$. Then $\alpha c^{p-1} - b < 0$. For otherwise, $\|T(1, 0)\| > 1$, which is a contradiction. Hence, $\lim_{k\to\infty} g(k) < 0$.

(i) For $g(0) < 0$, clearly g has no solution and we are done.

(ii) For $g(0) > 0$, i.e., $\alpha d^{p-1} - a > 0$, we have $\alpha b^{p-1} - c < 0$. Since otherwise, $\|T(0, 1)\| > 1$, a contradiction. Therefore, $\lim_{k\to\infty} f(k) < 0$, which implies that f has exactly one solution. Thus, there may be only one $(x, y) \in M_T$ such that $x < 0, y \geq 0$. Now, from $g(0) > 0$ and $\lim_{k\to\infty} g(k) < 0$, we get g has exactly one solution. Therefore, there may be only one $(x, y) \in M_T$ such that $x \geq 0, y > 0$. Hence, in this case, $|M_T| \leq 4$.

(iii) For $g(0) = 0$, from $\lim_{k\to\infty} g(k) < 0$, we get g has only one solution at $k = 0$. Therefore, there may be only one $(x, y) \in M_T$, which is $(0, 1)$, such that $x \geq 0, y > 0$. If $(0, 1) \in M_T$, then, from Eq. (3.5.7), we have $\alpha b^{p-1} - c = 0$, i.e., $\lim_{k\to\infty} f(k) = 0$. Therefore, f has only one solution. Thus, there may exist only one $(x, y) \in M_T$ such that $x < 0, y \geq 0$. If $(0, 1) \notin M_T$, then there does not exist any $(x, y) \in M_T$ such that $x \geq 0, y > 0$. Since f can have at most two solutions, there exist at most two $(x, y) \in M_T$ such that $x < 0, y \geq 0$. Therefore, $|M_T| \leq 4$.

(B) Suppose $f(0) = 0$, i.e., $\alpha a^{p-1} - d = 0$. Then $\alpha c^{p-1} - b \leq 0$, i.e., $\lim_{k\to\infty} g(k) \leq 0$. For otherwise, $\|T(1, 0)\| > 1$, which is a contradiction. If $\lim_{k\to\infty} f(k) > 0$, i.e., if $\alpha b^{p-1} - c > 0$, then taking $S = \begin{pmatrix} -1 & 0 \\ 0 & 1 \end{pmatrix} \begin{pmatrix} -a & b \\ c & d \end{pmatrix} \begin{pmatrix} 0 & 1 \\ 1 & 0 \end{pmatrix}$ $= \begin{pmatrix} -b & a \\ d & c \end{pmatrix}$ and applying (A), we have $|M_S| \leq 4$. Thus, from Lemma 3.5.3, we get $|M_S| = |M_T| \leq 4$. Therefore, we consider $\lim_{k\to\infty} f(k) \leq 0$. Note that, f has exactly one solution. Thus, there can be only one $(x, y) \in M_T$ such that $x < 0, y \geq 0$.

To get the number of $(x, y) \in M_T$ with $x \geq 0, y > 0$, we look for the number of solutions of the function g. Note that, if $g(0) > 0$, i.e., if $\alpha d^{p-1} - a > 0$, then taking $S = \begin{pmatrix} 0 & -1 \\ 1 & 0 \end{pmatrix} \begin{pmatrix} -a & b \\ c & d \end{pmatrix} \begin{pmatrix} 0 & -1 \\ 1 & 0 \end{pmatrix} = \begin{pmatrix} -d & c \\ b & a \end{pmatrix}$ and applying (A), we have $|M_S| \leq 4$, i.e., $|M_T| = |M_S| \leq 4$. Hence, we let $g(0) \leq 0$. Then, from $\lim_{k\to\infty} g(k) \leq 0$, we get that g has at most one solution. Thus, there can be only one $(x, y) \in M_T$ such that $x \geq 0, y > 0$. Therefore, $|M_T| \leq 4$ whenever $1 < p < 2$.

Next, assume that $2 < p < \infty$. Note that T^* is defined on ℓ_q^2, where $\frac{1}{p} + \frac{1}{q} = 1$ and $1 < q < 2$. Thus, $|M_{T^*}| \leq 4$. Using Lemma 3.5.6, we have $|M_T| = |M_{T*}| \leq 4$. Therefore, we get $|M_T| \leq 4$ for $1 < p < \infty$.

(II) Using similar arguments as (I), we can show that if $2 < p < \infty$, then $|m_T| \leq 4$. If $1 < p < 2$, we can use Lemma 3.5.5 to show that $|m_T| \leq 4$. Thus, for each $1 < p < \infty$, we get $|m_T| \leq 4$. □

Example 3.5.8 If we consider the operator $T = \begin{pmatrix} 1 & 1 \\ 1 & -1 \end{pmatrix}$ on $\ell_p^2(\mathbb{R})$ for $1 < p < \infty$, $p \neq 2$, then it can be checked that $|M_T| = |m_T| = 4$.

Now, combining Theorems 3.5.2 to 3.5.7 and Example 3.5.8, we can conclude the next theorem.

Theorem 3.5.9 ([89, Theorem 2.11]) *Suppose* $\mathbb{X} = \ell_p^2(\mathbb{R})$ $(1 < p < \infty)$. *Let* $T \in \mathbb{L}(\mathbb{X})$ *be such that* T *is not a scalar multiple of an isometry. Then* $|M_T| \leq 4$ *and* $|m_T| \leq 4$. *Moreover, these upper bounds are optimal.*

We end this chapter with the following question on operator norm attainment which seems to be unanswered.

Question 1 *Suppose* $\mathbb{X}$ *is a two-dimensional real strictly convex Banach space. Does there exist a linear operator* T *on* $\mathbb{X}$ *which is not a scalar multiple of an isometry such that* M_T *contains* $\left\{\frac{tx+(1-t)y}{\|tx+(1-t)y\|} : t \in [0, 1]\right\}$ *for some* $x, y \in S_{\mathbb{X}}$ *with* $x \neq y$?

The main objective of this chapter, as the reader has surely noticed, is to discuss the (minimal) norm attainment set of a bounded linear operator between Banach spaces. Apart from being important in its own right, we will observe that such optimization problems play fundamental roles in describing the geometry of operators. In the next chapter, we will further illustrate the importance of the operator norm attainment set (and natural generalizations of the same involving norming sequences of unit vectors) in understanding B–J orthogonality of operators. This in turn will help us in the subsequent chapters to discuss finer geometric properties of linear operators, including smoothness and the property of being extreme contractions. Considering all these, the importance of operator norm attainment set and its connections with the geometry of the underlying Banach spaces via semi-inner-products, can be hardly overstated.

Chapter 4
B–J Orthogonality of Operators

4.1 Introduction

Without a shadow of doubt, Birkhoff–James (B–J) orthogonality is the principal tool used in this monograph in studying the geometry of operator spaces. Therefore, understanding B–J orthogonality of operators is of paramount importance to us. In this chapter, we gradually build the theory of characterizing B–J orthogonality of operators between Banach (Hilbert) spaces, up to its fullest generality. We relate B–J orthogonality of operators with that of *certain special elements in the ground space, with which we are already familiar*. Indeed, as we will see, these special elements are nothing but those unit vectors in the domain space at which the concerned operator attains norm. By virtue of the theory developed in the previous chapter, the reader can now legitimately guess the importance of B–J orthogonality in the whole scheme of things. We will observe that two particular topological properties of the operator norm attainment set, namely, compactness and connectedness, play significant roles in determining the B–J orthogonality of operators between Banach spaces. It is elementary to observe that in the finite-dimensional case, the norm attainment set of an operator is trivially a non-empty compact subset of the unit sphere. However, as we have already observed in the previous chapter, it is not necessarily connected (not even the union of two connected subsets of the unit sphere). We will notice that whenever the norm attainment set of an operator is of the desired form with respect to the topological property of connectedness, B–J orthogonality of that operator with any other operator assumes a specially convenient form. Indeed, as will be illustrated in this chapter, the Banach space theoretic origin of the Bhatia–Šemrl-type properties of operators can be traced back to the topological properties of the operator norm attainment set. This observation allows us to gradually develop the theory of B–J orthogonality of operators, starting from the finite-dimensional case, and then proceeding to the more complicated infinite-dimensional case. Of course, in the latter scenario, the norm attainment set of an operator may very well be empty,

A. Mal et al., *Birkhoff–James Orthogonality and Geometry of Operator Spaces*,
Infosys Science Foundation Series, https://doi.org/10.1007/978-981-99-7111-4_4

and that forces us to further generalize the finite-dimensional theory so as to have a complete understanding of B–J orthogonality of operators between Banach spaces.

All the results of Sect. 4.2 hold for real as well as complex Hilbert spaces. It turns out that the interrelation is much more complicated for operators between Banach spaces, in comparison to operators between Hilbert spaces. Moreover, in the former case, in order to highlight the role of the underlying scalar field, we deal with the problems separately in the real case and the complex case. It turns out that in the latter case, a unified treatment for real and complex scalars seems more natural. In Sects. 4.3, 4.4 and 4.8 we work only with real Banach spaces. Section 4.5 is devoted to an analogous study on complex Banach spaces. The results of Sects. 4.6 and 4.7 hold for real as well as complex Banach spaces.

4.2 Orthogonality in $\mathbb{L}(\mathbb{H})$

The problem of characterizing B–J orthogonality of operators ("$T \perp_B A$") on Hilbert spaces was initiated independently by Bhatia and Šemrl [17] and Paul [99], at the near end of the last century. From the perspective of distance of an operator to one-dimensional subspaces of operators, the same (though the term orthogonality was not used) was also studied in [82]. The special cases when either T or A is the identity operator are of particular interest and were studied in [68, 142]. The following theorem which marked the beginning of a remarkably long and fruitful research expedition, turned out to be of particular interest in the whole story.

Theorem 4.2.1 ([17, 99]) *Let $\mathbb{H}$ be an infinite-dimensional Hilbert space. Let $T, A \in \mathbb{L}(\mathbb{H})$. Then $T \perp_B A$ if and only if there exists a sequence $\{x_n\} \subset S_{\mathbb{H}}$ such that $\|Tx_n\| \to \|T\|$ and $\langle Tx_n, Ax_n\rangle \to 0$ hold.*

In the finite-dimensional setting, Theorem 4.2.1 reduces to the following form which is perhaps even more elegant.

Theorem 4.2.2 ([17, Theorem 1.1]) *Let $\mathbb{H}$ be a finite-dimensional Hilbert space. Let $T, A \in \mathbb{L}(\mathbb{H})$. Then $T \perp_B A$ if and only if there exists $x \in M_T$ such that $\langle Tx, Ax\rangle = 0$ holds.*

Theorem 4.2.2, also known as the Bhatia–Šemrl Theorem, has been studied extensively by many authors in the progressively broader setting of C*-algebra and Hilbert C*-module (see [9–11, 13, 19, 152]). In [19] Bhattacharyya and Grover provided another proof of Theorems 4.2.1 and 4.2.2 using the tools of convex analysis. For a geometric approach towards the Bhatia–Šemrl Theorem in complex Banach spaces, we refer to the readers to [113, 114].

Following [122], an operator $T \in \mathbb{L}(\mathbb{H})$ is said to satisfy the Bhatia–Šemrl Property (B–Š Property) if for any $A \in \mathbb{L}(\mathbb{H})$, $T \perp_B A$ implies that there exists $x \in S_{\mathbb{H}}$ such that $Tx \perp_B Ax$. It seems natural to look for a natural extension of this powerful result to the infinite-dimensional setting. Note that if $\mathbb{H}$ is infinite-dimensional, then, for $T \in \mathbb{L}(\mathbb{H})$, M_T may be empty. Even if $M_T \neq \emptyset$, the following example shows that the above theorem can not be extended verbatim to bounded linear operators without any additional restriction on T.

Example 4.2.3 Consider $T \in \mathbb{L}(\ell_2)$ defined by $Te_1 = -e_1$, and $Te_n = (1 - 1/n)e_n$ for $n \geq 2$, where $\{e_n : n \in \mathbb{N}\}$ is the usual orthonormal basis for the Hilbert space ℓ_2. Clearly, T attains norm only at $\pm e_1$. Let $A = I$, the identity operator on ℓ_2. It is easy to check that $T \perp_B A$. Indeed, $\|(T + \lambda A)e_1\| \geq \|T\|$ for all $\lambda \leq 0$, and $\|(T + \lambda A)e_n\| \geq \|T\|$, for all $\lambda \geq 1/n$. But Te_1 is not orthogonal to Ae_1 in the sense of Birkhoff–James.

This shows that we require additional assumptions on the operator T so that it satisfies the B–Š Property. In the following theorem, we obtain a necessary and sufficient condition for a bounded linear operator T on an infinite-dimensional Hilbert space to satisfy B–Š property.

Theorem 4.2.4 ([105, Theorem 3.1]) *Let $T \in \mathbb{L}(\mathbb{H})$. Then, for any $A \in \mathbb{L}(\mathbb{H})$, $T \perp_B A \Leftrightarrow Tx_0 \perp Ax_0$ for some $x_0 \in M_T$ if and only if $M_T = S_{H_0}$, where H_0 is a finite-dimensional subspace of $\mathbb{H}$ and $\|T\|_{H_0^\perp} < \|T\|$.*

Proof Without loss of generality, we may and do assume that $\|T\| = 1$. We first prove the necessary part. From Theorem 3.3.5, it follows that M_T is either empty or is always a unit sphere of some subspace of H_0 of $\mathbb{H}$. Assume $M_T \neq \emptyset$. We show that the subspace H_0 is finite-dimensional. Suppose on the contrary that H_0 is infinite-dimensional. Then we can find a set $\{e_n : n \in \mathbb{N}\}$ of orthonormal vectors in H_0. Extend the set to a complete orthonormal basis $\mathcal{B} = \{e_\alpha : \alpha \in \Lambda \supset \mathbb{N}\}$ of $\mathbb{H}$. For each $e_\alpha \in H_0 \cap \mathcal{B}$, we have

$$\|T^*T\| = \|T\|^2 = \|Te_\alpha\|^2 = \langle T^*Te_\alpha, e_\alpha\rangle \leq \|T^*Te_\alpha\|\|e_\alpha\| \leq \|T^*T\|.$$

This implies that $T^*Te_\alpha = \lambda_\alpha e_\alpha$ for some scalar λ_α. Thus, $\{Te_\alpha : e_\alpha \in H_0 \cap \mathcal{B}\}$ is a set of orthonormal vectors in $\mathbb{H}$. Define $A : \mathcal{B} \longrightarrow \mathbb{H}$ as follows:

$$\begin{aligned} A(e_n) &= \frac{1}{n^2}Te_n,\ n \in \mathbb{N} \\ A(e_\alpha) &= Te_\alpha,\ e_\alpha \in H_0 \cap \mathcal{B} \setminus \{e_n : n \in \mathbb{N}\} \\ A(e_\alpha) &= 0,\ e_\alpha \in \mathcal{B} \setminus H_0 \cap \mathcal{B}. \end{aligned}$$

As $\{Te_\alpha : e_\alpha \in H_0 \cap \mathcal{B}\}$ is a set of orthonormal vectors in $\mathbb{H}$, it is easy to see that A can be extended as a bounded linear operator on $\mathbb{H}$. Now, for any scalar λ,

$$\|T + \lambda A\| \geq \|(T + \lambda A)e_n\| = \|(1 + \frac{\lambda}{n^2})Te_n\| = \left| 1 + \frac{\lambda}{n^2} \right| \|T\| \to \|T\|.$$

Thus, $T \perp_B A$.
We next prove that there exists no $x \in M_T$ such that $Tx \perp_B Ax$. Let $x = \sum_\alpha \langle x, e_\alpha\rangle e_\alpha \in M_T$. Then

$$\langle Tx, Ax\rangle = \sum_n \frac{1}{n^2}|\langle x, e_n\rangle|^2\|T\|^2 + \sum_{\alpha\notin\mathbb{N}} |\langle x, e_\alpha\rangle|^2\|T\|^2$$

and so $\langle Tx, Ax\rangle = 0$ if and only if $x = 0$. Thus, $T\perp_B A$ but there exists no $x \in M_T$ such that $Tx\perp_B Ax$. This contradiction proves that H_0 must be finite-dimensional. We next claim that $\|T\|_{{H_0}^\perp} < \|T\|$. Suppose $\|T\|_{{H_0}^\perp} = \|T\|$. As T does not attain its norm on ${H_0}^\perp$ and $\|T\| = \sup\{\|Tx\| : x \in S_{{H_0}^\perp}\}$, there exists $\{e_n\}$ in $S_{H_0^\perp}$ such that $\|Te_n\| \to \|T\|$. We have $\mathbb{H} = H_0 \oplus {H_0}^\perp$. Define $A : \mathbb{H} \longrightarrow \mathbb{H}$ as follows:

$$Az = Tx, \text{ where } z = x + y, x \in H_0, y \in {H_0}^\perp.$$

Then it is easy to see that A is a bounded linear operator on $\mathbb{H}$. Again, for any scalar λ, $\|T + \lambda A\| \geq \|(T + \lambda A)e_n\| = \|Te_n\|$ holds for each $n \in \mathbb{N}$. Then $\|T + \lambda A\| \geq \|T\|$ for all λ so that $T\perp_B A$. But there exists no $x \in M_T$ such that $\langle Tx, Ax\rangle = 0$. This contradiction completes the proof of the necessary part.

We next prove the sufficient part. If $\langle Tx_0, Ax_0\rangle = 0$ for some $x_0 \in M_T$, then $T\perp_B A$. Next, let $T\perp_B A$. Then, by Theorem 4.2.1, there exists $\{z_n\} \subset S_{\mathbb{H}}$ such that $\|Tz_n\| \to \|T\|$ and $\langle Tz_n, Az_n\rangle \to 0$. For each $n \in \mathbb{N}$, we have $z_n = x_n + y_n$, where $x_n \in H_0$, $y_n \in {H_0}^\perp$. Then $\|z_n\|^2 = 1 = \|x_n\|^2 + \|y_n\|^2$ and so $\|x_n\| \leq 1$, $\forall\, n \in \mathbb{N}$. Now $\{x_n\}$ is a bounded sequence in a finite-dimensional space H_0 and so has a convergent subsequence converging to some element of H_0. Without loss of generality, we may and do assume that $x_n \to x_0$ (say) in H_0 in norm. For each non-zero element $x \in H_0$, we have

$$\|T^*T\|\|x\|^2 \leq \|T\|^2\|x\|^2 = \|Tx\|^2 = \langle T^*Tx, x\rangle \leq \|T^*Tx\|\|x\| \leq \|T^*T\|\|x\|^2$$

and so $\langle T^*Tx, x\rangle = \|T^*Tx\|\|x\|$. Then, by the equality condition of Schwarz's inequality $T^*Tx = \lambda_x x$ for some λ_x. Now $\langle T^*Tx_n, y_n\rangle = \langle T^*Ty_n, x_n\rangle = 0$ and so

$$\begin{aligned}
&\langle T^*Tz_n, z_n\rangle = \langle T^*Tx_n, x_n\rangle + \langle T^*Tx_n, y_n\rangle + \langle T^*Ty_n, x_n\rangle + \langle T^*Ty_n, y_n\rangle \\
&\Rightarrow \lim \|Tz_n\|^2 = \lim \|Tx_n\|^2 + \lim \|Ty_n\|^2 \\
&\Rightarrow \lim \|Ty_n\|^2 = \|T\|^2(1 - \|x_0\|^2) \\
&\Rightarrow \lim \|Ty_n\|^2 = \|T\|^2 \lim \|y_n\|^2.
\end{aligned}$$

By hypothesis $\sup\{\|Ty\| : y \in {H_0}^\perp,\ \|y\| = 1\} < \|T\|$ and, so by the last equality, there does not exist any non-zero subsequence of $\{\|y_n\|\}$. So, we infer that $y_n = 0\,\forall\, n$ and $z_n = x_n$ for all n. Then $\langle Tz_n, Az_n\rangle \to 0 \Rightarrow \langle Tx_0, Ax_0\rangle = 0$. This completes the proof. □

Observe that in a strictly convex Banach space $\mathbb{X}$, for $x, y \in \mathbb{X}$ if $x \perp_B y$, then, for all non-zero scalars λ, $\|x + \lambda y\| > \|x\|$ hold. Following this motivation, the notion of strong B–J orthogonality was introduced in [104]. In an arbitrary Banach space $\mathbb{X}$, x is said to be strongly B–J orthogonal to y, written as $x \perp_{SB} y$, if for any non-zero scalar λ, $\|x + \lambda y\| > \|x\|$ holds. Note that $x \perp_{SB} y$ implies $x \perp_B y$, but the converse is not necessarily true. However, in a Hilbert space $\mathbb{H}$, the notion of B–J orthogonality and strong B–J orthogonality are equivalent, but this is not true in $\mathbb{L}(\mathbb{H})$. The next theorem provides a necessary condition for such equivalence in $\mathbb{L}(\mathbb{H})$. Here we recall that a scalar λ is said to be in the approximate point spectrum of an operator A, written as $\lambda \in \sigma_{app}(A)$, if there exists a sequence $\{x_n\} \subset S_{\mathbb{H}}$ such that $\|(A - \lambda I)x_n\| \to 0$ holds.

Theorem 4.2.5 ([126, Theorem 2.15]) *Let* $\mathbb{H}$ *be a Hilbert space,* $T, A \in \mathbb{L}(\mathbb{H})$ *and* $0 \notin \sigma_{app}(A)$. *Then* $T \perp_B A$ *and* $T \perp_{SB} A$ *are equivalent.*

Proof We only need to show that $T \perp_B A$ implies $T \perp_{SB} A$. If possible suppose that $T \perp_B A$ but $T \not\perp_{SB} A$. Then there exists $\lambda_0 \neq 0$ such that $\|T\| = \|T - \lambda_0 A\|$. Now, for all scalars λ, $\|T - \lambda_0 A\| \leq \|T + \lambda A\| = \|(T - \lambda_0 A) + (\lambda + \lambda_0)A\|$. Hence, $(T - \lambda_0 A) \perp_B A$. Therefore, it follows from Theorem 4.2.1 that there exists $\{x_n\}$ of unit vectors such that $\|(T - \lambda_0 A)x_n\| \to \|T - \lambda_0 A\|$ and $\langle (T - \lambda_0 A)x_n, Ax_n \rangle \to 0$. Now, passing through a subsequence, if necessary, and using the fact that $\|Ax_n\| \not\to 0$ (since $0 \notin \sigma_{app}(A)$), we get

$$\begin{aligned} \|T - \lambda_0 A\|^2 &= \lim_{n\to\infty} \|(T - \lambda_0 A)x_n\|^2 \\ &= \lim_{n\to\infty} \{\|Tx_n\|^2 + |\lambda_0|^2\|Ax_n\|^2 - \lambda_0\langle Ax_n, Tx_n\rangle - \overline{\lambda_0}\langle Tx_n, Ax_n\rangle\} \\ &- \lim_{n\to\infty} \{\|Tx_n\|^2 - |\lambda_0|^2\|Ax_n\|^2] \\ &< \|T\|^2. \end{aligned}$$

Therefore, $\|T - \lambda_0 A\| < \|T\|$, a contradiction. This proves the theorem. □

4.3 Orthogonality in $\mathbb{K}(\mathbb{X}, \mathbb{Y})$

Let us begin with the apparently simplest case of $\mathbb{X}$ being a finite-dimensional Banach space and $T, A \in \mathbb{L}(\mathbb{X})$. If there exists $x \in M_T$ such that $Tx \perp_B Ax$, then it is not difficult to see that $T \perp_B A$ holds. It follows from Theorem 4.2.2, that the converse is also true, whenever $\mathbb{X}$ is a Hilbert space. This tempted Bhatia and Šemrl to conjecture in [17] that $T \perp_B A \Rightarrow Tx \perp_B Ax$ for some $x \in M_T$, for any finite-dimensional Banach space $\mathbb{X}$. However, Li and Schneider presented an example in [75] to negate this conjecture. Let us once again recall from [122] that $T \in \mathbb{L}(\mathbb{X}, \mathbb{Y})$ is said to satisfy the B–Š Property if for any $A \in \mathbb{L}(\mathbb{X}, \mathbb{Y})$, $T \perp_B A$ implies that there exists $x \in M_T$ such that $Tx \perp_B Ax$. In [103], concrete examples are produced to illustrate the existence of operators satisfying as well as not satisfying the B–Š Property in the setting of $\|\cdot\|_\infty$ and $\|\cdot\|_1$ norm. Following similar idea, we can easily give an easy example of an operator not satisfying the B–Š Property.

Example 4.3.1 Consider $T, A \in \mathbb{L}(\ell_\infty^2)$ defined by $T(a, b) = (3a + b, -2a + 2b)$ and $A(a, b) = (a + 3b, 2a - 2b)$ for all $a, b \in \mathbb{R}$. Then $M_T = \{\pm(1, 1), \pm(1, -1)\}$

and $\|T\| = 4$. Now, for all $\lambda \geq 0$, $\|T + \lambda A\| \geq \|(T + \lambda A)(1, 1)\| = \|(4, 0) + \lambda(4, 0)\| \geq \|(4, 0)\| = \|T\|$. On the other hand, for all $\lambda \leq 0$, $\|T + \lambda A\| \geq \|(T + \lambda A)(1, -1)\| = \|(2, -4) + \lambda(-2, 4)\| \geq \|(2, -4)\| = \|T\|$. Therefore, $T \perp_B A$. However, $T(1, 1) \not\perp_B A(1, 1)$ and $T(1, -1) \not\perp_B A(1, -1)$.

Indeed, Benítez et al. proved the following theorem that essentially highlights the difficulty of working in the normed setting, as opposed to the case of Hilbert spaces.

Theorem 4.3.2 ([16]) *A finite-dimensional Banach space $\mathbb{X}$ is an inner product space if and only if for each $T, A \in \mathbb{L}(\mathbb{X})$, $T \perp_B A \Leftrightarrow Tx \perp_B Ax$ for some $x \in M_T$.*

Clearly, Theorem 4.3.2 indicates that to get an analogous result to Theorem 4.2.2 in case of an arbitrary Banach space, we need more restrictions on the operators. The first of such results, in the setting of finite-dimensional Banach spaces, was proved by Sain and Paul [120], motivated by the examples discussed in [103].

Theorem 4.3.3 ([120, Theorem 2.1]) *Let $\mathbb{X}$ be a finite-dimensional Banach space. Let $T \in \mathbb{L}(\mathbb{X})$ be such that T attains its norm at only $\pm D$, where D is a connected subset of $S_{\mathbb{X}}$. Then for $A \in \mathbb{L}(\mathbb{X})$ with $T \perp_B A$, there exists $x \in D$ such that $Tx \perp_B Ax$.*

Proof Suppose on the contrary that there does not exist any $x \in D$ such that $Tx \perp_B Ax$. We now obtain a contradiction in the following three steps to complete the proof of the theorem.
Step 1: In the first step, we show that $D = W_1 \cup W_2$, where

$$W_1 = \{x \in D : \|Tx + \lambda Ax\| > \|T\| \ \forall \ \lambda > 0\},$$

$$W_2 = \{x \in D : \|Tx + \lambda Ax\| > \|T\| \ \forall \ \lambda < 0\}.$$

Let $x_0 \in D$ be arbitrary. Since $Tx_0 \not\perp_B Ax_0$, we can find $0 \neq \lambda_0 \in \mathbb{R}$ such that $\|Tx_0 + \lambda_0 Ax_0\| < \|Tx_0\| = \|T\|$. Let us first assume that $\lambda_0 < 0$.

Now, for $\lambda > 0 \ \exists \ t \in (0, 1)$ such that

$$\begin{aligned}
Tx_0 &= t(Tx_0 + \lambda Ax_0) + (1 - t)(Tx_0 + \lambda_0 Ax_0) \\
\Rightarrow \|Tx_0\| &< t\|(Tx_0 + \lambda Ax_0)\| + (1 - t)\|Tx_0\| \\
\Rightarrow \|Tx_0\| &< \|(Tx_0 + \lambda Ax_0)\|.
\end{aligned}$$

$$\text{Therefore } \|Tx_0 + \lambda Ax_0\| > \|Tx_0\| = \|T\| \ \ \forall \ \lambda > 0.$$

Again, if we assume that $\lambda_0 > 0$ then similarly we can show that

$$\|Tx_0 + \lambda Ax_0\| > \|Tx_0\| = \|T\| \ \ \forall \ \lambda < 0.$$

Thus, for $x \in D$ either $\|Tx + \lambda Ax\| > \|T\| \, \forall \lambda > 0$ or $\|Tx + \lambda Ax\| > \|T\| \, \forall \lambda < 0$ and so $D = W_1 \cup W_2$.

Step 2: In step 2, we prove that $W_1 \neq \emptyset$ and $W_2 \neq \emptyset$.

If possible suppose that $W_1 = \emptyset$ i.e., for all $x \in D$, $\|Tx + \lambda Ax\| > \|Tx\| = \|T\|\ \forall\ \lambda < 0$. Since $Tx \not\perp_B Ax$ there exists $\lambda_0 > 0$ such that $\|Tx + \lambda_0 Ax\| < \|Tx\| = \|T\|$. By the convexity of the norm function we get,

$$\|Tx + \lambda Ax\| < \|Tx\| = \|T\| \quad \forall \quad \lambda \in (0, \lambda_0).$$

Choose λ_x such that $0 < \lambda_x < \min\{\lambda_0, 1\}$. Let us consider the continuous function $g : S_{\mathbb{X}} \times [-1, 1] \longrightarrow \mathbb{R}$ defined as

$$g(x, \lambda) = \|Tx + \lambda Ax\|.$$

We have $g(x, \lambda_x) = \|Tx + \lambda_x Ax\| < \|T\|$ and so by continuity of g there exist $r_x, \delta_x > 0$ such that

$$g(y, \lambda) < \|T\| \quad \forall \quad y \in B(x, r_x) \cap S_{\mathbb{X}} \text{ and } \forall\, \lambda \in (\lambda_x - \delta_x, \lambda_x + \delta_x).$$

Let $y \in B(x, r_x) \cap S_{\mathbb{X}}$. Then for $\lambda \in (0, \lambda_x)$ there exists $t \in (0, 1)$ such that

$$\begin{aligned} &Ty + \lambda Ay = t(Ty) + (1-t)(Ty + \lambda_x Ay) \\ \Rightarrow\ &\|Ty + \lambda Ay\| < t\|Ty\| + (1-t)\|Ty + \lambda_x Ay\| \\ \Rightarrow\ &\|Ty + \lambda Ay\| < \|T\|. \end{aligned}$$

Therefore, $g(y, \lambda) = \|Ty + \lambda Ay\| < \|T\|\ \forall\ y \in B(x, r_x) \cap S_{\mathbb{X}}$ and $\forall \lambda \in (0, \lambda_x)$.

Since $g(x, \lambda) = g(-x, \lambda)$, it follows that $\|Ty + \lambda Ay\| < \|T\|\ \forall\ y \in B(-x, r_x) \cap S_{\mathbb{X}}$ and $\forall \lambda \in (0, \lambda_x)$.

Next, for $z \in S_{\mathbb{X}}$ and $z \notin D \cup (-D)$, we have $g(z, 0) = \|Tz\| < \|T\|$. So by continuity of g there exist open balls $B(z, r_z) \cap S_{\mathbb{X}}$ and $(-\delta_z, \delta_z)$ such that $g(y, \lambda) = \|Ty + \lambda Ay\| < \|T\|\ \forall\ y \in B(z, r_z) \cap S_{\mathbb{X}}$ and $\forall \lambda \in (-\delta_z, \delta_z)$. Consider the open cover

$$\{B(x, r_x) \cap S_{\mathbb{X}},\ B(-x, r_x) \cap S_{\mathbb{X}} : x \in D\} \cup \{B(z, r_z) \cap S_{\mathbb{X}} : z \in S_{\mathbb{X}}, z \notin D \cup (-D)\}$$

of $S_{\mathbb{X}}$. Now, $\mathbb{X}$, being finite-dimensional, $S_{\mathbb{X}}$ is compact and so this cover has a finite subcover of the form

$$S_{\mathbb{X}} \subset \cup_{i=1}^{n_1} B(x_i, r_{x_i}) \cup_{i=1}^{n_1} B(-x_i, r_{x_i}) \cup_{k=1}^{n_2} B(z_k, r_{z_k}) \cap S_{\mathbb{X}},$$

for some positive integers n_1, n_2. Choose $\mu_0 \in (\cap_{i=1}^{n_1} (0, \lambda_{x_i})) \bigcap (\cap_{k=1}^{n_2} (-\delta_{z_k}, \delta_{z_k}))$. Now, $T + \mu_0 A$ being a linear operator on the finite-dimensional space $\mathbb{X}$, attains its norm at some $w_0 \in S_{\mathbb{X}}$. However, it follows from the choice of μ_0 that, $\|T + \mu_0 A\| = \|(T + \mu_0 A)w_0\| < \|T\|$ which contradicts that $T \perp_B A$. Thus, $W_1 \neq \emptyset$. Proceeding similarly we can show that $W_2 \neq \emptyset$.

Step 3: We finally show that W_1, W_2 form a separation of D.

Clearly, $\overline{W_1} \cap W_2 = \emptyset$ and $W_1 \cap \overline{W_2} = \emptyset$, otherwise we can find $x \in D$ such that $Tx \perp_B Ax$. As $D = W_1 \cup W_2$ and $\overline{W_1} \cap W_2 = \emptyset, \ W_1 \cap \overline{W_2} = \emptyset$ so we get a separation of D, this is a contradiction.

Thus, there exists some $x \in D$ such that $Tx \perp_B Ax$. □

Using Theorems 4.3.2 and 4.3.3, one can obtain the following characterization of Euclidean spaces.

Theorem 4.3.4 ([120, Theorem 2.2]) *A finite-dimensional real Banach space $\mathbb{X}$ is a Hilbert space if and only if for any linear operator T on $\mathbb{X}$, M_T is the unit sphere of some subspace of $\mathbb{X}$.*

Proof The necessary part follows from Theorem 3.3.5. For the sufficient part, suppose $\mathbb{X}$ is a finite-dimensional real Banach space such that M_T is the unit sphere of some subspace of $\mathbb{X}$. Then clearly T attains its norm only at $\pm D$ where D is a connected subset of $S_{\mathbb{X}}$. So from Theorem 4.3.3, it follows that given any $T, A \in \mathbb{L}(\mathbb{X})$ with $T \perp_B A$ there exists $x \in S_{\mathbb{X}}$ such that $\|Tx\| = \|T\|$ and $Tx \perp_B Ax$. Applying Theorem 4.3.2 we conclude that $\mathbb{X}$ is a Hilbert space. □

Theorem 4.3.3 exhibits a nice connection between orthogonality of operators and that in the ground space in terms of the norm attainment set. A more general result in this direction is given below.

Theorem 4.3.5 ([105, Lemma 2.1]) *Let $T \in \mathbb{L}(\mathbb{X}, \mathbb{Y})$ and let $M_T = D \cup (-D)$, where D is a non-empty compact connected subset of $S_{\mathbb{X}}$. Then for any $A \in \mathbb{L}(\mathbb{X}, \mathbb{Y})$, either there exists $x \in M_T$ such that $Tx \perp_B Ax$ or there exists $\lambda_0 \neq 0$ such that $\|Tx + \lambda_0 Ax\| < \|Tx\| \ \forall \ x \in M_T$.*

Proof Suppose that there exists no $x \in M_T$ such that $Tx \perp_B Ax$. Let

$$W_1 = \{x \in D : \|Tx + \lambda_x Ax\| < \|T\| \text{ for some } \lambda_x > 0\}$$

$$\text{and } W_2 = \{x \in D : \|Tx + \lambda_x Ax\| < \|T\| \text{ for some } \lambda_x < 0\}.$$

Clearly both W_1, W_2 are open sets in D and $D = W_1 \cup W_2$. The connectedness of D ensures that either $D = W_1$ or $D = W_2$. Consider the case $D = W_1$. Then for each $x \in D$, there exists $\lambda_x \in (0, 1)$ such that $\|Tx + \lambda_x Ax\| < \|Tx\| = \|T\|$. By the convexity of the norm function we get,

$$\|Tx + \lambda Ax\| < \|Tx\| = \|T\| \ \forall \ \lambda \in (0, \lambda_x).$$

Let us consider the continuous function $g : S_{\mathbb{X}} \times [-1, 1] \longrightarrow \mathbb{R}$ defined as

$$g(x, \lambda) = \|Tx + \lambda Ax\|.$$

We have $g(x, \lambda_x) = \|Tx + \lambda_x Ax\| < \|T\|$ and so by the continuity of g, there exist $r_x, \delta_x > 0$ such that $g(y, \lambda) < \|T\| \;\; \forall \; y \in B(x, r_x) \cap S_{\mathbb{X}}$ and $\forall \, \lambda \in (\lambda_x - \delta_x, \lambda_x + \delta_x)$. Let $y \in B(x, r_x) \cap D$. Then for any $\lambda \in (0, \lambda_x)$ we get

$$Ty + \lambda Ay = \Big(1 - \frac{\lambda}{\lambda_x}\Big)Ty + \frac{\lambda}{\lambda_x}(Ty + \lambda_x Ay)$$
$$\Rightarrow \|Ty + \lambda Ay\| < (1 - \frac{\lambda}{\lambda_x})\|T\| + \frac{\lambda}{\lambda_x}\|T\|$$
$$\Rightarrow \|Ty + \lambda Ay\| < \|T\|.$$

Therefore, $g(y, \lambda) < \|T\| \quad \forall \quad y \in B(x, r_x) \cap D$ and $\forall \, \lambda \in (0, \lambda_x)$. Consider the open cover $\{B(x, r_x) \cap D : x \in D\}$ of D. By the compactness of D, this cover has a finite subcover $\{B(x_i, r_{x_i}) \cap D : i = 1, 2, \ldots, n\}$ so that

$$D \subset \cup_{i=1}^{n} B(x_i, r_{x_i}).$$

Choose $\lambda_0 \in \cap_{i=1}^{n}(0, \lambda_{x_i})$. Then for any $x \in M_T$, $\|Tx + \lambda_0 Ax\| < \|T\|$. If $D = W_2$ then proceeding similarly we can find $\lambda_0 < 0$ such that for any $x \in M_T$, $\|Tx + \lambda_0 Ax\| < \|T\|$. □

Remark 4.3.6 The above result is significant in the sense that given $T \in \mathbb{L}(\mathbb{X}, \mathbb{Y})$ with $M_T = D \cup (-D)$ (D is as given above), then, for any arbitrary $A \in \mathbb{L}(\mathbb{X}, \mathbb{Y})$ either of the following always holds
(i) there exists $x \in M_T$ such that $Tx \perp_B Ax$.
(ii) $Ax \in (Tx)^+$ for each $x \in M_T$.
(iii) $Ax \in (Tx)^-$ for each $x \subset M_T$.
This highlights the role of the norm attainment set in the study of B–J orthogonality.

We next provide a complete characterization of $T \perp_B A$ for compact operators on a reflexive Banach space, without any restriction on the norm attainment set M_T.

Theorem 4.3.7 ([126, Theorem 2.1]) *Let $\mathbb{X}$ be a reflexive Banach space and let $\mathbb{Y}$ be any Banach space. Then for any $T, A \in \mathbb{K}(\mathbb{X}, \mathbb{Y})$, $T \perp_B A$ if and only if there exist $x, y \in M_T$ such that $Ax \in (Tx)^+$ and $Ay \in (Ty)^-$.*

Proof We first prove the sufficient part. Since $Ax \in (Tx)^+$, $\|T + \lambda A\| \geq \|Tx + \lambda Ax\| \geq \|Tx\| = \|T\|$ for all $\lambda \geq 0$. Similarly $Ay \in (Ty)^-$ implies that $\|T + \lambda A\| \geq \|Ty + \lambda Ay\| \geq \|Ty\| = \|T\|$ for all $\lambda \leq 0$. This completes the proof of the sufficient part.

We next prove the necessary part. For each $n \in \mathbb{N}$, $T + \frac{1}{n}A$ is a compact linear operator on the reflexive Banach space $\mathbb{X}$ and so there exists $x_n \in S_{\mathbb{X}}$ such that $\|(T + \frac{1}{n}A)x_n\| = \|T + \frac{1}{n}A\|$. Since $\mathbb{X}$ is reflexive, $B_{\mathbb{X}}$ is weakly compact. Therefore, $\{x_n\}$ has a weakly convergent subsequence. Without loss of generality, we assume that $\{x_n\}$ weakly converges to x (say). Since T and A are compact linear operators, $Tx_n \to Tx$ and $Ax_n \to Ax$. Since $T \perp_B A$, $\|T + \frac{1}{n}A\| \geq \|T\|$ for all $n \in \mathbb{N}$. Therefore, for all $n \in \mathbb{N}$,

$$\|\left(T+\frac{1}{n}A\right)x_n\| = \|T+\frac{1}{n}A\| \geq \|T\| \geq \|Tx_n\|.$$

Letting $n \to \infty$, we see that, $\|Tx\| \geq \|T\| \geq \|Tx\|$. This proves that $\|Tx\| = \|T\|$, i.e., $x \in M_T$. Now we show that $Ax \in (Tx)^+$. For any $\lambda \geq \frac{1}{n}$, we claim that $\|Tx_n + \lambda Ax_n\| \geq \|Tx_n\|$. Otherwise, $Tx_n + \frac{1}{n}Ax_n = (1-\frac{1}{n\lambda})Tx_n + \frac{1}{n\lambda}(Tx_n + \lambda Ax_n)$ gives that,

$$\begin{aligned}
\|T+\frac{1}{n}A\| &= \|Tx_n + \frac{1}{n}Ax_n\| \\
&\leq (1-\frac{1}{n\lambda})\|Tx_n\| + \frac{1}{n\lambda}\|(Tx_n + \lambda Ax_n)\| \\
&< (1-\frac{1}{n\lambda})\|Tx_n\| + \frac{1}{n\lambda}\|Tx_n\| \\
&= \|Tx_n\| \leq \|T\|,
\end{aligned}$$

a contradiction. This completes the proof of our claim. Now for any $\lambda > 0$, there exists $n_0 \in \mathbb{N}$ such that $\lambda > \frac{1}{n_0}$. So for all $n \geq n_0$, $\|Tx_n + \lambda Ax_n\| \geq \|Tx_n\|$. Therefore, letting $n \to \infty$, we have $\|Tx + \lambda Ax\| \geq \|Tx\|$. This completes the proof of the fact that $Ax \in (Tx)^+$. Considering the sequence of compact operators $\{T - \frac{1}{n}A\}$ and proceeding similarly we can show that there exists $y \in M_T$ such that $Ay \in (Ty)^-$. This completes the proof. □

Keeping in mind that each finite-dimensional Banach space is reflexive and each bounded linear operator on a finite-dimensional Banach space is compact, we immediately get the following corollary, which was first proved in [135, Theorem 2.2] assuming $\mathbb{X} = \mathbb{Y}$.

Corollary 4.3.8 *Let $\mathbb{X}$, $\mathbb{Y}$ be finite-dimensional Banach spaces. Then for any $T, A \in \mathbb{L}(\mathbb{X}, \mathbb{Y})$, $T \perp_B A$ if and only if there exist $x, y \in M_T$ such that $Ax \in (Tx)^+$ and $Ay \in (Ty)^-$.*

The next theorem deals with the B–Š Property of compact operators on a reflexive Banach space, under some additional condition on the norm attainment set. We need the following easy proposition to get the desired result.

Proposition 4.3.9 ([119, Proposition 2.1]) *Let $\mathbb{X}$ be a Banach space. Let A be a closed subset of $\mathbb{X}$ such that $A = B \cup C$, where B and C are connected. If A is not connected, then B and C are closed.*

Theorem 4.3.10 ([119, Theorem 2.2]) *Let $\mathbb{X}$ be a reflexive Banach space and let $\mathbb{Y}$ be any Banach space. Let $T, A \in \mathbb{K}(\mathbb{X}, \mathbb{Y})$. Suppose either* (i) *or* (ii) *holds.*

(i) *M_T is a connected subset of $S_{\mathbb{X}}$.*

(ii) *M_T is not connected but $M_T = D \cup (-D)$, where D is a non-empty connected subset of $S_{\mathbb{X}}$.*

Then $T \perp_B A$ if and only if there exists $x \in M_T$ such that $Tx \perp_B Ax$.

Proof The sufficient part follows trivially. We only prove the necessary part. Let $T \perp_B A$.

First let (i) be true. Then consider the sets

$$W_1 = \{x \in M_T : Ax \in (Tx)^+\},$$

$$W_2 = \{x \in M_T : Ax \in (Tx)^-\}.$$

Then by Theorem 4.3.7, $W_1 \neq \emptyset,\ W_2 \neq \emptyset$. It is easy to check that W_1 and W_2 are closed. Applying Proposition 3.2.2, we get $M_T = W_1 \cup W_2$. Since M_T is connected, $W_1 \cap W_2 \neq \emptyset$. Let $x \in W_1 \cap W_2$. Then $Ax \in (Tx)^+$ and $Ax \in (Tx)^-$. Therefore, $Tx \perp_B Ax$.

Next, suppose that (ii) is true. Then by Proposition 4.3.9, D is closed. Again considering

$$W_1 = \{x \in D : Ax \in (Tx)^+\},$$

$$W_2 = \{x \in D : Ax \in (Tx)^-\}$$

and proceeding as (i) we can find $x \in D$ such that $Tx \perp_B Ax$. This completes the proof. □

We would like to remark that a result analogous to the previous theorem was obtained by Wójcik in [151] with additional assumptions of smoothness and strict convexity of the range space. Now, applying Theorem 4.3.10, we get the following theorem which characterizes Euclidean spaces among finite-dimensional Banach spaces.

Theorem 4.3.11 ([119, Theorem 2.3]) *Let $\mathbb{X}$ be a finite-dimensional Banach space. Then the following statements are equivalent:*
(i) Given any $T \in \mathbb{L}(\mathbb{X})$, we have M_T is the unit sphere of some subspace of $\mathbb{X}$.
(ii) Given any $T \in \mathbb{L}(\mathbb{X})$, we have $M_T = D_T \cup (-D_T)$, where D_T is a connected subset of $S_{\mathbb{X}}$.
(iii) $\mathbb{X}$ is an Euclidean space.

Proof $(i) \Rightarrow (ii)$ is obvious.

$(ii) \Rightarrow (iii)$. Let $T, A \in \mathbb{L}(\mathbb{X})$ be such that $T \perp_B A$. Since $M_T = D_T \cup (-D_T)$, where D_T is a connected subset of $S_{\mathbb{X}}$, applying Theorem 4.3.10, we can find $x \in M_T$ such that $Tx \perp_B Ax$. Hence, by Theorem 4.3.2, $\mathbb{X}$ is an Euclidean space.

$(iii) \Rightarrow (i)$. Follows from Theorem 3.3.5. □

The converse of Theorem 4.3.10 was proved in [122, Theorem 2.3] for two-dimensional Banach spaces.

Theorem 4.3.12 *Let $\mathbb{X}$ be a two-dimensional Banach space and let $T \in \mathbb{L}(\mathbb{X})$. If M_T has more than two components, then T does not satisfy B–Š property, i.e., there exists a linear operator $A \in \mathbb{L}(\mathbb{X})$ such that $T \perp_B A$ but $Tx \not\perp_B Ax$ for any $x \in M_T$.*

Combining Theorems 4.3.10 and 4.3.12, we immediately get the following characterization of operators satisfying the B–Š Property.

Theorem 4.3.13 ([122, Theorem 2.4]) *Let $\mathbb{X}$ be a two-dimensional Banach space and let $T \in \mathbb{L}(\mathbb{X})$. Then T satisfies the B–Š Property if and only if M_T has at most two components.*

The proof of Theorem 4.3.12 heavily depends on the geometry of the concerned two-dimensional Banach space. We do not know whether Theorem 4.3.12 can be generalized for finite-dimensional Banach spaces of dimension more than two. In view of this, the following conjecture makes sense.

Conjecture 1 Let $\mathbb{X}$ be an n-dimensional ($n > 2$) Banach space and let $T \in \mathbb{L}(\mathbb{X})$. Then T satisfies the B–Š Property if and only if M_T has at most two components.

Recall that in [58], James proved that each bounded linear functional on a reflexive Banach space must attain norm. Therefore, as a consequence of Theorem 4.3.7, we get a characterization of B–J orthogonality in any reflexive dual Banach space.

Theorem 4.3.14 ([126, Theorem 2.13]) *Let $\mathbb{X}$ be a reflexive Banach space. Then for any $f, g \in \mathbb{X}^*$, $f \perp_B g$ if and only if there exist $x, y \in M_f$ such that $f(x).g(x) \geq 0$ and $f(y).g(y) \leq 0$. In addition, if $\mathbb{X}$ is strictly convex then $f \perp_B g$ if and only if there exists $x \in M_f$ such that $f(x).g(x) = 0$.*

4.4 Orthogonality in $\mathbb{L}(\mathbb{X}, \mathbb{Y})$

In this section, our goal is to explore B–J orthogonality in $\mathbb{L}(\mathbb{X}, \mathbb{Y})$. Although arbitrary bounded linear operator on a reflexive Banach space may not attain its norm, Wójcik in [150, Lemma 3.1] proved that if the concerned bounded linear operators is not far from the space of compact operators and if the space of compact linear operators satisfies a special property, then the operator must attain its norm. In this case, [150, Lemma 3.1] also provides a structure of extreme points of $J(T)$, where $J(T)$ is the set of all norming linear functionals of T. To study $Ext\,J(T)$ we need the following two well known theorems, which are worth stating in their own right.

Theorem 4.4.1 ([117, Theorem 1.3]) *Let $\mathbb{X}, \mathbb{Y}$ be Banach spaces over $\mathcal{K}(= \mathbb{R}$ or $\mathbb{C})$. Then*

$$Ext(B_{\mathbb{K}(\mathbb{X},\mathbb{Y})^*}) = \{x^{**} \otimes y^* \in \mathbb{K}(\mathbb{X}, \mathbb{Y})^* : x^{**} \in Ext(B_{\mathbb{X}^{**}}),\ y^* \in Ext(B_{\mathbb{Y}^*})\},$$

where $x^{**} \otimes y^*(T) = x^{**}(T^* y^*)$ for every $T \in \mathbb{K}(\mathbb{X}, \mathbb{Y})$.

Theorem 4.4.2 *Let $\mathbb{X}, \mathbb{Y}$ be Banach spaces over $\mathcal{K}$ such that $\mathbb{X}$ is reflexive. Then*

$$Ext(B_{\mathbb{K}(\mathbb{X},\mathbb{Y})^*}) = \{y^* \otimes x \in \mathbb{K}(\mathbb{X}, \mathbb{Y})^* : x \in Ext(B_{\mathbb{X}}),\ y^* \in Ext(B_{\mathbb{Y}^*})\},$$

where $y^* \otimes x(T) = y^*(Tx)$ for every $T \in \mathbb{K}(\mathbb{X}, \mathbb{Y})$.

We are now in a position to describe the structure of $Ext J(T)$.

Lemma 4.4.3 ([150, Lemma 3.1]) *Let $\mathbb{X}$ be a reflexive Banach space over the field $\mathcal{K}(= \mathbb{R}$ or $\mathbb{C})$ and let $\mathbb{Y}$ be a Banach space over the field $\mathcal{K}$. Suppose that $\mathbb{K}(\mathbb{X}, \mathbb{Y})$ is an $M-$ideal in $\mathbb{L}(\mathbb{X}, \mathbb{Y})$. Let $T \in \mathbb{L}(\mathbb{X}, \mathbb{Y})$, $\|T\| = 1$ and $dist(T, \mathbb{K}(\mathbb{X}, \mathbb{Y})) < 1$. Then $M_T \cap Ext(B_{\mathbb{X}}) \neq \emptyset$ and*

$$Ext\ J(T) = \{y^* \otimes x \in \mathbb{K}(\mathbb{X}, \mathbb{Y})^* : x \in M_T \cap Ext(B_{\mathbb{X}}),\ y^* \in Ext\ J(Tx)\},$$

where $y^* \otimes x : \mathbb{L}(\mathbb{X}, \mathbb{Y}) \to \mathcal{K}$ is defined by $y^* \otimes x(S) = y^*(Sx)$ for every $S \in \mathbb{L}(\mathbb{X}, \mathbb{Y})$.

Proof Clearly $J(T)$ is a non-empty weak*-compact convex subset of $S_{\mathbb{L}(\mathbb{X},\mathbb{Y})^*}$. It is easy to observe that $Ext J(T) \neq \emptyset$. Since $J(T)$ is an extremal subset of $B_{\mathbb{L}(\mathbb{X},\mathbb{Y})^*}$, we get $Ext J(T) \subset Ext(B_{\mathbb{L}(\mathbb{X},\mathbb{Y})^*})$. By the hypothesis $\mathbb{K}(\mathbb{X}, \mathbb{Y})$ is an $M-$ideal in $\mathbb{L}(\mathbb{X}, \mathbb{Y})$ and so we have

$$\mathbb{L}(\mathbb{X}, \mathbb{Y})^* = \mathbb{K}(\mathbb{X}, \mathbb{Y})^* \oplus_1 \mathbb{K}(\mathbb{X}, \mathbb{Y})^\dagger,$$

where $\mathbb{K}(\mathbb{X}, \mathbb{Y})^\dagger = \{F \in \mathbb{L}(\mathbb{X}, \mathbb{Y})^* : F(S) = 0\ \forall\ S \in \mathbb{K}(\mathbb{X}, \mathbb{Y})\}$, the annihilator of $\mathbb{K}(\mathbb{X}, \mathbb{Y})$. Then

$$Ext J(T) \subset Ext(B_{(\mathbb{K}(\mathbb{X},\mathbb{Y})^* \oplus_1 \mathbb{K}(\mathbb{X},\mathbb{Y})^\dagger)}).$$

So extreme points of $J(T)$ are either extreme points of $B_{\mathbb{K}(\mathbb{X},\mathbb{Y})^*}$ or extreme points of $B_{\mathbb{K}(\mathbb{X},\mathbb{Y})^\dagger}$. We claim that $Ext J(T) \subset B_{\mathbb{K}(\mathbb{X},\mathbb{Y})^*}$. Let $\phi \in Ext J(T)$. If possible, let $\phi \in B_{\mathbb{K}(\mathbb{X},\mathbb{Y})^\dagger}$. Since $\text{dist}(T, \mathbb{K}(\mathbb{X}, \mathbb{Y})) < 1$, so there exists $K \in \mathbb{K}(\mathbb{X}, \mathbb{Y})$ such that $\|T - K\| < 1$. Then $\phi(T - K) \leq \|T - K\| < 1$ and again $\phi(T - K) = \phi(T) - \phi(K) = \|T\| - 0 = 1$, which are contradictory. So we must have $Ext J(T) \subset B_{\mathbb{K}(\mathbb{X},\mathbb{Y})^*}$ and consequently

$$Ext J(T) \subset Ext(B_{\mathbb{K}(\mathbb{X},\mathbb{Y})^*}). \tag{4.4.1}$$

Using Theorem 4.4.2 we can write $\phi = y^* \otimes x$, where $x \in Ext(B_{\mathbb{X}})$, $y^* \in Ext(B_{\mathbb{Y}^*})$. Now, $\|T\| = \phi(T) = y^*(Tx) \leq \|Tx\| \leq \|T\|$. This shows that $y^*(Tx) = \|Tx\| = \|T\|$. Thus, $x \in M_T$ and $y^* \in J(Tx)$. So $M_T \cap Ext(B_{\mathbb{X}}) \neq \emptyset$. We next show the equality

$$Ext\ J(T) = \{y^* \otimes x \in \mathbb{K}(\mathbb{X}, \mathbb{Y})^* : x \in M_T \cap Ext(B_{\mathbb{X}}),\ y^* \in Ext\ J(Tx)\}.$$

It is easy to see that

$$\{y^* \otimes x \in \mathbb{K}(\mathbb{X}, \mathbb{Y})^* : x \in M_T \cap Ext(B_{\mathbb{X}}),\ y^* \in Ext\ J(Tx)\} \subset J(T). \tag{4.4.2}$$

Observe that $J(Tx)$ is an extremal subset of $B_{\mathbb{Y}^*}$ and so $Ext J(Tx) \subset Ext(B_{\mathbb{Y}^*})$. Thus, we get,

$$\{y^* \otimes x \in \mathbb{K}(\mathbb{X}, \mathbb{Y})^* : x \in M_T \cap Ext(B_{\mathbb{X}}), y^* \in Ext\ J(Tx)\} \subset Ext(B_{\mathbb{K}(\mathbb{X},\mathbb{Y})^*}). \tag{4.4.3}$$

From (4.4.1), (4.4.2) and (4.4.3) we get,

$$\{y^* \otimes x \in \mathbb{K}(\mathbb{X}, \mathbb{Y})^* : x \in M_T \cap Ext(B_{\mathbb{X}}), y^* \in Ext\ J(Tx)\} \subset Ext\ J(T).$$

The converse inclusion holds easily. This completes the proof. □

We recall another fundamental theorem concerning B–J orthogonality by Singer.

Theorem 4.4.4 ([141, Theorem 1.1, pp. 170]) *Let $\mathbb{X}$ be a Banach space. Then for $x, y \in \mathbb{X}$, $x \perp_B y \Leftrightarrow \exists\ \phi, \psi \in Ext(B_{\mathbb{X}^*})$ and $\alpha \in [0, 1]$ such that $\phi(x) = \psi(x) = \|x\|$ and $\alpha\phi(y) + (1 - \alpha)\psi(y) = 0$.*

Above two results help us to characterize B–J orthogonality of bounded linear operators under certain conditions.

Theorem 4.4.5 ([87, Theorem 3.2]) (i) *Let $\mathbb{X}$ be a reflexive Banach space and let $\mathbb{Y}$ be a Banach space. Suppose that $\mathbb{K}(\mathbb{X}, \mathbb{Y})$ is an $M-$ideal in $\mathbb{L}(\mathbb{X}, \mathbb{Y})$. Let $T \in \mathbb{L}(\mathbb{X}, \mathbb{Y})$ be such that $\|T\| = 1$ and $dist(T, \mathbb{K}(\mathbb{X}, \mathbb{Y})) < 1$. Then for any $A \in \mathbb{L}(\mathbb{X}, \mathbb{Y})$, $T \perp_B A$ if and only if there exist $x_1, x_2 \in M_T \cap Ext(B_{\mathbb{X}})$ such that $Ax_1 \in (Tx_1)^+$ and $Ax_2 \in (Tx_2)^-$.*

(ii) *In addition, if $M_T = D \cup (-D)$, where D is a closed connected subset of $S_{\mathbb{X}}$, then $T \perp_B A$ if and only if there exists $x \in M_T$ such that $Tx \perp_B Ax$.*

Proof (i) The proof of the sufficient part follows trivially. We only prove the necessary part. Let $T \perp_B A$. Then using Theorem 4.4.4, we can find $\phi, \psi \in Ext(B_{\mathbb{L}(\mathbb{X},\mathbb{Y})^*}) \cap J(T)$ and $t \in [0, 1]$ such that $t\phi(A) + (1 - t)\psi(A) = 0$. Clearly, $\phi, \psi \in Ext J(T)$. Therefore, by Lemma 4.4.3, there exist $x_i \in M_T \cap Ext(B_{\mathbb{X}})$ and $y_i^* \in Ext J(Tx_i)$ for $i = 1, 2$ such that $\phi = y_1^* \otimes x_1$ and $\psi = y_2^* \otimes x_2$. Now,

$$\begin{aligned} t\phi(A) + (1 - t)\psi(A) = 0 \\ \Rightarrow ty_1^* \otimes x_1(A) + (1 - t)y_2^* \otimes x_2(A) = 0 \\ \Rightarrow ty_1^*(Ax_1) + (1 - t)y_2^*(Ax_2) = 0. \end{aligned}$$

Thus, clearly $y_1^*(Ax_1)$ and $y_2^*(Ax_2)$ are of opposite signs. Without loss of generality, we assume that $y_1^*(Ax_1) \geq 0$ and $y_2^*(Ax_2) \leq 0$. Thus, for any $\lambda \geq 0$, $\|Tx_1 + \lambda Ax_1\| \geq |y_1^*(Tx_1 + \lambda Ax_1)| \geq \|Tx_1\|$, since $y_1^* \in Ext J(Tx_1)$. Hence, $Ax_1 \in (Tx_1)^+$. Similarly, $y_2^*(Ax_2) \leq 0$ and $y_2^* \in Ext J(Tx_2)$ gives that $Ax_2 \in (Tx_2)^-$.
(ii) The proof of the sufficient part follows trivially. We only prove the necessary part. Let $T \perp_B A$ and $M_T = D \cup (-D)$, where D is a closed connected subset of $S_{\mathbb{X}}$. Let

$$W_1 = \{x \in D : Ax \in (Tx)^+\},$$

$$W_2 = \{x \in D : Ax \in (Tx)^-\}.$$

Then by (i), $W_1 \neq \emptyset$, $W_2 \neq \emptyset$. Observe that both W_1 and W_2 are closed. Applying Proposition 3.2.2, we get $D = W_1 \cup W_2$. Since D is connected, $W_1 \cap W_2 \neq \emptyset$. Let $x \in W_1 \cap W_2$. Then from Proposition 3.2.2, it follows that $Tx \perp_B Ax$. This completes the proof. □

We would like to mention that if $\mathbb{X}, \mathbb{Y}$ are arbitrary Banach spaces and if $T \in \mathbb{L}(\mathbb{X}, \mathbb{Y})$ is such that $M_T = D \cup (-D)$, where D is a closed connected subset of $S_{\mathbb{X}}$, then, for any $A \in \mathbb{L}(\mathbb{X}, \mathbb{Y})$, $T \perp_B A$ may not necessarily imply that there exists some $x \in M_T$ such that $Tx \perp_B Ax$ holds. We exhibit the following example in support of this claim.

Example 4.4.6 Consider the operators $T, A : \ell_1 \to \ell_1$ defined as follows:

$$Te_1 = e_1, Te_n = \left(1 - \frac{1}{n^2}\right) e_n, \ n \geq 2 \text{ and}$$

$$Ae_n = \frac{1}{n^2} e_n, \ n \geq 1,$$

where $e_i = (0, 0, \ldots, 0, 1, 0, \ldots)$, 1 at the i-th coordinate. It is straightforward to check that $M_T = \{\pm e_1\}$, $\|T\| = 1$ and $Te_1 \not\perp_B Ae_1$. However, for all scalars λ,

$$\|T + \lambda A\| \geq \|(T + \lambda A)e_n\| = |1 - \frac{1}{n^2} + \frac{\lambda}{n^2}| \to 1 = \|T\|.$$

Therefore, $T \perp_B A$.

The previous example proves that, for $T \in \mathbb{L}(\mathbb{X}, \mathbb{Y})$, $M_T = \{\pm x_0\}$ is not sufficient to conclude $Tx_0 \perp_B Ax_0$, whenever $T \perp_B A$ holds. Therefore, for such a result to hold true, we require additional conditions. The following theorem is a case in point.

Theorem 4.4.7 ([126, Theorem 3.1]) *Let $\mathbb{X}, \mathbb{Y}$ be Banach spaces. Let $T \in \mathbb{L}(\mathbb{X}, \mathbb{Y})$ and $M_T = \{\pm x_0\}$ for some $x_0 \in S_{\mathbb{X}}$ and T further satisfies the following property:*

Given any $\delta > 0$, if $\{H_\alpha : \alpha \in \Lambda\}$ is the collection of all hyperspaces such that $d(x_0, H_\alpha) > \delta$ then $\sup\{\|Tx\| : x \in (\cup_\alpha H_\alpha) \cap S_{\mathbb{X}}\} < \|T\|$.

Then for any $A \in \mathbb{L}(\mathbb{X}, \mathbb{Y})$, $T \perp_B A$ if and only if $Tx_0 \perp_B Ax_0$.

Proof We only show that $T \perp_B A$ implies $Tx_0 \perp_B Ax_0$, The converse part is trivial. If possible let $T \perp_B A$ but $Tx_0 \not\perp_B Ax_0$. Then there exists a scalar $\lambda_0 \neq 0$ such that $\|Tx_0 + \lambda_0 Ax_0\| < \|T\|$. So we can find an $\epsilon_1 > 0$ such that

$$\|Tx_0 + \lambda_0 Ax_0\| < \|T\| - \epsilon_1.$$

Without loss of generality, we may and do assume that $\lambda_0 > 0$. By continuity of the function $T + \lambda_0 A$ at the point x_0, we can find an open ball $B(x_0, r_0)$ such that

$$\|Tx + \lambda_0 Ax\| < \|T\| - \epsilon_1 \ \forall x \in B(x_0, r_0).$$

Let $\lambda \in (0, \lambda_0)$. Then for all $z \in B(x_0, r_0) \cap S_{\mathbb{X}}$,

$$\begin{aligned} Tz + \lambda Az &= (1 - \frac{\lambda}{\lambda_0})Tz + \frac{\lambda}{\lambda_0}(Tz + \lambda_0 Az) \\ \Rightarrow \|Tz + \lambda Az\| &\leq (1 - \frac{\lambda}{\lambda_0})\|Tz\| + \frac{\lambda}{\lambda_0}\|(Tz + \lambda_0 Az)\| \\ \Rightarrow \|Tz + \lambda Az\| &< (1 - \frac{\lambda}{\lambda_0})\|T\| + \frac{\lambda}{\lambda_0}(\|T\| - \epsilon_1) = \|T\| - \frac{\lambda}{\lambda_0}\epsilon_1. \end{aligned}$$

Since $\|(T + \lambda A)z\| = \|(T + \lambda A)(-z)\|$, we get for all $z \in (B(x_0, r_0) \cup B(-x_0, r_0)) \cap S_{\mathbb{X}}$,

$$\|Tz + \lambda Az\| < \|T\| - \frac{\lambda}{\lambda_0}\epsilon_1, \forall \lambda \in (0, \lambda_0).$$

Consider $C = S_{\mathbb{X}} \setminus (B(x_0, r_0) \cup B(-x_0, r_0))$. For any $z \in C$, x_0 and z being linearly independent, from Theorem 2.1.13, there exists $\beta_z \in \mathbb{R}$ such that $x_0 + \beta_z z \perp_B z$. Therefore, $\|x_0 + \lambda z\| \geq \|x_0 + \beta_z z\| = c_z$(say), for all $\lambda \in \mathbb{R}$. Observe that $c_z > 0$ as $\{x_0, z\}$ is linearly independent. Define a function $f_z : \text{span}\{x_0, z\} \to \mathbb{R}$ by $f_z(ax_0 + bz) = a$. Note that f_z is a bounded linear functional on $\text{span}\{x_0, z\}$ and so by the Hahn–Banach theorem there exists a norm preserving extension, say, f of f_z on $\mathbb{X}$. Clearly $\|f\| = \|f_z\| \leq \frac{1}{c_z}$. Let $H_z = \ker(f)$. Then for all $y \in H_z$, $1 = |f(x_0 - y)| \leq \|f\|\|x_0 - y\| \leq \frac{1}{c_z}\|x_0 - y\|$. So $\|x_0 - y\| \geq c_z$. Therefore, $d(x_0, H_z) \geq c_z$. Thus, for each $z \in C$, there exists a hyperspace H_z of $\mathbb{X}$ containing z such that $d(x_0, H_z) \geq c_z$. So $C \subset (\cup_{z \in C} H_z) \cap S_{\mathbb{X}}$. Observe that there exists a $\delta > 0$ such that $d(x_0, H_z) > \delta$ for all $z \in C$. Now, by the hypothesis $\sup\{\|Tx\| : x \in (\cup_{z \in C} H_z) \cap S_{\mathbb{X}}\} < \|T\|$ and so $\sup\{\|Tx\| : x \in C\} < \|T\|$. Hence, there exists $\epsilon_2 > 0$ such that $\sup\{\|Tx\| : x \in C\} < \|T\| - \epsilon_2$.

Choose $0 < \widetilde{\lambda} < \min\{\lambda_0, \frac{\epsilon_2}{2\|A\|}\}$. Then for all $z \in C$, we get

$$\begin{aligned} \|Tz + \widetilde{\lambda}Az\| &\leq \|Tz\| + |\widetilde{\lambda}|\|Az\| \\ &< \|T\| - \epsilon_2 + |\widetilde{\lambda}|\|A\| \\ &< \|T\| - \frac{1}{2}\epsilon_2. \end{aligned}$$

Choose $\epsilon = \min\{\frac{1}{2}\epsilon_2, \frac{\widetilde{\lambda}}{\lambda_0}\epsilon_1\}$. Thus, for all $x \in S_{\mathbb{X}}$ we get,

$$\|Tx + \widetilde{\lambda}Ax\| < \|T\| - \epsilon.$$

This shows that $\|T + \widetilde{\lambda}A\| < \|T\|$, which contradicts the fact that $T \perp_B A$. This completes the proof. □

We next provide another instance where the conclusion of Theorem 4.4.7 holds true. Since suitable modification of Theorem 4.4.7 suffices to prove the next result, we conveniently skip the proof.

Theorem 4.4.8 ([105, Theorem 2.2]) *Let $\mathbb{X}$ be a Banach space, $T \in \mathbb{L}(\mathbb{X}, \mathbb{Y})$, $M_T = D \cup (-D)$ (D is a non empty compact connected subset of $S_{\mathbb{X}}$). Suppose* $\sup\{\|Tx\| : x \in C\} < \|T\|$ *for all closed subset C of $S_{\mathbb{X}}$ with $d(M_T, C) > 0$. Then for any $A \in \mathbb{L}(\mathbb{X}, \mathbb{Y})$, $T \perp_B A$ if and only if there exists $z \in M_T$ such that $Tz \perp_B Az$.*

Since bounded linear operators on Banach spaces may not attain norm, characterizing B–J orthogonality in $\mathbb{L}(\mathbb{X}, \mathbb{Y})$ is far more difficult. For this purpose, we require the notion of norming sequence. We say that a sequence $\{x_n\}$ of $S_{\mathbb{X}}$ is a norming sequence for T if $\|Tx_n\| \to \|T\|$. We first obtain an easy sufficient condition for $T \perp_B A$, where T, A are bounded linear operators on arbitrary Banach spaces.

Proposition 4.4.9 ([126, Proposition 2.3]) *Let $\mathbb{X}$ and $\mathbb{Y}$ be Banach spaces. Let T, $A \in \mathbb{L}(\mathbb{X}, \mathbb{Y})$. Suppose there exist two norming sequences $\{x_n\}$ and $\{y_n\}$ for T such that $Ax_n \in (Tx_n)^+$ and $Ay_n \in (Ty_n)^-$ for all $n \in \mathbb{N}$. Then $T \perp_B A$.*

Proof We first claim that $A \in T^+$. Let $\lambda \geq 0$. Since $Ax_n \in (Tx_n)^+$, we have $\|T + \lambda A\| \geq \|Tx_n + \lambda Ax_n\| \geq \|Tx_n\|$ for all $n \in \mathbb{N}$. Therefore, letting $n \to \infty$, and noting that $\{x_n\}$ is a norming sequence for T, we get $\|T + \lambda A\| \geq \|T\|$. Thus, $A \in T^+$. Similarly considering the norming sequence $\{y_n\}$ and the fact that $Ay_n \in (Ty_n)^-$ we can show that $A \in T^-$. Thus, from $A \in T^+$ and $A \in T^-$ we conclude that $T \perp_B A$. □

The following example shows that the condition in Proposition 4.4.9, is not necessary for $T \perp_B A$.

Example 4.4.10 Define $T, A : \ell_1 \to \ell_1$ by $Te_n = (1 - \frac{1}{n+1})e_n$, $n \geq 1$, and $Ae_n = \frac{1}{n+1}e_n$, $n \geq 1$, where $e_n = (0, 0, \ldots, 0, 1, 0, \ldots)$, with 1 in the n-th position and 0 elsewhere. Let $x = \sum_{n=1}^{\infty} a_n e_n \in \ell_1$ where $a_n \in \mathbb{R}$ for all n. Then

$$\|Tx\| = \sum_{n=1}^{\infty} |a_n||1 - \frac{1}{n+1}| \leq \sum_{n=1}^{\infty} |a_n| = \|x\|.$$

So $\|T\| \leq 1$. Also, $\|Te_n\| = (1 - \frac{1}{n+1}) \to 1$ as $n \to \infty$. Hence, $\|T\| = 1$. First we show that $T \perp_B A$. For any scalar λ, $\|T + \lambda A\| \geq \|(T + \lambda A)e_n\| = |1 - \frac{(1-\lambda)}{n+1}| \to 1$, so that $\|T + \lambda A\| \geq \|T\|$ for all λ. Now for $x \in \ell_1 \setminus \{0\}$, we get

$$\|(T - A)x\| = \sum_{n=1}^{\infty} |a_n||1 - \frac{2}{n+1}| < \sum_{n=1}^{\infty} |a_n||1 - \frac{1}{n+1}| = \|Tx\|,$$

which implies that $\|(T - A)x\| < \|Tx\|$ for all $x \in \ell_1 \setminus \{0\}$. In particular, we must have, $Ax \notin (Tx)^-$ for all $x \in \ell_1 \setminus \{0\}$. Thus, we observe that in this example, although $T \perp_B A$, it is not possible to find a norming sequence $\{y_n\}$ for T such that $Ay_n \in (Ty_n)^-$ for all $n \in \mathbb{N}$.

To continue the investigation of B–J orthogonality of bounded linear operators, we need the following two notions introduced in [126]. Following [126], for $x, y \in \mathbb{X}$ and $\epsilon \in [0, 1)$, we say that $y \in x^{+\epsilon}$, if $\|x + \lambda y\| \geq \sqrt{1-\epsilon^2}\|x\|$ holds for each scalar $\lambda \geq 0$. Similarly, we say that $y \in x^{-\epsilon}$, if $\|x + \lambda y\| \geq \sqrt{1-\epsilon^2}\|x\|$ holds for each scalar $\lambda \leq 0$.

In the following theorem, we study $T \perp_B A$, where T is a rotund point. Recall that $x \in S_{\mathbb{X}}$ is said to be a rotund point of $B_{\mathbb{X}}$ if $\|y\| = \frac{\|x+y\|}{2} = 1$ implies that $x = y$. To get our desired result, we use the next lemma which gives a characterization of rotund points in terms of B–J orthogonality. Note that the proof of the lemma can be obtained similarly as the proof of [123, Theorem 2.4].

Lemma 4.4.11 *Let $\mathbb{X}$ be a Banach space.Then $x \in S_{\mathbb{X}}$ is a rotund point if and only if $x \perp_B y \Rightarrow x \perp_{SB} y$ for any $y \in \mathbb{X} \setminus \{\theta\}$.*

Theorem 4.4.12 ([126, Theorem 2.6]) *Let $\mathbb{X}$ and $\mathbb{Y}$ be Banach spaces. Let T be a rotund point of $\mathbb{L}(\mathbb{X}, \mathbb{Y})$. Then for any $A \in \mathbb{L}(\mathbb{X}, \mathbb{Y})$, $T \perp_B A$ if and only if there exist two norming sequences $\{x_n\}$, $\{y_n\}$ for T and two sequences of positive real numbers $\{\epsilon_n\}$, $\{\delta_n\}$ such that*

(i) $\epsilon_n \to 0$, $\delta_n \to 0$ *as* $n \to \infty$.

(ii) $Ax_n \in (Tx_n)^{+\epsilon_n}$ *and* $Ay_n \in (Ty_n)^{-\delta_n}$ *for all* $n \in \mathbb{N}$.

Proof We first prove the sufficient part. We claim that $A \in T^+$. Let $\lambda \geq 0$. Since $Ax_n \in (Tx_n)^{+\epsilon_n}$ for all $n \in \mathbb{N}$, so, $\|Tx_n + \lambda Ax_n\| \geq \sqrt{1-\epsilon_n^2}\ \|Tx_n\|$. Therefore, $\|T + \lambda A\| \geq \|(T + \lambda A)x_n\| \geq \sqrt{1-\epsilon_n^2}\|Tx_n\|$. Since $\{x_n\}$ is a norming sequence for T and $\epsilon_n \to 0$, so we get, $\|T + \lambda A\| \geq \|T\|$. Thus, $A \in T^+$. Similarly we can show that $A \in T^-$. Hence, $T \perp_B A$. This completes the proof of the sufficient part.

We next prove the necessary part. Since $T \in \mathbb{L}(\mathbb{X}, \mathbb{Y})$ is a rotund point, clearly T is non-zero. Also, by Lemma 4.4.11, $T \perp_B A \Rightarrow T \perp_{SB} A$ for any non-zero $A \in \mathbb{L}(\mathbb{X}, \mathbb{Y})$.

Therefore, for each $n \in \mathbb{N}$,

$$\|T + \frac{1}{n}A\| > \|T\|.$$

Therefore, for each $n \in \mathbb{N}$, there exists $x_n \in S_{\mathbb{X}}$ such that $\|(T + \frac{1}{n}A)x_n\| > \|T\|$. We claim that $\{x_n\}$ is a norming sequence for T, i.e., $\|Tx_n\| \to \|T\|$. Indeed,

$$\|Tx_n\| = \|(T + \frac{1}{n}A)x_n - \frac{1}{n}Ax_n\| \geq \|(T + \frac{1}{n}A)x_n\| - \frac{1}{n}\|Ax_n\| > \|T\| - \frac{1}{n}\|A\| \to \|T\|$$

as $n \to \infty$. Since $x_n \in S_{\mathbb{X}}$, $\|Tx_n\| \leq \|T\|$. This establishes our claim. Since $\|Tx_n\| \to \|T\| > 0$, there exists $n_1 \in \mathbb{N}$ such that $\|Tx_n\| > \frac{\|T\|}{2} > 0$ for all $n \geq n_1$. Choose $n_2 \in \mathbb{N}$ such that $n_2 > \frac{2\|A\|}{\|T\|}$. Let $n_0 = \max\{n_1, n_2\}$. Then for all $n \geq n_0$, $0 < \frac{\|A\|}{n\|Tx_n\|} < \frac{2\|A\|}{n\|T\|} < 1$, which implies that, for all $n \geq n_0$, $0 < 1 - \frac{\|A\|}{n\|Tx_n\|} < 1$.

Choose $\epsilon_n = \sqrt{1 - (1 - \frac{\|A\|}{n\|Tx_n\|})^2}$. Then clearly $\epsilon_n \to 0$ as $n \to \infty$. We claim that $Ax_n \in (Tx_n)^{+\epsilon_n}$ for all $n \geq n_0$. Let $n \geq n_0$. Then for $0 < \lambda \leq \frac{1}{n}$,

$$\|Tx_n + \lambda Ax_n\| \geq \|Tx_n\| - |\lambda|\|Ax_n\| \geq \|Tx_n\| - \frac{1}{n}\|A\| = \sqrt{1 - \epsilon_n^2}\|Tx_n\|.$$

For $\lambda > \frac{1}{n} > 0$, we claim that $\|Tx_n + \lambda Ax_n\| \geq \|Tx_n\|$. Suppose on the contrary, we have $\|Tx_n + \lambda Ax_n\| < \|Tx_n\|$ for some $\lambda > \frac{1}{n}$. Now, there exists $t \in (0, 1)$ such that $Tx_n + \frac{1}{n}Ax_n = t(Tx_n) + (1-t)(Tx_n + \lambda Ax_n)$. This implies that,

$$\|Tx_n + \frac{1}{n}Ax_n\| \leq t\|Tx_n\| + (1-t)\|Tx_n + \lambda Ax_n\| < t\|Tx_n\| + (1-t)\|Tx_n\| = \|Tx_n\| \leq \|T\|,$$

a contradiction. Therefore, for all $\lambda \geq 0$, $\|Tx_n + \lambda Ax_n\| \geq \|Tx_n\| - \frac{1}{n}\|A\| = \sqrt{1-\epsilon_n^2}\ \|Tx_n\|$. This completes the proof of our claim. Similarly, considering $\|T - \frac{1}{n}A\| > \|T\|$ for each $n \in \mathbb{N}$, we can find the desired norming sequence $\{y_n\}$ for T and a sequence $\{\delta_n\}$ in $\mathbb{R}^+$ such that (i) and (ii) are satisfied. □

Observe that the sufficient part of Theorem 4.4.12 holds true even if T is not rotund. However, we now exhibit an example to show that if T is not rotund then the conditions (i) and (ii) of Theorem 4.4.12 are not necessary for $T \perp_B A$.

Example 4.4.13 Define $T, A : \ell_1 \to \ell_1$ by $Te_1 = \frac{1}{2}e_1,\ Te_n = (1 - \frac{1}{n^4})e_n, n \geq 2$, and $Ae_1 = \frac{1}{2}e_1,\ Ae_n = \frac{1}{n^2}(\frac{1}{n^2} - 1)e_n,\ n \geq 2$, where $e_n = (0, 0, \ldots, 0, 1, 0, \ldots)$, with 1 in the n-th position and 0 elsewhere. Then both T and A are bounded linear operators on ℓ_1 with $\|T\| = 1$. Also $T \perp_B A$, since for any scalar λ,

$$\|T + \lambda A\| \geq \|(T + \lambda A)e_n\| = \|(1 - \frac{1}{n^4})e_n + \lambda\frac{1}{n^2}(\frac{1}{n^2} - 1)e_n\| = |1 - \frac{1}{n^4} + \frac{\lambda}{n^4} - \frac{\lambda}{n^2}| \to 1.$$

Now for each $x = (a_1, a_2, \ldots, \ldots) \in S_{\ell_1}$, we have

$$\begin{aligned}
\|Tx + Ax\| &= \|(a_1, (1 - \frac{1}{2^2})a_2, \ldots, (1 - \frac{1}{n^2})a_n, \ldots)\| \\
&= |a_1| + |a_2|(1 - \frac{1}{2^2}) + \ldots + |a_n|(1 - \frac{1}{n^2}) + \ldots \\
&= |\frac{a_1}{2}| + |a_2|(1 - \frac{1}{2^4}) + \ldots + |a_n|(1 - \frac{1}{n^4}) + \ldots \\
&\quad + |\frac{a_1}{2}| + |a_2|(\frac{1}{2^4} - \frac{1}{2^2}) + \ldots + |a_n|(\frac{1}{n^4} - \frac{1}{n^2}) + \ldots \\
&\leq \|Tx\| + \frac{1}{2} + \sum_{n=2}^{\infty}(\frac{1}{n^4} - \frac{1}{n^2}) \\
&< \|Tx\| - 0.06 \\
&\leq \|T\| - 0.06.
\end{aligned}$$

This shows that there does not exist $\{\epsilon_n\}$ with $\epsilon_n \to 0$ and $\{x_n\} \subset S_{\ell_1}$ such that $\|Tx_n\| \to \|T\|$ and $Ax_n \in (Tx_n)^{+\epsilon_n}$, for otherwise

$$\sqrt{1 - \epsilon_n{}^2}\, \|Tx_n\| \leq \|Tx_n + Ax_n\| < \|Tx_n\| - 0.06,$$

and so letting $n \to \infty$ we get a contradiction. Thus, $T \perp_B A$ but T does not satisfy hypothesis of Theorem 4.4.12.

Characterization of B–J orthogonality of operators in its most general form is obtained in the next theorem.

Theorem 4.4.14 ([126, Theorem 2.8]) *Let $\mathbb{X}$ and $\mathbb{Y}$ be Banach spaces. Let $T \in \mathbb{L}(\mathbb{X}, \mathbb{Y})$ be non-zero. Then for any $A \in \mathbb{L}(\mathbb{X}, \mathbb{Y})$, $T \perp_B A$ if and only if either* (a) *or* (b) *holds*

(a) *There exists a norming sequence $\{x_n\}$ for T such that $\|Ax_n\| \to 0$ as $n \to \infty$.*

(b) *There exist two norming sequences $\{x_n\}$, $\{y_n\}$ for T and two sequences of positive real numbers $\{\epsilon_n\}$, $\{\delta_n\}$ such that*

(i) *$\epsilon_n \to 0$, $\delta_n \to 0$ as $n \to \infty$.*

(ii) *$Ax_n \in (Tx_n)^{+\epsilon_n}$ and $Ay_n \in (Ty_n)^{-\delta_n}$ for all $n \in \mathbb{N}$.*

Proof We first prove the sufficient part. Suppose (a) holds. For any scalar λ, $\|T + \lambda A\| \geq \|Tx_n + \lambda Ax_n\| \geq \|Tx_n\| - |\lambda|\|Ax_n\| \to \|T\|$ as $n \to \infty$. Therefore, $T \perp_B A$. Next, suppose (b) holds. Then following exactly the same line of arguments as in the proof of the sufficient part of Theorem 4.4.12, we obtain $T \perp_B A$. This completes the proof of the sufficient part.

Let us now prove the comparatively involved necessary part.

Suppose (a) does not hold. Without loss of generality, we may and do assume that $\|A\| \leq 1$. Since $T \perp_B A$, for any non-zero scalar λ, $\|T + \lambda A\| \geq \|T\|$. In particular, for each $n \in \mathbb{N}$,

$$\|T + \frac{1}{n} A\| > \|T\| - \frac{1}{n^3}.$$

Therefore, for each $n \in \mathbb{N}$, there exists a sequence $\{x_n\}$ in $S_{\mathbb{X}}$ such that $\|(T + \frac{1}{n}A)x_n\| > \|T\| - \frac{1}{n^3} \geq \|Tx_n\| - \frac{1}{n^3}$. We claim that $\|Tx_n\| \to \|T\|$. Indeed,

$$\|Tx_n\| = \|(T + \frac{1}{n}A)x_n - \frac{1}{n}Ax_n\| \geq \|(T + \frac{1}{n}A)x_n\| - \frac{1}{n}\|Ax_n\| > \|T\| - \frac{1}{n^3} - \frac{1}{n}\|A\| \to \|T\|$$

as $n \to \infty$. Since $x_n \in S_{\mathbb{X}}$, $\|Tx_n\| \leq \|T\|$. This proves our claim.

Since (a) does not hold, we assume that, $\inf_{n \in \mathbb{N}} \|Ax_n\| = c > 0$. Choose $n_1 \in \mathbb{N}$ such that $n_1 > \frac{2\|T\|}{c}$. Since $\|Tx_n\| \to \|T\| > 0$, there exists $n_2 \in \mathbb{N}$ such that $\|Tx_n\| > \frac{\|T\|}{2} > 0$ for all $n \geq n_2$. Choose $n_3 \in \mathbb{N}$ such that $n_3 > \frac{2}{\|T\|}$. Let $n_0 = \max\{n_1, n_2, n_3\}$. Then for all $n \geq n_0$, $0 < \frac{1}{n\|Tx_n\|} < \frac{2}{n\|T\|} < 1$, which implies that, for all $n \geq n_0$, $0 < 1 - \frac{1}{n\|Tx_n\|} < 1$.

Choose $\epsilon_n = \sqrt{1 - (1 - \frac{1}{n\|Tx_n\|})^2}$. Then clearly $\epsilon_n \to 0$ as $n \to \infty$.

We claim that $Ax_n \in (Tx_n)^{+\epsilon_n}$ for all $n \geq n_0$. Let $n \geq n_0$. Then for $0 \leq \lambda < \frac{1}{n}$,

$$\|Tx_n + \lambda Ax_n\| \geq \|Tx_n\| - \lambda\|Ax_n\| \geq \|Tx_n\| - \frac{1}{n}.$$

For $\frac{1}{n} \leq \lambda \leq n$, we claim that $\|Tx_n + \lambda Ax_n\| \geq \|Tx_n\| - \frac{1}{n}$. Suppose on the contrary, we have $\|Tx_n + \lambda Ax_n\| < \|Tx_n\| - \frac{1}{n}$ for some $\frac{1}{n} \leq \lambda \leq n$. Now, $Tx_n + \frac{1}{n}Ax_n = (1 - \frac{1}{n\lambda})Tx_n + \frac{1}{n\lambda}(Tx_n + \lambda Ax_n)$. Therefore, we get

$$\begin{aligned}
\|Tx_n\| - \frac{1}{n^3} &< \|Tx_n + \frac{1}{n}Ax_n\| \\
&\leq (1 - \frac{1}{n\lambda})\|Tx_n\| + \frac{1}{n\lambda}\|(Tx_n + \lambda Ax_n)\| \\
&< (1 - \frac{1}{n\lambda})\|Tx_n\| + \frac{1}{n\lambda}(\|Tx_n\| - \frac{1}{n}) \\
&= \|Tx_n\| - \frac{1}{n^2\lambda}.
\end{aligned}$$

This shows that $\lambda > n$, which is not possible. Thus, for $0 \leq \lambda \leq n$, $\|Tx_n + \lambda Ax_n\| \geq \|Tx_n\| - \frac{1}{n}$. Hence, $0 \leq \lambda \leq \frac{2\|T\|}{c}$ gives that $\|Tx_n + \lambda Ax_n\| \geq \|Tx_n\| - \frac{1}{n}$. Now, for $\lambda > \frac{2\|T\|}{c}$,

$$\|Tx_n + \lambda Ax_n\| \geq \lambda\|Ax_n\| - \|Tx_n\| \geq \lambda c - \|Tx_n\| > 2\|T\| - \|Tx_n\| \geq \|Tx_n\| - \frac{1}{n}.$$

Therefore, for all $\lambda \geq 0$, $\|Tx_n + \lambda Ax_n\| \geq \|Tx_n\| - \frac{1}{n} = \sqrt{1 - c_n^2}\,\|Tx_n\|$. This completes the proof of our claim.

Similarly, considering $\|T - \frac{1}{n}A\| > \|T\| - \frac{1}{n^3}$ for each $n \in \mathbb{N}$, we can find the desired norming sequence $\{y_n\}$ for T and a sequence $\{\delta_n\}$ in $\mathbb{R}^+$ such that all the conditions of (b) are satisfied. This completes the proof. □

As an immediate consequence of Theorems 4.4.12 and 4.4.14, we obtain a characterization of B–J orthogonality in the dual space of a Banach space. To get the desired results, we simply need to observe the following fact regarding real numbers.

Proposition 4.4.15 ([126, Proposition 2.9]) *For any two real numbers x, y and for any $\epsilon \in [0, 1)$, $y \in x^{+\epsilon}$ if and only if $xy \geq 0$. Similarly, $y \in x^{-(\epsilon)}$ if and only if $xy \leq 0$.*

B–J orthogonality in $\mathbb{X}^*$ assumes the following form if $\mathbb{X}^*$ is strictly convex.

Theorem 4.4.16 ([126, Theorem 2.10]) *Let $\mathbb{X}$ be a Banach space such that $\mathbb{X}^*$ is strictly convex. Then for any $f, g \in \mathbb{X}^*$, $f \perp_B g$ if and only if there exist $\{x_n\}$, $\{y_n\}$ in $S_{\mathbb{X}}$ such that*

(i) $|f(x_n)| \to \|f\|$ and $|f(y_n)| \to \|f\|$ as $n \to \infty$

(ii) $f(x_n).g(x_n) \geq 0$ and $f(y_n).g(y_n) \leq 0$ for all $n \in \mathbb{N}$.

Proof Since $\mathbb{X}^*$ is strictly convex, f, g are rotund points of $\mathbb{X}^*$. The rest of the proof follows directly from Theorem 4.4.12 and Proposition 4.4.15. □

If $\mathbb{X}^*$ is not necessarily strictly convex, then we get the following theorem using Theorem 4.4.14 and Proposition 4.4.15.

Theorem 4.4.17 ([126, Theorem 2.11]) *Let $\mathbb{X}$ be a Banach space. Then for any $f, g \in \mathbb{X}^*$, $f \perp_B g$ if and only if either* (a) *or* (b) *holds*
(a) *there exists $\{x_n\}$ in $S_{\mathbb{X}}$ such that $|f(x_n)| \to \|f\|$ and $g(x_n) \to 0$.*
(b) *there exist $\{x_n\}, \{y_n\}$ in $S_{\mathbb{X}}$ such that*
(i) $|f(x_n)| \to \|f\|$ *and* $|f(y_n)| \to \|f\|$ *as* $n \to \infty$
(ii) $f(x_n).g(x_n) \geq 0$ *and* $f(y_n).g(y_n) \leq 0$ *for all* $n \in \mathbb{N}$.

4.5 Orthogonality of Operators on Complex Banach Spaces

In the previous two sections, we have studied B–J orthogonality for real Banach spaces only. In this section, our aim is to continue the study for complex Banach spaces. We begin this section with a generalization of Theorem 4.4.14. The proof follows using similar arguments as in Theorem 4.4.14. For the sake of completeness, we give the details of the proof here.

Theorem 4.5.1 *Let $\mathbb{X}$ and $\mathbb{Y}$ be Banach spaces. Let $T \in \mathbb{L}(\mathbb{X}, \mathbb{Y})$ be non-zero. Then for any $A \in \mathbb{L}(\mathbb{X}, \mathbb{Y})$, $T \perp_B A$ if and only if either* (a) *or* (b) *holds*
(a) *There exists a norming sequence $\{x_n\}$ for T such that $\|Ax_n\| \to 0$ as $n \to \infty$.*
(b) *For each $\alpha \in U$, there exist two norming sequences $\{x_n = x_n(\alpha)\}$, $\{y_n = y_n(\alpha)\}$ for T and two sequences of positive real numbers $\{\epsilon_n = \epsilon_n(\alpha)\}$, $\{\delta_n = \delta_n(\alpha)\}$ such that*
(i) $\epsilon_n \to 0$, $\delta_n \to 0$ *as* $n \to \infty$.
(ii) $Ax_n \in (Tx_n)_\alpha^{+\epsilon_n}$ *and* $Ay_n \in (Ty_n)_\alpha^{-\delta_n}$ *for all* $n \in \mathbb{N}$.

Proof We first prove the easier sufficient part.

Suppose (a) holds. Now, for any $\lambda \in \mathbb{C}$, $\|T + \lambda A\| \geq \|Tx_n + \lambda Ax_n\| \geq \|Tx_n\| - |\lambda|\|Ax_n\| \to \|T\|$ as $n \to \infty$. Therefore, $T \perp_B A$.

Now, suppose (b) holds. Let $\lambda \in \mathbb{C}$. Then $\lambda = t\alpha$ for some $\alpha \in U$ and $t \in \mathbb{R}$. If $t \geq 0$ then $Ax_n \in (Tx_n)_\alpha^{+\epsilon_n}$ for all $n \in \mathbb{N}$, gives that $\|Tx_n + \lambda Ax_n\| = \|Tx_n + t\alpha Ax_n\| \geq \sqrt{1 - \epsilon_n^2}\ \|Tx_n\|$. This implies, $\|T + \lambda A\| \geq \|(T + \lambda A)x_n\| = \|Tx_n + t\alpha Ax_n\| \geq \sqrt{1 - \epsilon_n^2}\|Tx_n\|$. Since $\epsilon_n \to 0$ and $\|Tx_n\| \to \|T\|$ as $n \to \infty$, we obtain, $\|T + \lambda A\| \geq \|T\|$. Similarly, if $t \leq 0$ then $Ay_n \in (Ty_n)_\alpha^{-\delta_n}$ for all $n \in \mathbb{N}$ and so $\|T + \lambda A\| \geq \|T\|$. This completes the proof of the sufficient part.

Let us now prove the comparatively trickier necessary part.

Suppose (a) does not hold. Without loss of generality, let us assume that $\|A\| \leq 1$. Since $T \perp_B A$, for any non-zero scalar λ, $\|T + \lambda A\| \geq \|T\|$. In particular, for $\alpha \in U$ and for each $n \in \mathbb{N}$,

$$\|T + \frac{\alpha}{n} A\| > \|T\| - \frac{1}{n^3}.$$

Therefore, for each $n \in \mathbb{N}$, there exists a sequence $\{x_n\}$ in $S_{\mathbb{X}}$ such that

$$\|(T + \frac{\alpha}{n} A)x_n\| > \|T\| - \frac{1}{n^3} \geq \|Tx_n\| - \frac{1}{n^3}.$$

We claim that $\|Tx_n\| \to \|T\|$. Indeed,

$$\begin{aligned}\|Tx_n\| &= \|(T + \frac{\alpha}{n} A)x_n - \frac{\alpha}{n} Ax_n\| \\ &\geq \|(T + \frac{\alpha}{n} A)x_n\| - |\frac{\alpha}{n}|\|Ax_n\| \\ &> \|T\| - \frac{1}{n^3} - \frac{1}{n}\|A\| \to \|T\| \text{ as } n \to \infty.\end{aligned}$$

Since $x_n \in S_{\mathbb{X}}$, $\|Tx_n\| \leq \|T\|$. This proves our claim.

Since (a) does not hold, we assume that, $\inf_{n\in\mathbb{N}} \|Ax_n\| = c > 0$. Choose $n_1 \in \mathbb{N}$ such that $n_1 > \frac{2\|T\|}{c}$. Since $\|Tx_n\| \to \|T\| > 0$, there exists $n_2 \in \mathbb{N}$ such that $\|Tx_n\| > \frac{\|T\|}{2} > 0$ for all $n \geq n_2$. Choose $n_3 \in \mathbb{N}$ such that $n_3 > \frac{2}{\|T\|}$. Let $n_0 = \max\{n_1, n_2, n_3\}$. Then for all $n \geq n_0$, $0 < \frac{1}{n\|Tx_n\|} < \frac{2}{n\|T\|} < 1$, which implies that, for all $n \geq n_0$, $0 < 1 - \frac{1}{n\|Tx_n\|} < 1$.

Choose $\epsilon_n = \sqrt{1 - (1 - \frac{1}{n\|Tx_n\|})^2}$. Then clearly $\epsilon_n \to 0$ as $n \to \infty$.

We claim that $Ax_n \in (Tx_n)_{\alpha}^{+\epsilon_n}$ for all $n \geq n_0$. Let $n \geq n_0$. Then for $0 \leq t < \frac{1}{n}$,

$$\|Tx_n + t\alpha Ax_n\| \geq \|Tx_n\| - t\|Ax_n\| \geq \|Tx_n\| - \frac{1}{n}.$$

For $\frac{1}{n} \leq t \leq n$, we claim that $\|Tx_n + t\alpha Ax_n\| \geq \|Tx_n\| - \frac{1}{n}$. Suppose on the contrary, we have $\|Tx_n + t\alpha Ax_n\| < \|Tx_n\| - \frac{1}{n}$ for some $\frac{1}{n} \leq t \leq n$. Now, $Tx_n + \frac{\alpha}{n} Ax_n = (1 - \frac{1}{nt})Tx_n + \frac{1}{nt}(Tx_n + t\alpha Ax_n)$. Therefore,

$$\begin{aligned}\|Tx_n\| - \frac{1}{n^3} &< \|Tx_n + \frac{\alpha}{n} Ax_n\| \\ &\leq (1 - \frac{1}{nt})\|Tx_n\| + \frac{1}{nt}\|(Tx_n + t\alpha Ax_n)\| \\ &< (1 - \frac{1}{nt})\|Tx_n\| + \frac{1}{nt}(\|Tx_n\| - \frac{1}{n}) \\ &= \|Tx_n\| - \frac{1}{n^2 t}.\end{aligned}$$

This implies that $t > n$, a contradiction. Thus, for $0 \le t \le n$, $\|Tx_n + t\alpha Ax_n\| \ge \|Tx_n\| - \frac{1}{n}$. Therefore, $0 \le t \le \frac{2\|T\|}{c}$ gives that $\|Tx_n + t\alpha Ax_n\| \ge \|Tx_n\| - \frac{1}{n}$. Now, for $t > \frac{2\|T\|}{c}$,

$$\|Tx_n + t\alpha Ax_n\| \ge t\|Ax_n\| - \|Tx_n\| \ge tc - \|Tx_n\| > 2\|T\| - \|Tx_n\| \ge \|Tx_n\| - \frac{1}{n}.$$

Therefore, for all $t \ge 0$, $\|Tx_n + t\alpha Ax_n\| \ge \|Tx_n\| - \frac{1}{n} = \sqrt{1 - \epsilon_n^2}\ \|Tx_n\|$. This completes the proof of our claim.

Similarly, considering $\|T - \frac{\alpha}{n} A\| > \|T\| - \frac{1}{n^3}$ for each $n \in \mathbb{N}$, we can find the desired norming sequences $\{y_n\}$ for T and $\{\delta_n\}$ in $\mathbb{R}^+$ such that all the conditions of (b) are satisfied. This completes the proof of the necessary part of the theorem. □

For compact operators on reflexive Banach spaces, Theorem 4.5.1 assumes the following form:

Theorem 4.5.2 ([107, Theorem 2.3]) *Let $\mathbb{X}$ be a reflexive Banach space and let $\mathbb{Y}$ be any Banach space. Suppose $T, A \in \mathbb{K}(\mathbb{X}, \mathbb{Y})$. Then $T \perp_B A$ if and only if for each $\alpha \in U$ there exist $x = x(\alpha), y = y(\alpha) \in M_T$ such that $Ax \in (Tx)_\alpha^+$ and $Ay \in (Ty)_\alpha^-$.*

Proof Suppose $T \perp_B A$. Then either (a) or (b) of Theorem 4.5.1 holds. If (a) holds then there exists a sequence $\{x_n\}$ such that $\|Tx_n\| \to \|T\|$ and $\|Ax_n\|$ converges to 0. Then since $\mathbb{X}$ is reflexive, $\{x_n\}$ has a weakly convergent subsequence, say $\{x_{n_k}\}$ converging weakly to x (say). Since T, A are compact operators, $\{Tx_{n_k}\}$, $\{Ax_{n_k}\}$ converges to Tx, Ax respectively. Hence, $x \in M_T$ and $Ax = 0$. Thus, $Ax \in (Tx)_\alpha^+$ and $Ax \in (Tx)_\alpha^-$ for each $\alpha \in U$. If (b) holds, then by using similar arguments as in case (a), we can find $x = x(\alpha), y = (\alpha) \in M_T$ such that $Ax \in (Tx)_\alpha^+$ and $Ay \in (Ty)_\alpha^-$. □

As an immediate consequence of Theorem 4.5.2, we get the characterization of B–J orthogonality of operators between finite-dimensional complex Banach spaces.

Corollary 4.5.3 ([107, Corollary 2.4]) *Let $\mathbb{X}$, $\mathbb{Y}$ be finite-dimensional Banach spaces. Let $T, A \in \mathbb{L}(\mathbb{X}, \mathbb{Y})$. Then $T \perp_B A$ if and only if for each $\alpha \in U$ there exist $x = x(\alpha), y = y(\alpha) \in M_T$ such that $Ax \in (Tx)_\alpha^+$ and $Ay \in (Ty)_\alpha^-$.*

Proof Since every finite-dimensional Banach space is reflexive and every bounded linear operator on a finite-dimensional Banach space is compact, the proof of the corollary follows from Theorem 4.5.2. □

In Theorem 4.3.10, we have seen that if T, A are compact operators on real reflexive Banach space such that M_T is a connected set, then $T \perp_B A$ implies that $Tx \perp_B Ax$ for some $x \in M_T$. Now, we obtain a complex version of Theorem 4.3.10.

Theorem 4.5.4 *Let $\mathbb{X}$ be a reflexive Banach space and let $\mathbb{Y}$ be any Banach space. Suppose $T, A \in \mathbb{L}(\mathbb{X}, \mathbb{Y})$ are such that M_T is a closed connected subset of $S_{\mathbb{X}}$. Then $T \perp_B A$ if and only if for each $\alpha \in U$, there exists $x = x(\alpha) \in M_T$ such that $Tx \perp_B^\alpha Ax$.*

Proof The proof of the sufficient part of the theorem is trivial. We only prove the necessary part. Since $T \perp_B A$, applying Theorem 4.5.2, it follows that, for each $\alpha \in U$, there exist $x = x(\alpha)$, $y = y(\alpha) \in M_T$ such that $Ax \in (Tx)^+_\alpha$ and $Ay \in (Ty)^-_\alpha$. Now, considering the line passing through α and $-\alpha$ in the complex plane and following the same arguments as in Theorem 4.3.10, it can be concluded that there exists $u \in M_T$ such that $Au \in (Tu)^+_\alpha$ and $Au \in (Tu)^-_\alpha$, by using the connectedness of M_T. This implies that $Tu \perp^\alpha_B Au$. This establishes the theorem. □

We would like to mention that Theorem 4.5.4 was initially proved in [107, Theorem 2.9] for $\mathbb{L}(\mathbb{X})$, where $\mathbb{X}$ is a finite-dimensional Banach space.

4.6 Strong B–J Orthogonality in $\mathbb{L}(\mathbb{X}, \mathbb{Y})$

In Theorem 4.2.5, we have proved that, for $T, A \in \mathbb{L}(\mathbb{H})$, if $0 \notin \sigma_{app}(A)$, then $T \perp_B A$ and $T \perp_{SB} A$ are equivalent. The result fails to hold, in general, if the operators are defined on arbitrary Banach spaces. As for example, consider $T, A \in \mathbb{L}(\ell^2_\infty)$ defined by $T(a, b) = (a + b, 0)$ and $A(a, b) = (a - b, a + b)$ for all $a, b \in \mathbb{R}$. It is straightforward to check the following

$$0 \notin \sigma_{app}(A),\ M_T = \{\pm(1, 1)\},\ T(1, 1) \perp_B A(1, 1) \text{ and } \|T + A\| = \|T\|.$$

Thus, $T \perp_B A$ but $T \not\perp_{SB} A$. This example clearly shows that, for $T, A \in \mathbb{L}(\mathbb{X}, \mathbb{Y})$, even if $0 \notin \sigma_{app}(A)$ holds $T \perp_B A$ may not be equivalent to $T \perp_{SB} A$. However, in the next theorem, we show that such case does not arise if the range space is uniformly convex. Let us first recall the definition of uniformly convex Banach space.

Definition 4.6.1 A Banach space $\mathbb{X}$ is said to be uniformly convex if to each ϵ, $0 < \epsilon \leq 2$, there corresponds a $\delta(\epsilon) > 0$ such that the conditions $\|x\| = \|y\| = 1$ and $\|x - y\| > \epsilon$ imply $\frac{\|x+y\|}{2} \leq 1 - \delta(\epsilon)$.

We prove the following theorem for real Banach spaces only. The result can also be proved for complex Banach spaces analogously. In the complex case, we only need to use Theorem 4.5.1, instead of using Theorem 4.4.14.

Theorem 4.6.2 ([126, Theorem 2.16]) *Let $\mathbb{X}$, $\mathbb{Y}$ be Banach spaces such that $\mathbb{Y}$ is uniformly convex. Let $T, A \in \mathbb{L}(\mathbb{X}, \mathbb{Y})$. If $0 \notin \sigma_{app}(A)$ then $T \perp_B A$ and $T \perp_{SB} A$ are equivalent.*

Proof The proof of the part $T \perp_{SB} A \Rightarrow T \perp_B A$ is trivial. We show that $T \perp_B A \Rightarrow T \perp_{SB} A$. Suppose on the contrary that $T \perp_B A$ but $T \not\perp_{SB} A$. Without loss of generality, we assume that $\|T\| = 1$. Applying Theorem 4.4.14, we get norming sequences $\{x_n\}$, $\{y_n\}$ for T and real sequences $\{\epsilon_n\}$, $\{\delta_n\} \subseteq [0, 1)$ such that $Ax_n \in (Tx_n)^{+\epsilon_n}$, $Ay_n \in (Ty_n)^{-\delta_n}$ and $\epsilon_n \to 0$, $\delta_n \to 0$. Since $T \not\perp_{SB} A$, there exists a non-zero scalar λ_0 such that $\|T + \lambda_0 A\| = \|T\|$. Without loss of generality, we assume that $\lambda_0 > 0$. Then for any $\lambda \in (0, \lambda_0)$, $T + \lambda A = (1 - \frac{\lambda}{\lambda_0})T + \frac{\lambda}{\lambda_0}(T + \lambda_0 A)$. Thus, from $T \perp_B A$, it follows that

$$\|T + \lambda A\| \leq (1 - \frac{\lambda}{\lambda_0})\|T\| + \frac{\lambda}{\lambda_0}\|(T + \lambda_0 A)\| = (1 - \frac{\lambda}{\lambda_0})\|T\| + \frac{\lambda}{\lambda_0}\|T\| = \|T\| \leq \|T + \lambda A\|.$$

Therefore, $\|T + \lambda A\| = \|T\|$ for all $0 \leq \lambda \leq \lambda_0$. Thus, for all $\lambda \in [0, \lambda_0]$, we have $\|T\| = \|T + \lambda A\| \geq \|(T + \lambda A)x_n\| \geq \sqrt{1 - \epsilon_n{}^2}\|Tx_n\| \to \|T\|$. Hence

$$\|(T + \lambda A)x_n\| \to \|T\| = 1 \;\forall\; \lambda \in [0, \lambda_0]. \tag{4.6.1}$$

Let $\inf \|Ax_n\| = c > 0$, since $0 \notin \sigma_{app}(A)$. Take $\lambda_1, \lambda_2 \in [0, \lambda_0]$ such that $\lambda_1 \neq \lambda_2$ and $|\lambda_1 - \lambda_2| \leq \frac{2}{c}$. Now, choose $0 < \epsilon < c|\lambda_1 - \lambda_2|$. Then

$$\|(Tx_n + \lambda_1 Ax_n) - (Tx_n + \lambda_2 Ax_n)\| = |\lambda_1 - \lambda_2|\|Ax_n\| \geq |\lambda_1 - \lambda_2|c > \epsilon.$$

On the other hand, $\|Tx_n + \lambda_i Ax_n\| \leq \|T\| = 1 \Rightarrow Tx_n + \lambda_i Ax_n \in B_{\mathbb{Y}}$ for $i = 1, 2$. Since $\mathbb{Y}$ is uniformly convex, clearly $\|\frac{Tx_n + \lambda_1 Ax_n}{2} + \frac{Tx_n + \lambda_2 Ax_n}{2}\| \leq 1 - \delta_{\mathbb{Y}}(\epsilon)$, where $\delta_{\mathbb{Y}}(\epsilon) > 0$ is the modulus of convexity of $\mathbb{Y}$. This implies that $\|Tx_n + \frac{\lambda_1 + \lambda_2}{2}Ax_n\| \leq 1 - \delta_{\mathbb{Y}}(\epsilon)$. Thus, $\lim_{n\to\infty} \|Tx_n + \frac{\lambda_1 + \lambda_2}{2}Ax_n\| \leq 1 - \delta_{\mathbb{Y}}(\epsilon) < 1$, which contradicts (4.6.1). This completes the proof. □

Theorem 4.6.2 indicates that when $\mathbb{Y}$ is not uniformly convex, it is interesting to study B–J orthogonality in $\mathbb{L}(\mathbb{X}, \mathbb{Y})$ by classifying them into two exhaustive categories, namely "$T \perp_{SB} A$" and "$T \perp_B A$ but $T \not\perp_{SB} A$." We next proceed in this particular direction. At first, we explore strong B–J orthogonality in $\mathbb{L}(\mathbb{X}, \mathbb{Y})$.

Theorem 4.6.3 ([88, Theorem 3.5]) *Let $\mathbb{X}$, $\mathbb{Y}$ be finite-dimensional Banach spaces. Let $T, A \in \mathbb{L}(\mathbb{X}, \mathbb{Y})$. Then $T \perp_{SB} A$ if and only if for any $\epsilon > 0$, there exists $\lambda_\epsilon > 0$ such that, for each $|\lambda| < \lambda_\epsilon$, there exists $y_\lambda \in (\cup_{x \in M_T} B(x, \epsilon)) \cap S_{\mathbb{X}}$ such that $\|Ty_\lambda + \lambda Ay_\lambda\| > \|T\|$.*

Proof We first prove the sufficient part. Let $\epsilon > 0$ be given. Then there exists $\lambda_\epsilon > 0$ such that, for each $|\lambda| < \lambda_\epsilon$, there exists $y_\lambda \in (\cup_{x \in M_T} B(x, \epsilon)) \cap S_{\mathbb{X}}$ such that $\|Ty_\lambda + \lambda Ay_\lambda\| > \|T\|$. Therefore, $\|T + \lambda A\| \geq \|Ty_\lambda + \lambda Ay_\lambda\| > \|T\|$. Now, by the convexity of the norm function, it is easy to observe that $T \perp_{SB} A$.

We next prove the necessary part. Suppose that $T \perp_{SB} A$. Let $\epsilon > 0$ be given. Now, using the compactness of $S_{\mathbb{X}}$, it is easy to observe that $\sup\{\|Tz\| : z \in S_{\mathbb{X}} \setminus (\cup_{x \in M_T} B(x, \epsilon))\} < \|T\| - \delta$ for some $\delta > 0$. Let $z \in S_{\mathbb{X}} \setminus (\cup_{x \in M_T} B(x, \epsilon))$. Then $\|Tz\| < \|T\| - \delta$. Therefore,

$$\begin{aligned}\|Tz + \lambda Az\| &\leq \|Tz\| + |\lambda|\|Az\| \\ &< \|T\| - \delta + |\lambda|\|A\| \\ &< \|T\| \quad (\text{for some } |\lambda| < \lambda_\epsilon).\end{aligned}$$

Now, since $\|T + \lambda A\| > \|T\|$, $(T + \lambda A)$ does not attain its norm in $S_{\mathbb{X}} \setminus (\cup_{x \in M_T} B(x, \epsilon)$ for each $|\lambda| < \lambda_\epsilon$. Hence, there exists $y_\lambda \in S_{\mathbb{X}} \cap (\cup_{x \in M_T} B(x, \epsilon))$ such that $\|Ty_\lambda + \lambda A y_\lambda\| = \|T + \lambda A\| > \|T\|$. This completes the proof. □

Now, we consider the other case, i.e., "$T \perp_B A$ but $T \not\perp_{SB} A$." To study this case for compact operators, we need the following lemma, which is a reformulation of [88, Theorem 2.3]

Lemma 4.6.4 *Let $\mathbb{X}$ be a reflexive Banach space and let $\mathbb{Y}$ be any Banach space. Let $T, A \in \mathbb{K}(\mathbb{X}, \mathbb{Y})$. Let λ be a scalar with $|\lambda| = 1$. Then $\|T + \lambda A\| = \|T\| + \|A\|$ if and only if there exists $x \in M_T \cap M_A$ such that $\|Tx + \lambda Ax\| = \|Tx\| + \|Ax\|$.*

Proof First we prove the necessary part. Let λ be a scalar with $|\lambda| = 1$ and $\|T + \lambda A\| = \|T\| + \|A\|$. The operator $T + \lambda A$ is a compact operator on a reflexive Banach space and so it attains its norm. Therefore, there exists $x \in S_{\mathbb{X}}$ such that $\|T + \lambda A\| = \|(T + \lambda A)x\|$. Thus,

$$\begin{aligned} \|T\| + \|A\| &= \|T + \lambda A\| \\ &= \|(T + \lambda A)x\| \\ &\leq \|Tx\| + \|Ax\| \\ &\leq \|Tx\| + \|A\| \\ &\leq \|T\| + \|A\|. \end{aligned}$$

This implies that $\|Tx + \lambda Ax\| = \|Tx\| + \|Ax\|$ and $\|Tx\| = \|T\|$. Similarly, $\|Ax\| = \|A\|$. Therefore, $x \in M_T \cap M_A$. This completes the proof of the necessary part.

For the sufficient part, let $x \in M_T \cap M_A$ be such that $\|Tx + \lambda Ax\| = \|Tx\| + \|Ax\|$. Therefore,

$$\|T\| + \|A\| \geq \|T + \lambda A\| \geq \|(T + \lambda A)x\| = \|Tx\| + \|Ax\| = \|T\| + \|A\|.$$

Thus, $\|T + \lambda A\| = \|T\| + \|A\|$. This completes the proof. □

In Theorem 4.3.10, we have seen that if T, A are compact operators on a reflexive Banach space such that $T \perp_B A$ holds, then $M_T = D \cup (-D)$ (D is a non-empty connected subset) is sufficient to get $x \in M_T$ such that $Tx \perp_B Ax$ holds. In the next theorem, we show that $T \not\perp_{SB} A$ is another sufficient condition for the same conclusion.

Theorem 4.6.5 ([88, Theorem 3.6]) *Let $\mathbb{X}$ be a reflexive Banach space and let $\mathbb{Y}$ be any Banach space. Let $T, A \in \mathbb{K}(\mathbb{X}, \mathbb{Y})$ be such that $T \perp_B A$ but $T \not\perp_{SB} A$. Then there exists $x \in M_T$ such that $Tx \perp_B Ax$.*

Proof Without loss of generality, we assume that $\|T\| = 1$. Let $T \perp_B A$ but $T \not\perp_{SB} A$. Then there exists a scalar $\mu \neq 0$ such that $\|T + \mu A\| = \|T\| = 1$. Consider $B = -\mu A - T$. Then $\|B\| = 1$. Since B–J orthogonality is homogeneous, we have $T \perp_B$

$-\mu A$, i.e., $T \perp_B (B + T)$. Therefore, from Theorem 2.1.15, it follows that there exists $\Phi \in \mathbb{K}(\mathbb{X}, \mathbb{Y})^*$ such that $\|\Phi\| = 1$, $\Phi(T) = \|T\|$ and $\Phi(T + B) = 0$. Clearly, $\Phi(-B) = \|T\| = 1 = \|-B\|$. Therefore,

$$\|T\| + \|-B\| = \Phi(T) + \Phi(-B) = \Phi(T - B) \leq \|\Phi\|\|T - B\| = \|T - B\| \leq \|T\| + \|-B\|.$$

Thus, $\|T - B\| = \|T\| + \|-B\|$. Applying Lemma 4.6.4, we get $x \in M_T \cap M_B$ such that $\|Tx - Bx\| = \|Tx\| + \|-Bx\|$. We note that there exists $g \in S_{\mathbb{X}^*}$ such that $g(Tx - Bx) = \|Tx - Bx\|$. Therefore,

$$g(Tx) + g(-Bx) = g(Tx - Bx) = \|Tx - Bx\| = \|Tx\| + \|-Bx\|.$$

On the other hand, $g(Tx) \leq \|Tx\|$ and $g(-Bx) \leq \|-Bx\|$ gives that $g(Tx) = \|Tx\| = 1$ and $g(-Bx) = \|-Bx\| = 1$. This implies that $g(Tx + Bx) = 0$. Therefore, from Theorem 2.1.15, it follows that $Tx \perp_B (Tx + Bx)$, i.e., $Tx \perp_B (-\mu Ax)$. Hence, $Tx \perp_B Ax$. This completes the proof. □

Combining Theorems 4.6.5 and 4.3.12 we gain a considerable insight on the number of components of the norm attainment set of an operator on two-dimensional Banach spaces.

Corollary 4.6.6 ([88, Corollary 3.8]) *Let $\mathbb{X}$ be a two-dimensional Banach space and let $T \in \mathbb{L}(\mathbb{X})$. If for any $A \in \mathbb{L}(\mathbb{X})$, $T \perp_B A$ implies that $T \not\perp_{SB} A$ then M_T cannot have more than two components.*

Proof Let $T \in \mathbb{L}(\mathbb{X})$. Suppose for any $A \in \mathbb{L}(\mathbb{X})$, $T \perp_B A$ implies that $T \not\perp_{SB} A$. Suppose on the contrary that M_T has more than two components. Then by Theorem 4.3.12, there exists $A \in \mathbb{L}(\mathbb{X})$ such that $T \perp_B A$ but $Tx \not\perp_B Ax$ for any $x \in M_T$. Following Theorem 4.6.5, we get $T \perp_{SB} A$, which contradicts the hypothesis. Thus, M_T cannot have more than two components. □

However, we exhibit an easy example to show that the converse of Corollary 4.6.6 is not true.

Example 4.6.7 Consider $T, A \in \mathbb{L}(\ell_\infty^2)$ defined by $T(x, y) = (x, x)$ and $A(x, y) = (-x, x)$ respectively. Clearly, M_T has only two components but $T \perp_{SB} A$, since $\|T + \lambda A\| \geq \|(T + \lambda A)(1, 1)\| > \|T(1, 1)\| = \|T\|$ for all $\lambda \in \mathbb{R} \setminus \{0\}$.

To generalize Theorem 4.6.5 for bounded linear operators on arbitrary Banach spaces, we need the following two lemmas. The first lemma can be regarded as a generalization of Lemma 4.6.4 and a reformulation of [88, Theorem 2.6].

Lemma 4.6.8 *Let $\mathbb{X}, \mathbb{Y}$ be Banach spaces and let $T, A \in \mathbb{L}(\mathbb{X}, \mathbb{Y})$. Suppose λ is a scalar with $|\lambda| = 1$. Then $\|T + \lambda A\| = \|T\| + \|A\|$ if and only if there exists a sequence $\{x_n\}$ in $S_{\mathbb{X}}$ such that*

$$\lim_{n\to\infty} \|Tx_n\| = \|T\|, \quad \lim_{n\to\infty} \|Ax_n\| = \|A\|$$

and

$$\lim_{n\to\infty} \|Tx_n + \lambda A x_n\| = \|T\| + \|A\|.$$

Proof First we prove the necessary part. Let $\|T + \lambda A\| = \|T\| + \|A\|$. Then there exists a sequence $\{x_n\}$ in $S_{\mathbb{X}}$ such that $\lim_{n\to\infty} \|(T + \lambda A)x_n\| = \|T + \lambda A\|$. Both $\{\|Tx_n\|\}$ and $\{\|Ax_n\|\}$ are bounded sequences of real numbers and so they have convergent subsequences. Without loss of generality, we assume that $\lim_{n\to\infty} \|Tx_n\|$ and $\lim_{n\to\infty} \|Ax_n\|$ exists. Therefore,

$$\begin{aligned}
\|T\| + \|A\| &= \|T + \lambda A\| \\
&= \lim_{n\to\infty} \|(T + \lambda A)x_n\| \\
&\le \lim_{n\to\infty} \|Tx_n\| + \lim_{n\to\infty} \|Ax_n\| \\
&\le \lim_{n\to\infty} \|Tx_n\| + \|A\| \\
&\le \|T\| + \|A\|.
\end{aligned}$$

This implies that $\lim_{n\to\infty} \|Tx_n + \lambda A x_n\| = \lim_{n\to\infty} \|Tx_n\| + \lim_{n\to\infty} \|Ax_n\|$ and $\lim_{n\to\infty} \|Tx_n\| = \|T\|$. Similarly, $\lim_{n\to\infty} \|Ax_n\| = \|A\|$. This completes the proof of the necessary part.
We next prove the sufficient part. Assume that there exists a sequence $\{x_n\}$ in $S_{\mathbb{X}}$ such that $\lim_{n\to\infty} \|Tx_n\| = \|T\|$, $\lim_{n\to\infty} \|Ax_n\| = \|A\|$ and $\lim_{n\to\infty} \|Tx_n + \lambda A x_n\| = \|T\| + \|A\|$. Therefore,

$$\begin{aligned}
\|T\| + \|A\| &\ge \|T + \lambda A\| \\
&\ge \lim_{n\to\infty} \|(T + \lambda A)x_n\| \\
&= \|T\| + \|A\|.
\end{aligned}$$

Thus, $\|T + \lambda A\| = \|T\| + \|A\|$. This completes the proof. □

Lemma 4.6.9 ([88, Lemma 3.10]) *Let $\mathbb{X}$ be a Banach space. Suppose $\{x_n\}$, $\{y_n\}$ are two bounded sequences of $\mathbb{X}$ such that*

$$\lim_{n\to\infty} \|x_n + y_n\| = \lim_{n\to\infty} \|x_n\| + \lim_{n\to\infty} \|y_n\|.$$

Then for each $k \in \mathbb{N}$, there exists $f_{n_k} \in S_{\mathbb{X}^*}$ such that

$$\lim_{k\to\infty} f_{n_k}(x_{n_k}) = \lim_{k\to\infty} \|x_{n_k}\|$$

and

$$\lim_{k\to\infty} f_{n_k}(y_{n_k}) = \lim_{k\to\infty} \|y_{n_k}\|.$$

Proof By the Hahn–Banach Theorem, for each $n \in \mathbb{N}$, there exists $f_n \in S_{\mathbb{X}^*}$ such that $f_n(x_n + y_n) = \|x_n + y_n\|$. Therefore, $\lim_{n\to\infty} f_n(x_n + y_n) = \lim_{n\to\infty} \|x_n + y_n\|$.

Since $\{x_n\}$, $\{y_n\}$ are bounded sequences of $\mathbb{X}$, $\{f_n(x_n)\}$, $\{f_n(y_n)\}$ are bounded sequences of $\mathbb{R}$. Therefore, $\{f_n(x_n)\}$, $\{f_n(y_n)\}$ have convergent subsequences, say, $\{f_{n_k}(x_{n_k})\}$, $\{f_{n_k}(y_{n_k})\}$ respectively. Then $\lim_{k\to\infty} f_{n_k}(x_{n_k})$ and $\lim_{k\to\infty} f_{n_k}(y_{n_k})$ exist and so

$$\begin{aligned}\lim_{k\to\infty} f_{n_k}(x_{n_k}) + \lim_{k\to\infty} f_{n_k}(y_{n_k}) &= \lim_{k\to\infty} f_{n_k}(x_{n_k} + y_{n_k})\\ &= \lim_{k\to\infty} \|x_{n_k} + y_{n_k}\|\\ &= \lim_{k\to\infty} \|x_{n_k}\| + \lim_{k\to\infty} \|y_{n_k}\|.\end{aligned}$$

On the other hand, from $\lim_{k\to\infty} f_{n_k}(x_{n_k}) \leq \lim_{k\to\infty} \|x_{n_k}\|$ and $\lim_{k\to\infty} f_{n_k}(y_{n_k}) \leq \lim_{k\to\infty} \|y_{n_k}\|$, it follows that

$$\lim_{k\to\infty} f_{n_k}(x_{n_k}) = \lim_{k\to\infty} \|x_{n_k}\|$$

and

$$\lim_{k\to\infty} f_{n_k}(y_{n_k}) = \lim_{k\to\infty} \|y_{n_k}\|.$$

This completes the proof. □

Now, we are ready to generalize Theorem 4.6.5 for bounded linear operators.

Theorem 4.6.10 ([88, Theorem 3.11]) *Let $\mathbb{X}$, $\mathbb{Y}$ be Banach spaces and let $T, A \in \mathbb{L}(\mathbb{X}, \mathbb{Y})$. Suppose $T \perp_B A$ but $T \not\perp_{SB} A$. Then either there exists a norming sequence $\{x_n\}$ for T such that $Ax_n \to 0$ or there exists a norming sequence $\{x_n\}$ for T and a sequence $\{\epsilon_n\}$ in $\mathbb{R}^+$ such that $\epsilon_n \to 0$ and $Tx_n \perp_D^{\epsilon_n} Ax_n$.*

Proof Without loss of generality, we assume that $\|T\| = 1$. Let $T \perp_B A$ but $T \not\perp_{SB} A$. Then there exists a scalar $\mu \neq 0$ such that $\|T + \mu A\| = \|T\| = 1$. Let $B = -\mu A - T$. Then $\|B\| = 1$. Now, $T \perp_B A$ gives that $T \perp_B (B + T)$. Therefore, by Theorem 2.1.15, there exists $\Phi \in \mathbb{L}(\mathbb{X}, \mathbb{Y})^*$ such that $\|\Phi\| = 1$, $\Phi(T) = \|T\|$ and $\Phi(T + B) = 0$. Clearly, $\Phi(-B) = \|T\| = 1 = \|-B\|$. Therefore,

$$\|T\| + \|-B\| = \Phi(T) + \Phi(-B) = \Phi(T - B) \leq \|\Phi\|\|T - B\| = \|T - B\| \leq \|T\| + \|-B\|.$$

Thus, $\|T - B\| = \|T\| + \|-B\|$. From Lemma 4.6.8, it follows that there exists a sequence $\{x_n\}$ in $S_{\mathbb{X}}$ such that

$$\lim_{n\to\infty} \|Tx_n\| = \|T\|, \ \lim_{n\to\infty} \|Bx_n\| = \|B\|$$

and

$$\lim_{n\to\infty} \|Tx_n - Bx_n\| = \lim_{n\to\infty} \|Tx_n\| + \lim_{n\to\infty} \|-Bx_n\|.$$

Now, by Lemma 4.6.9, without loss of generality, we may assume that, for each $n \in \mathbb{N}$ there exists $f_n \in S_{\mathbb{Y}^*}$ such that

$$\lim_{n\to\infty} f_n(Tx_n) = \lim_{n\to\infty} \|Tx_n\| = \|T\| = 1$$

and

$$\lim_{n\to\infty} f_n(-Bx_n) = \lim_{n\to\infty} \|-Bx_n\| = \|B\| = 1.$$

Clearly, $f_n(Tx_n) = r_n$ and $f_n(-Bx_n) = s_n$, where $|r_n| \le 1$, $|s_n| \le 1$, $r_n \to 1$ and $s_n \to 1$. Therefore, $f_n(Tx_n + Bx_n) = r_n - s_n$, which converges to 0 as $n \to \infty$. If $\{Ax_n\}$ has a subsequence converging to 0 then we are done. So assume that $\{Ax_n\}$ has no subsequence converging to 0. Without loss of generality, we may assume that $\inf_n \|Ax_n\| = c > 0$. Now, observe that $1 > 1 - (|r_n| - \frac{2}{c|\mu|}|r_n - s_n|)^2 \ge 0$, since $|r_n| \to 1$, $|r_n| \le 1$ and $|r_n - s_n| \to 0$. Let $\epsilon_n^2 = 1 - (|r_n| - \frac{2}{c|\mu|}|r_n - s_n|)^2$ for each $n \in \mathbb{N}$. Then clearly $\epsilon_n \in [0, 1)$ for each $n \in \mathbb{N}$ and $\epsilon_n \to 0$. We show that $Tx_n \perp_D^{\epsilon_n} Ax_n$. First, let $|\lambda| \ge \frac{2}{c}\|Tx_n\| \ge \frac{2}{\|Ax_n\|}\|Tx_n\|$. Then

$$\|Tx_n + \lambda Ax_n\| \ge |\lambda|\|Ax_n\| - \|Tx_n\| \ge \|Tx_n\| \ge \sqrt{1 - \epsilon_n^2}\|Tx_n\|.$$

Let $|\lambda| < \frac{2}{c}\|Tx_n\| \le \frac{2}{c}$. Then

$$\begin{aligned}
\|Tx_n + \lambda Ax_n\| &= \|Tx_n - \frac{\lambda}{\mu}(Tx_n + Bx_n)\| \\
&\ge |f_n\Big(Tx_n - \frac{\lambda}{\mu}(Tx_n + Bx_n)\Big)| \\
&= |r_n - \frac{\lambda}{\mu}(r_n - s_n)| \\
&\ge |r_n| - \frac{|\lambda|}{|\mu|}|r_n - s_n| \\
&> |r_n| - \frac{2}{c|\mu|}|r_n - s_n| \\
&= \sqrt{1 - \epsilon_n^2} \\
&\ge \sqrt{1 - \epsilon_n^2}\|Tx_n\|.
\end{aligned}$$

Hence, $Tx_n \perp_D^{\epsilon_n} Ax_n$. This completes the proof. □

We end this section with an example which shows that if $T \perp_B A$ then the conditions of Theorem 4.6.10 may not be sufficient for $T \not\perp_{SB} A$.

Example 4.6.11 Let $\mathbb{X} = \ell_\infty(\mathbb{R})$. Define linear operators $T, A : \mathbb{X} \to \mathbb{X}$ as follows:

$$T(x_1, x_2, x_3, \ldots) = (x_1, x_1, x_1, \ldots)$$
$$A(x_1, x_2, x_3, \ldots) = (-x_2, x_2, x_2, \ldots).$$

Clearly $T, A \in \mathbb{L}(\mathbb{X})$ with $\|T\| = \|A\| = 1$. Now, for $\lambda \neq 0$, $\|T + \lambda A\| \geq \|(T + \lambda A)(1, 1, 0, 0, 0, \ldots)\| = \|(1 - \lambda, 1 + \lambda, 1 + \lambda, \ldots)\| = 1 + |\lambda| > 1 = \|T\|$.

Therefore, $T \perp_{SB} A$. But $(1, 0, 0, \ldots) \in M_T$ and choosing $y_n = (1, 0, 0, \ldots)$ we get $\|Ty_n\| \to \|T\|$, $\|Ay_n\| \to 0$.

Similarly, defining $B : \mathbb{X} \longrightarrow \mathbb{X}$ by $B(x_1, x_2, x_3, \ldots) = (-x_1, x_1, x_1, \ldots)$, we see that $T \perp_{SB} B$ although there exist $y_n = (1, 0, 0, \ldots) \in M_T$, $\epsilon_n = 0$ such that $\|Ty_n\| \to \|T\|$ and $Ty_n \perp_D^{\epsilon_n} By_n$.

4.7 Orthogonality to a Subspace of Operators

Suppose that $\mathcal{W}$ is a closed subspace of $\mathbb{L}(\mathbb{X}, \mathbb{Y})$ and $T \in \mathbb{L}(\mathbb{X}, \mathbb{Y})$. Till now, we have restricted our study to the case $T \perp_B \mathcal{W}$, where $\dim(\mathcal{W}) = 1$. In this section, we consider the case $\dim(\mathcal{W}) \geq 1$. Throughout the section, we use the following classical result by Singer.

Theorem 4.7.1 ([141, Theorem 1.1, pp. 170]) *Let $\mathbb{X}$ be a Banach space. Let $x \in \mathbb{X}$ and $\mathcal{W}$ be a subspace of $\mathbb{X}$ such that $\dim(\mathcal{W}) = n$ and $x \notin \mathcal{W}$. Then $x \perp_B \mathcal{W}$ if and only if there exist h extreme points $f_1, f_2, \ldots, f_h \in Ext(B_{\mathbb{X}^*})$, ($h \leq n + 1$ if the scalars are real and $h \leq 2n + 1$ if the scalars are complex), h numbers $\lambda_1, \lambda_2, \ldots, \lambda_h > 0$ such that $\sum_{i=1}^{h} \lambda_i = 1$, $\sum_{i=1}^{h} \lambda_i f_i(w) = 0$ for all $w \in \mathcal{W}$ and $f_i(x) = \|x\|$.*

We first characterize $T \perp_B \mathcal{W}$, where T is a compact operator and $\mathcal{W}$ is a finite-dimensional subspace of compact operators. We note that for a non-zero $T \in \mathbb{L}(\mathbb{X}, \mathbb{Y})$, $J(T) = \{\Phi \in \mathbb{L}(\mathbb{X}, \mathbb{Y})^* : \|\Phi\| = 1, \Phi(T) = \|T\|\}$ is a non-empty weak* compact convex set. Clearly in Theorem 4.7.1, $f_i \in Ext(B_{\mathbb{X}^*}) \cap J(x) = Ext(J(x))$. Now, we are ready to prove our desired result, which was first proved in [84, Theorem 2.4] assuming that $\mathbb{Y}$ is finite-dimensional.

Theorem 4.7.2 *Let $\mathbb{X}$ be a reflexive Banach space and let $\mathbb{Y}$ be any Banach space. Suppose $\mathcal{W}$ is a subspace of $\mathbb{K}(\mathbb{X}, \mathbb{Y})$ with $\dim(\mathcal{W}) = n$. Then for any $T \in \mathbb{K}(\mathbb{X}, \mathbb{Y})$, $T \perp_B \mathcal{W}$ if and only if there exist $x_i \in M_T \cap Ext(B_{\mathbb{X}})$, $y_i^* \in Ext(J(Tx_i))$ and $\lambda_i > 0$ for $1 \leq i \leq h$, ($h \in \mathbb{N}$, $h \leq n + 1$ if the scalars are real and $h \leq 2n + 1$ if the scalars are complex) such that $\sum_{i=1}^{h} \lambda_i = 1$ and $\sum_{i=1}^{h} \lambda_i y_i^*(Ax_i) = 0$ for all $A \in \mathcal{W}$.*

Proof We first prove the sufficient part. From $y_i^* \otimes x_i(T) = y_i^*(Tx_i) = \|Tx_i\| = \|T\|$, it follows that $y_i^* \otimes x_i \in J(T)$. Let $\Phi = \sum_{i=1}^{h} \lambda_i y_i^* \otimes x_i$. Then $\Phi \in J(T)$, since $J(T)$ is a convex set. Suppose that $A \in \mathcal{W}$. Now, $\Phi(A) = \sum_{i=1}^{h} \lambda_i y_i^* \otimes x_i(A) = \sum_{i=1}^{h} \lambda_i y_i^*(Ax_i) = 0$. Therefore,

$$\|T + A\| \geq |\Phi(T + A)| = |\Phi(T)| = \|T\|,$$

i.e., $T \perp_B A \Rightarrow T \perp_B \mathcal{W}$. This completes the proof of the sufficient part.
For the necessary part, suppose that $T \perp_B \mathcal{W}$. From Theorem 4.7.1, it follows that

there exist h numbers $\lambda_1, \ldots, \lambda_h > 0$ ($h \leq n+1$ if the scalars are real and $h \leq 2n+1$ if the scalars are complex) and $\Phi_1, \ldots \Phi_h \in Ext J(T)$ such that $\sum_{i=1}^{h} \lambda_i = 1$ and $\sum_{i=1}^{h} \lambda_i \Phi_i(A) = 0$ for all $A \in \mathcal{W}$. Now, $Ext J(T) \subseteq Ext(B_{\mathbb{K}(\mathbb{X},\mathbb{Y})^*})$. From Theorem 4.4.2, we have

$$Ext(B_{\mathbb{K}(\mathbb{X},\mathbb{Y})^*}) = \{y^* \otimes x \in \mathbb{K}(\mathbb{X},\mathbb{Y})^* : y^* \in Ext(B_{\mathbb{Y}^*}), x \in Ext(B_{\mathbb{X}})\}.$$

Therefore, for each $1 \leq i \leq h$, $\Phi_i = y_i^* \otimes x_i$, where $y_i^* \in Ext(B_{\mathbb{Y}^*}), x_i \in Ext(B_{\mathbb{X}})$. Now,

$$\|T\| = \Phi_i(T) = y_i^* \otimes x_i(T) = y_i^*(Tx_i) \leq \|Tx_i\| \leq \|T\|$$

implies that $\|Tx_i\| = \|T\|$, i.e., $x_i \in M_T \cap Ext(B_{\mathbb{X}})$ and $y_i^*(Tx_i) = \|Tx_i\|$, i.e., $y_i^* \in J(Tx_i) \cap Ext(B_{\mathbb{Y}^*}) = Ext J(Tx_i)$. Moreover, for all $A \in \mathcal{W}$,

$$\sum_{i=1}^{h} \lambda_i \Phi_i(A) = 0 \Rightarrow \sum_{i=1}^{h} \lambda_i y_i^* \otimes x_i(A) = 0 \Rightarrow \sum_{i=1}^{h} \lambda_i y_i^*(Ax_i) = 0.$$

This completes the proof. □

Recall that in a smooth Banach space, there exists a unique s.i.p. compatible with the norm. Now, if we assume that $\mathbb{Y}$ to be smooth, then Theorem 4.7.2 assumes the following form which also generalizes [84, Corollary 2.5].

Corollary 4.7.3 *Suppose $\mathbb{X}, \mathbb{Y}$ are Banach spaces such that $\mathbb{X}$ is reflexive and $\mathbb{Y}$ is smooth. Let $T \in \mathbb{K}(\mathbb{X},\mathbb{Y})$ be non-zero and let $\mathcal{W}$ be a subspace of $\mathbb{K}(\mathbb{X},\mathbb{Y})$ with* $\dim(\mathcal{W}) = n$. *Let $[\ ,\ ]$ be the unique s.i.p. on $\mathbb{Y}$. Then $T \perp_B \mathcal{W}$ if and only if there exist $x_i \in M_T \cap Ext(B_{\mathbb{X}})$ and $\lambda_i > 0$ for $1 \leq i \leq h$ ($h \in \mathbb{N}$, $h \leq n+1$ if the scalars are real and $h \leq 2n+1$ if the scalars are complex) such that $\sum_{i=1}^{h} \lambda_i = 1$ and $\sum_{i=1}^{h} \lambda_i [Ax_i, Tx_i] = 0$ for all $A \in \mathcal{W}$.*

Proof By Theorem 4.7.2, $T \perp_B \mathcal{W}$ if and only if there exist $x_i \in M_T \cap Ext(B_{\mathbb{X}})$, $y_i^* \in Ext J(Tx_i)$ and $\lambda_i > 0$ for $1 \leq i \leq h$ such that $\sum_{i=1}^{h} \lambda_i = 1$ and $\sum_{i=1}^{h} \lambda_i y_i^*$ $(Ax_i) = 0$ for all $A \in \mathcal{W}$. Since $\mathbb{Y}$ is smooth, and $y_i^* \in Ext J(Tx_i)$, we have $y_i^*(z) = [z, Tx_i]$ for all $z \in \mathbb{Y}$. Therefore, $\sum_{i=1}^{h} \lambda_i y_i^*(Ax_i) = 0$ gives that $\sum_{i=1}^{h} \lambda_i [Ax_i, Tx_i] = 0$ for all $A \in \mathcal{W}$. This proves the result. □

Till now in this section, we have considered only compact operators. Our next goal is to generalize Theorem 4.7.2 for bounded linear operators. In the following theorem, we will see that in some spaces, if for a bounded linear operator T, $dist(T, \mathbb{K}(\mathbb{X},\mathbb{Y})) < \|T\|$ then we may consider finite-dimensional subspace $\mathcal{W}$ of bounded linear operators to get an analogous result of Theorem 4.7.2. The proof follows using Lemma 4.4.3 and using similar line of arguments as in Theorem 4.7.2. Therefore, we only state the result here.

Theorem 4.7.4 *Let $\mathbb{X}$ be a reflexive Banach space and let $\mathbb{Y}$ be any Banach space. Suppose $\mathbb{K}(\mathbb{X},\mathbb{Y})$ is an M-ideal in $\mathbb{L}(\mathbb{X},\mathbb{Y})$. Suppose $T \in \mathbb{L}(\mathbb{X},\mathbb{Y})$ is a non-zero* *operator with $dist(T, \mathbb{K}(\mathbb{X}, \mathbb{Y})) < \|T\|$. Let $\mathcal{W}$ be a subspace of $\mathbb{L}(\mathbb{X}, \mathbb{Y})$ such that $\dim(\mathcal{W}) = n$. Then $T \perp_B \mathcal{W}$ if and only if there exist $x_i \in M_T \cap Ext(B_{\mathbb{X}})$, $y_i^* \in Ext J(Tx_i)$ and $\lambda_i > 0$ for $1 \leq i \leq h$, ($h \in \mathbb{N}$, $h \leq n+1$ if the scalars are real and $h \leq 2n+1$ if the scalars are complex) such that $\sum_{i=1}^{h} \lambda_i = 1$ and $\sum_{i=1}^{h} \lambda_i y_i^*(Ax_i) = 0$ for all $A \in \mathcal{W}$.*

As an immediate consequence of Theorem 4.7.4, we get the following corollary analogous to Corollary 4.7.3.

Corollary 4.7.5 *Let $\mathbb{X}$ be a reflexive Banach space and let $\mathbb{Y}$ be a smooth Banach space. Suppose $\mathbb{K}(\mathbb{X}, \mathbb{Y})$ is an M-ideal in $\mathbb{L}(\mathbb{X}, \mathbb{Y})$. Suppose $T \in \mathbb{L}(\mathbb{X}, \mathbb{Y})$ is a non-zero operator with $dist(T, \mathbb{K}(\mathbb{X}, \mathbb{Y})) < \|T\|$. Let $\mathcal{W}$ be a subspace of $\mathbb{L}(\mathbb{X}, \mathbb{Y})$ such that $\dim(\mathcal{W}) = n$. Let $[\ ,\ ]$ be the unique s.i.p. on $\mathbb{Y}$. Then $T \perp_B \mathcal{W}$ if and only if there exist $x_i \in M_T \cap Ext(B_{\mathbb{X}})$ and $\lambda_i > 0$ for $1 \leq i \leq h$ ($h \in \mathbb{N}$, $h \leq n+1$ if the scalars are real and $h \leq 2n+1$ if the scalars are complex) such that $\sum_{i=1}^{h} \lambda_i = 1$ and $\sum_{i=1}^{h} \lambda_i [Ax_i, Tx_i] = 0$ for all $A \in \mathcal{W}$.*

The case where $\mathcal{W}$ is a one-dimensional subspace of $\mathbb{L}(\mathbb{X}, \mathbb{Y})$ is of particular interest. In the following theorem, we see what happens for complex scalars.

Theorem 4.7.6 *Let $\mathbb{X}$ be a reflexive complex Banach space and let $\mathbb{Y}$ be a complex Banach space. Suppose $\mathbb{K}(\mathbb{X}, \mathbb{Y})$ is an M-ideal in $\mathbb{L}(\mathbb{X}, \mathbb{Y})$. Suppose $T \in \mathbb{L}(\mathbb{X}, \mathbb{Y})$ is a non-zero operator with $dist(T, \mathbb{K}(\mathbb{X}, \mathbb{Y})) < \|T\|$. Let $A \in \mathbb{L}(\mathbb{X}, \mathbb{Y})$.*
(i) *Then $T \perp_B A$ if and only if there exist $x_i \in M_T \cap Ext(B_{\mathbb{X}})$, $y_i^* \in Ext J(Tx_i)$ and $\lambda_i \geq 0$ for $1 \leq i \leq 3$, such that $\sum_{i=1}^{3} \lambda_i = 1$ and $\sum_{i=1}^{3} \lambda_i y_i^*(Ax_i) = 0$.*
(ii) In particular, if $M_T = \{\alpha x_0 : |\alpha| = 1\}$ holds for some $x_0 \in S_{\mathbb{X}}$, then $T \perp_B A$ if and only if $Tx_0 \perp_B Ax_0$.

Proof The proof of (i) follows from Theorem 4.7.4. We only prove (ii). If $Tx_0 \perp_B Ax_0$, then clearly $T \perp_B A$. On the other hand, suppose that $T \perp_B A$. Then using (i), we get for $1 \leq i \leq 3$, scalars α_i with $|\alpha_i| = 1$, $\lambda_i \geq 0$, $y_i^* \in Ext J(\alpha_i Tx_0)$ such that $\lambda_1 + \lambda_2 + \lambda_3 = 1$ and $\sum_{i=1}^{3} \alpha_i \lambda_i y_i^*(Ax_0) = 0$. Suppose that $f = \sum_{i=1}^{3} \alpha_i \lambda_i y_i^*$. Then $f(Ax_0) = 0$. Moreover, it is easy to check that $f(Tx_0) = \|Tx_0\|$ and $\|f\| = 1$. Therefore, $Tx_0 \perp_B Ax_0$. This completes the proof. □

We can reformulate Corollary 4.7.3 in the following way, if we assume $\mathbb{X} = \mathbb{Y} = \mathcal{K}^n$, where $\mathcal{K}^n$ is equipped with the usual inner product. In this case, we can denote $\mathbb{L}(\mathcal{K}, \mathcal{K})$ by $M_n(\mathcal{K})$, the collection of all $n \times n$ matrices over $\mathcal{K}$.

Corollary 4.7.7 ([84, Corollary 2.6]) *Let $\mathcal{W}$ be a subspace of $M_n(\mathcal{K})$ and let $T \in M_n(\mathcal{K})$ such that $\|T\| = 1$. Then $T \perp_B \mathcal{W}$ if and only if there exists a density matrix $P \in M_n(\mathcal{K})$ such that $T^*TP = P$ and $tr((TP)^*A) = 0$ for all $A \in \mathcal{W}$ (note that a positive semidefinite matrix with trace 1 is called a density matrix).*

Proof We first prove the necessary part. Let $T \perp_B \mathcal{W}$. Then by Corollary 4.7.3, there exist $x_i \in M_T \cap Ext(B_{\mathbb{X}})$ and $\lambda_i > 0$ for $1 \leq i \leq h$ such that $\sum_{i=1}^{h} \lambda_i = 1$

and $\sum_{i=1}^{h} \lambda_i [Ax_i, Tx_i] = 0$ for all $A \in \mathcal{W}$. Since in an inner product space, the inner product is the only s.i.p., we have

$$\sum_{i=1}^{h} \lambda_i \langle Ax_i, Tx_i \rangle = 0 \Rightarrow \sum_{i=1}^{h} \lambda_i \langle Tx_i x_i^*, A \rangle = 0 \Rightarrow \langle TP, A \rangle = 0,$$

where $P = \sum_{i=1}^{h} \lambda_i x_i x_i^*$. Clearly, P is a density matrix. It then follows that $tr((TP)^* A) = 0$. For each $x_i \in M_T$, $T^*Tx_i = x_i$. Therefore,

$$T^*TP = T^*T(\sum_{i=1}^{h} \lambda_i x_i x_i^*) = \sum_{i=1}^{h} \lambda_i T^*Tx_i x_i^* = \sum_{i=1}^{h} \lambda_i x_i x_i^* = P.$$

This completes the proof of the necessary part.
For the sufficient part, suppose there exists a density matrix $P \in M_n(\mathcal{K})$ such that $T^*TP = P$ and $tr((TP)^* A) = 0$ for all $A \in \mathcal{W}$. Since P is a density matrix, there exist $\lambda_i > 0$ and $x_i \in \mathcal{K}^n$, $\|x_i\| = 1$ for some $1 \leq i \leq h$ such that $\sum_{i=1}^{h} \lambda_i = 1$ and $P = \sum_{i=1}^{h} \lambda_i x_i x_i^*$. Note that λ_i is an eigenvalue of P corresponding to the eigenvector x_i. Now, $T^*TP = P$ implies that $x_i \in M_T$. Also $tr((TP)^* A) = 0$ implies that $\sum_{i=1}^{h} \lambda_i \langle Ax_i, Tx_i \rangle = 0$. Therefore, by Corollary 4.7.3, we get $T \perp_B \mathcal{W}$. This completes the proof. □

We would like to mention that Corollary 4.7.7 was first proved by Grover [41, Theorem 1] using a different technique. To generalize this result, we need the following lemma.

Lemma 4.7.8 ([141, Lemma 1.1, pp. 166]) *Let $\mathbb{X}$ be a Banach space of finite dimension k and let $f \in S_{\mathbb{X}^*}$. Then there exist $f_1, f_2, \ldots, f_h \in Ext(B_{\mathbb{X}^*})$, where $1 \leq h \leq k$ if the scalars are real and $1 \leq h \leq 2k - 1$ if the scalars are complex and scalars $\lambda_1, \lambda_2, \ldots, \lambda_h > 0$ such that $\sum_{i=1}^{h} \lambda_i = 1$ and $f = \sum_{i=1}^{h} \lambda_i f_i$.*

Now, the promised generalization of Corollary 4.7.7 in $\mathbb{L}(\mathbb{H})$ for an arbitrary Hilbert space $\mathbb{H}$:

Theorem 4.7.9 ([84, Theorem 2.8]) *Let $\mathbb{H}$ be a Hilbert space. Suppose $T \in \mathbb{L}(\mathbb{H})$ is such that $\|T\| = 1$, $M_T = S_{H_0}$, where H_0 is a finite-dimensional subspace of $\mathbb{H}$ and $\|T\|_{H_0^\perp} < 1$. Then for any subspace $\mathcal{W}$ of $\mathbb{L}(\mathbb{H})$, $T \perp_B \mathcal{W}$ if and only if there exist $x_1, x_2, \ldots, x_n \in M_T$ and $\lambda_1, \lambda_2, \ldots, \lambda_n > 0$ such that $\sum_{i=1}^{n} \lambda_i = 1$ and $\sum_{i=1}^{n} \lambda_i \langle Ax_i, Tx_i \rangle = 0$ for all $A \in \mathcal{W}$.*

Proof We prove the theorem for a complex Hilbert space. For a real Hilbert space, the proof follows similarly. We first prove the necessary part. We claim that $dist(T, \mathbb{K}(\mathbb{H})) < 1$. If possible, suppose that $dist(T, \mathbb{K}(\mathbb{H})) \geq 1$. Then for every $S \in \mathbb{K}(\mathbb{H})$, $dist(T, \text{span}\{S\}) \geq dist(T, \mathbb{K}(\mathbb{H})) \geq 1$. Then for each scalar λ, $\|T - \lambda S\| \geq dist(T, \text{span}\{S\}) \geq 1$. Therefore, $T \perp_B S$. Let $z \in \mathbb{H}$. Then z can be written as $z = x + y$, $x \in H_0$, $y \in H_0^\perp$. Define $S : \mathbb{H} \to \mathbb{H}$ by $Sz = Tx$. Clearly,

$S \in \mathbb{K}(\mathbb{H})$, since H_0 is finite-dimensional. Hence, $T \perp_B S$. By Theorem 4.2.4, there exists $x \in M_T = S_{H_0}$ such that $Tx \perp Sx$, i.e., $Tx \perp Tx$, a contradiction. This proves the claim.

Now, we show that span $Ext J(T)$ is finite-dimensional. Since $\mathbb{K}(\mathbb{H})$ is an $M-$ideal in $\mathbb{L}(\mathbb{H})$ and $dist(T, \mathbb{K}(\mathbb{H})) < 1$, by Lemma 4.4.3,

$$Ext J(T) = \{y^* \otimes x : x \in M_T, y^* \in Ext J(Tx)\},$$

where $y^* \otimes x(S) = y^*(Sx) = \langle Sx, Tx \rangle$ for every $S \in \mathbb{L}(\mathbb{H})$. So we can write $Ext J(T) = \{x \otimes Tx : x \in M_T\}$, where $x \otimes Tx(S) = \langle Sx, Tx \rangle$ for every $S \in \mathbb{L}(\mathbb{H})$. Suppose that $\{e_1, e_2, \ldots, e_k\}$ is an orthonormal basis of H_0. Then

$$\begin{aligned}
&\text{span } Ext J(T) \\
&= \text{span } \{x \otimes Tx : x \in M_T\} \\
&= \text{span } \{\sum_{i,j=1}^{k} a_i \overline{a_j} e_i \otimes Te_j : \sum_{i=1}^{k} |a_i|^2 = 1\} \\
&= \text{span } \{e_i \otimes Te_j : 1 \leq i, j \leq k\}.
\end{aligned}$$

Therefore, span$Ext J(T)$ is finite-dimensional.

Next, we construct $\Phi \in J(T)$ such that $\Phi(A) = 0$ for all $A \in \mathcal{W}$. Let $\mathcal{Z} =$ span$\{T, \mathcal{W}\}$. Define $\phi : \mathcal{Z} \to \mathbb{C}$ by $\phi(\alpha T + A) = \alpha$, where $\alpha \in \mathbb{C}$ and $A \in \mathcal{W}$. Since $T \perp_B \mathcal{W}$, $\phi \in S_{\mathcal{Z}^*}$. By the Hahn–Banach Theorem, there exists $\Phi \in \mathbb{L}(\mathbb{H})^*$ such that $\|\Phi\| = \|\phi\| = 1$ and $\Phi|_{\mathcal{Z}} = \phi$. Therefore, $\Phi(T) = 1$ implies that $\Phi \in J(T)$ and $\Phi(A) = 0$ for all $A \in \mathcal{W}$. Since $J(T)$ is weak* compact convex subset of $\mathbb{L}(\mathbb{H})^*$, using the Krein-Milman Theorem, we get,

$$\begin{aligned}
J(T) &= \overline{\text{conv}}^{w*}\, Ext J(T) \subseteq \overline{\text{span } Ext J(T)}^{w*} \\
&= \overline{\text{span } Ext J(T)} = \text{span } Ext J(T),
\end{aligned}$$

since span $Ext J(T)$ is finite-dimensional. Now, using Lemma 4.7.8, we get extreme points $\Phi_1, \Phi_2, \ldots, \Phi_n$ of the unit ball of span $Ext J(T)$ and $\lambda_1, \lambda_2, \ldots, \lambda_n > 0$ such that $\sum_{i=1}^n \lambda_i = 1$ and $\Phi = \sum_{i=1}^n \lambda_i \Phi_i$. Observe that $\Phi_i \in Ext J(T)$ for all $1 \leq i \leq n$. So for $1 \leq i \leq n$, there exists $x_i \in M_T$ such that $\Phi_i = x_i \otimes Tx_i$. Now, for all $A \in \mathcal{W}$,

$$\Phi(A) = 0 \Rightarrow \sum_{i=1}^{n} \lambda_i \Phi_i(A) = 0 \Rightarrow \sum_{i=1}^{n} \lambda_i x_i \otimes Tx_i(A) = 0 \Rightarrow \sum_{i=1}^{n} \lambda_i \langle Ax_i, Tx_i \rangle = 0.$$

This completes the proof of the necessary part.

For the sufficient part, suppose that $\sum_{i=1}^n \lambda_i \langle Ax_i, Tx_i \rangle = 0$ for all $A \in \mathcal{W}$, where $\sum_{i=1}^n \lambda_i = 1$ and $\lambda_i > 0$ for $1 \leq i \leq n$. Then $\sum_{i=1}^n \lambda_i x_i \otimes Tx_i(A) = 0$. Now, $x_i \otimes Tx_i \in J(T)$ and $J(T)$ is convex, so $\sum_{i=1}^n \lambda_i x_i \otimes Tx_i \in J(T)$. Let $\Phi = \sum_{i=1}^n \lambda_i x_i \otimes$

Tx_i. Then for every $A \in \mathcal{W}$, $\|T + A\| \geq |\Phi(T + A)| = |\Phi(T)| = 1$. Thus, $T \perp_B \mathcal{W}$. This completes the proof of the theorem. □

Observe that, in Theorem 4.7.2, we have studied $T \perp_B \mathcal{W}$, where $\mathcal{W}$ is a finite-dimensional subspace of compact operators. Instead of considering finite-dimensional subspace, if we consider an arbitrary subspace $\mathcal{W}$ of bounded linear operators, then to get analogous result of Theorem 4.7.2, we need further restrictions on the operator T. Let us recall that, if $J(T)$ contains exactly k-linearly independent functionals, then T is called k-smooth operator. In this case, dimension of span $Ext J(T)$ will be k.

Theorem 4.7.10 *Suppose $\mathbb{X}, \mathbb{Y}$ are Banach spaces such that $\mathbb{X}$ is reflexive. Let $\mathbb{K}(\mathbb{X}, \mathbb{Y})$ be an M-ideal in $\mathbb{L}(\mathbb{X}, \mathbb{Y})$. Let $T \in \mathbb{L}(\mathbb{X}, \mathbb{Y})$ be such that the following conditions hold:*
(i) $\|T\| = 1$ and $\text{dist}(T, \mathbb{K}(\mathbb{X}, \mathbb{Y})) < 1$,
(ii) T is k-smooth for some $k < \infty$.
Then for a subspace $\mathcal{W}$ of $\mathbb{L}(\mathbb{X}, \mathbb{Y})$, $T \perp_B \mathcal{W}$ if and only if there exist $x_i \in M_T \cap Ext(B_{\mathbb{X}})$, $y_i^* \in Ext J(Tx_i)$ and $\lambda_i > 0$ for $1 \leq i \leq n$, $(n \in \mathbb{N})$ such that $\sum_{i=1}^{n} \lambda_i = 1$ and $\sum_{i=1}^{n} \lambda_i y_i^*(Ax_i) = 0$ for all $A \in \mathcal{W}$.

Proof The sufficient part follows similarly as in Theorem 4.7.2. We only prove the necessary part of the theorem. It follows from (ii) that span $Ext J(T)$ is finite-dimensional. Let $\mathcal{Z} = span\{T, \mathcal{W}.\}$ Now, as in Theorem 4.7.9, we define a linear functional ϕ on $\mathcal{Z}$ by $\phi(\alpha T + A) = \alpha$, where $\alpha \in \mathcal{K}$ and $A \in \mathcal{W}$. Using $T \perp_B \mathcal{W}$, we can show that $\|\phi\| = 1$. Therefore, by the Hahn–Banach Theorem, there exists $\Phi \in \mathbb{L}(\mathbb{X}, \mathbb{Y})^*$ such that $\|\Phi\| = 1$ and $\Phi|_{\mathcal{Z}} = \phi$. Clearly, $\Phi \in J(T)$. Since $J(T)$ is weak* compact, convex subset of $\mathbb{L}(\mathbb{X}, \mathbb{Y})^*$, using the Krein-Milman Theorem, we get,

$$\begin{aligned} J(T) &= \overline{\text{conv}}^{w*}(Ext J(T)) \subseteq \overline{\text{span}\, Ext J(T)}^{w*} \\ &= \overline{\text{span}\, Ext J(T)} = \text{span}\, Ext J(T), \end{aligned}$$

since span $Ext J(T)$ is finite-dimensional. Now, proceeding similarly as in Theorem 4.7.9, we get $\Phi_1, \Phi_2, \ldots, \Phi_n \in Ext J(T)$ and $\lambda_1, \lambda_2, \ldots, \lambda_n > 0$ such that $\sum_{i=1}^{n} \lambda_i = 1$ and $\Phi = \sum_{i=1}^{n} \lambda_i \Phi_i$. By Lemma 4.4.3, $\Phi_i = y_i^* \otimes x_i$, where $x_i \in M_T \cap Ext(B_{\mathbb{X}})$ and $y_i^* \in Ext J(Tx_i)$ for each $1 \leq i \leq n$. Now, for each $A \in \mathcal{W}$, $\Phi(A) = \phi(A) = 0$. Thus,

$$\sum_{i=1}^{n} \lambda_i \Phi_i(A) = 0 \Rightarrow \sum_{i=1}^{n} \lambda_i y_i^* \otimes x_i(A) = 0 \Rightarrow \sum_{i=1}^{n} \lambda_i y_i^*(Ax_i) = 0.$$

This completes the proof of the theorem. □

We would like to remark here that Theorem 4.7.10 is a generalization of [84, Theorem 2.9]. Now, we concentrate on a special type of subspace $\mathcal{W}$ of $\mathbb{L}(\mathbb{X}, \mathbb{Y})$ to study $T \perp_B \mathcal{W}$. Consider a closed subspace Z of $\mathbb{X}$. Suppose $L_Z(\mathbb{X}, \mathbb{Y}) := \{T \in \mathbb{L}(\mathbb{X}, \mathbb{Y}) : T|_Z = 0\}$. Considering a closed hyperspace Z of $\mathbb{X}$, Wójcik in [148, Theorem 2.1(a)] studied $T \perp_B L_Z(\mathbb{X}, \mathbb{Y})$ for real scalars and assuming $Ext(B_{\mathbb{Y}^*})$ to be closed. However, as we prove next, [148, Theorem 2.1(a)] can be generalized for complex scalars also and the restriction on $Ext(B_{\mathbb{Y}^*})$ is redundant.

Theorem 4.7.11 ([84, Theorem 2.7]) *Suppose $\mathbb{X}$, $\mathbb{Y}$ are Banach spaces such that $\mathbb{X}$ is reflexive and $\dim \mathbb{Y} = n$. Let $f \in S_{\mathbb{X}^*}$ and let $Z = ker(f)$. Suppose $A \in \mathbb{L}(\mathbb{X}, \mathbb{Y})$, $T_0 \in L_Z(\mathbb{X}, \mathbb{Y})$. Let $A_0 = A - T_0$ do not attain norm on Z, i.e., $M_{A_0} \cap Z = \emptyset$. Let*

$$F_{A_0} = \{y^* \in Ext(B_{\mathbb{Y}^*}) : \exists x \in M_{A_0} \cap Ext(B_{\mathbb{X}}) \text{ with } f(x) > 0, y^*(A_0x) = \|A_0\|\}.$$

Then $A_0 \perp_B L_Z(\mathbb{X}, \mathbb{Y})$ if and only if $0 \in conv(F_{A_0})$.

Proof Let $\mathcal{W} = L_Z(\mathbb{X}, \mathbb{Y})$. Since $\mathbb{Y}$ is finite-dimensional, it is easy to see that $\mathcal{W}$ is finite-dimensional. Suppose that $A_0 \perp_B \mathcal{W}$. Without loss of generality, we may and do assume that $\|A_0\| = 1$. Then by Theorem 4.7.2, there exist $x_i \in M_{A_0} \cap Ext(B_{\mathbb{X}})$, $y_i^* \in ExtJ(A_0x_i)$ and $\lambda_i > 0$ for $1 \leq i \leq h$, $(h \in \mathbb{N})$ such that $\sum_{i=1}^h \lambda_i = 1$ and $\sum_{i=1}^h \lambda_i y_i^*(Sx_i) = 0$ for all $S \in \mathcal{W}$. Assume that $x \in M_f$ and $f(x) = 1$. Then for each $1 \leq i \leq h$, $x_i = a_i x + z_i$, where $a_i \in K$ and $z_i \in Z$. Without loss of generality, we may assume that $a_i > 0$ for all $1 \leq i \leq h$. Then $f(x_i) > 0$ and so $y_i^* \in F_{A_0}$ for all $1 \leq i \leq h$. Now, for all $S \in \mathcal{W}$,

$$\sum_{i=1}^{h} \lambda_i y_i^*(Sx_i) = 0 \Rightarrow \sum_{i=1}^{h} \lambda_i a_i y_i^*(Sx) = 0.$$

Let $\{y_1, y_2, \ldots, y_n\}$ be a basis of $\mathbb{Y}$. For each $1 \leq j \leq n$, define $S_j \in \mathcal{W}$ by $S_j(ax + z) = ay_j$, where $a \in \mathcal{K}$, $z \in Z$. Now,

$$\sum_{i=1}^{h} \lambda_i a_i y_i^*(S_j x) = 0 \Rightarrow \sum_{i=1}^{h} \lambda_i a_i y_i^*(y_j) = 0 \text{ for all } 1 \leq j \leq n.$$

Therefore, $\sum_{i=1}^h \lambda_i a_i y_i^* = 0$. Let $\alpha = \sum_{i=1}^h \lambda_i a_i > 0$. Then from $\sum_{i=1}^h \frac{\lambda_i a_i}{\alpha} y_i^* = 0$ we get, $0 \in \text{conv}(F_{A_0})$.

Conversely, suppose that $0 \in \text{conv}(F_{A_0})$. Then $\sum_{i=1}^h \mu_i y_i^* = 0$, where $\mu_i > 0$, $\sum_{i=1}^h \mu_i = 1$ and $y_i^* \in F_{A_0}$. Clearly, for each $1 \leq i \leq h$, there exists $x_i \in M_{A_0} \cap Ext(B_{\mathbb{X}})$ such that $f(x_i) > 0$ and $y_i^*(A_0x_i) = \|A_0\| = \|A_0x_i\|$. Therefore, $y_i^* \in ExtJ(A_0x_i)$. Now, let $x \in M_f$ and $f(x) = 1$. Let $x_i = a_i x + z_i$, where $a_i \in \mathcal{K}$, $z_i \in Z$. Then from $f(x_i) > 0$, we have $a_i > 0$. Let $\alpha = \sum_{i=1}^h \frac{\mu_i}{a_i} > 0$ and $\lambda_i = \frac{\mu_i}{a_i \alpha} > 0$ Then for all $S \in \mathcal{W}$,

$$\sum_{i=1}^{h} \lambda_i y_i^*(Sx_i) = \sum_{i=1}^{h} \frac{\mu_i}{a_i \alpha} y_i^*(Sx_i) = \sum_{i=1}^{h} \frac{\mu_i}{\alpha} y_i^*(Sx) = \frac{1}{\alpha}(\sum_{i=1}^{h} \mu_i y_i^*)(Sx) = 0.$$

Therefore, by Theorem 4.7.2, $T \perp_B \mathcal{W}$. This completes the proof of the theorem. □

4.8 Norm of an Operator Through Orthogonality

Distance of an operator from a subspace of operators is an important and interesting area of study in approximation theory. The problem has attracted the attention of many accomplished and active researchers [8, 17, 18, 36, 100, 131, 153]. Note that if $T \in \mathbb{L}(\mathbb{H})$ for a finite-dimensional Hilbert space $\mathbb{H}$, and $T_0(= T - a_0 I) \perp_B I$, then $dist(T, \text{span}\{I\}) = \|T_0\|$. Ando in [8] and Fuji and Nakamoto in [36] proved that

$$dist(T, \text{span}\{I\}) = \|T_0\| = \max\{|\langle Tx, y\rangle| : \|x\| = \|y\| = 1, \langle x, y\rangle = 0\}. \tag{4.8.1}$$

In [100, Theorem 5], Paul et al. generalized (4.8.1) by considering span$\{A\}$ instead of span$\{I\}$, where A is an arbitrary operator in $\mathbb{L}(\mathbb{H})$ such that $0 \notin \sigma_{app}(A)$. If $T_0(= T - a_0 A) \perp_B A$, then $dist(T, \text{span}\{A\}) = \|T_0\|$ and by [100, Theorem 5], if $0 \notin \sigma_{app}(A)$, then

$$\begin{aligned} dist(T, \text{span}\{A\}) &= \|T_0\| && (4.8.2)\\ &= \max\{|\langle T_0 x, y\rangle| : \|x\| = \|y\| = 1, \langle Ax, y\rangle = 0\} && (4.8.3)\\ &= \max\{|\langle Tx, y\rangle| : \|x\| = \|y\| = 1, \langle Ax, y\rangle = 0\}. && (4.8.4) \end{aligned}$$

The aim of this section is to apply the results of the previous sections to further generalize the above distance formula in the setting of Banach spaces. We also derive various lower bounds for the distance of an operator from a one-dimensional subspace. The notion of s.i.p. is an integral part of this section. Therefore, at first we relate the notion of s.i.p. with the notions x^+ and x^- to successfully apply the results of previous sections.

Theorem 4.8.1 ([119, Theorem 2.4]) *Let $\mathbb{X}$ be a Banach space and let $x, y \in \mathbb{X}$. Then the following are true:*
(i) $y \in x^+$ if and only if there exists a s.i.p. $[\ ,\ ]$ on $\mathbb{X}$ such that $[y, x] \geq 0$.
(ii) $y \in x^-$ if and only if there exists a s.i.p. $[\ ,\ ]$ on $\mathbb{X}$ such that $[y, x] \leq 0$.

Proof (i) We first prove the sufficient part. Suppose there exists a s.i.p. $[\ ,\ ]$ on $\mathbb{X}$ such that $[y, x] \geq 0$. Then for any $\lambda \geq 0$,

$$\|x + \lambda y\|\|x\| \geq |[x + \lambda y, x]| = |\|x\|^2 + \lambda[y, x]| = \|x\|^2 + \lambda[y, x] \geq \|x\|^2.$$

This shows that $y \in x^+$.

Conversely, let $y \in x^+$. Then from Proposition 3.2.2, it follows that either $x \perp_B y$ or $y \in x^+ \setminus x^-$. If $x \perp_B y$, then by [33, Proposition 5.3], there exists a s.i.p $[\ ,\ ]$ on $\mathbb{X}$ such that $[y, x] = 0$. So assume that $y \in x^+ \setminus x^-$. Without loss of generality, we assume that $x \in S_{\mathbb{X}}$. Now, we consider two possible cases of $\{x, y\}$ is linearly dependent and linearly independent. Suppose $\{x, y\}$ is linearly dependent. Then $y = \alpha x$, for some scalar $\alpha \geq 0$, as $y \in x^+$. By virtue of the Hahn–Banach Theorem, for each $v \in S_{\mathbb{X}}$, there exists at least one linear functional $f_v \in S_{\mathbb{X}^*}$ such that $f_v(v) = \|v\| = 1$. We select exactly one such f_v for each $v \in S_{\mathbb{X}}$. For $\lambda \in \mathbb{R}$, we choose $f_{\lambda v} \in \mathbb{X}^*$ such that $f_{\lambda v} = \lambda f_v$. Now, define a mapping $[\ ,\ ] : \mathbb{X} \times \mathbb{X} \longrightarrow \mathbb{R}$ by $[u, v] = f_v(u)$ for all $u,\ v \in \mathbb{X}$. This mapping is clearly a s.i.p. on $\mathbb{X}$. Moreover, $[y, x] = f_x(y) = f_x(\alpha x) = \alpha\|x\| \geq 0$, as $\alpha \geq 0$.
Next, suppose $\{x, y\}$ is linearly independent. Let $Z = \text{span}\{x, y\}$. Then by Theorem 2.1.12, there exists $z \in Z \setminus \{0\}$ such that $x \perp_B z$. Hence, $y = ax + bz$ for some $a, b \in \mathbb{R}$. Applying Theorem 2.1.15 we can find a linear functional $g \in S_{\mathbb{X}^*}$ such that $g(x) = \|x\| = 1$ and $g(z) = 0$. As before, for each $v (\neq x) \in S_{\mathbb{X}}$, choose exactly one $f_v \in S_{\mathbb{X}^*}$ such that $f_v(v) = \|v\| = 1$ and for x, choose $f_x = g$. For $\lambda \in \mathbb{R}$, choose $f_{\lambda v} = \lambda f_v$ for all $v \in S_{\mathbb{X}}$. Then the mapping $[\ ,\ ] : \mathbb{X} \times \mathbb{X} \longrightarrow \mathbb{R}$ defined by $[u, v] = f_v(u)$ for all $u,\ v \in \mathbb{X}$, is a s.i.p. on $\mathbb{X}$. Moreover, $[y, x] = f_x(y) = f_x(ax + bz) = a$. We claim that $a \geq 0$. For if $a < 0$, then, for all $\lambda \leq 0$,

$$\|x + \lambda y\| = \|x + \lambda y\|\|f_x\| \geq f_x(x + \lambda y) = f_x((1 + \lambda a)x + \lambda bz) = (1 + \lambda a)\|x\| \geq \|x\|.$$

Therefore, $y \in x^-$, a contradiction. Thus, for any $y \in x^+$ there exists a s.i.p. $[\ ,\ ]$ on $\mathbb{X}$ such that $[y, x] \geq 0$. This completes the proof of (i).
(ii) Follows similarly as (i). □

Wójcik in [151, Lemma 5.1] first generalized (4.8.2) in the setting of Banach spaces, assuming several restrictions on the spaces and the operators.

Theorem 4.8.2 ([151, Lemma 5.1]) *Let $\mathbb{X}, \mathbb{Y}$ be reflexive Banach spaces. Further assume that $\mathbb{Y}$ is smooth and strictly convex. Let $T, A \in \mathbb{K}(\mathbb{X}, \mathbb{Y})$ and let M_T be either a connected subset of $S_{\mathbb{X}}$ or $|M_T| = 2$. Then $T \perp_B A$ implies that*

$$\|T\| = \sup\{[Tx, y] : x \in S_{\mathbb{X}}, y \in S_{\mathbb{Y}}, [Ax, y] = 0\}.$$

The above theorem was then generalized in [119, Theorem 2.5] in the following way.

Theorem 4.8.3 ([119, Theorem 2.5]) *Let $\mathbb{X}$ be a reflexive Banach space and let $\mathbb{Y}$ be any Banach space. Suppose $T, A \in \mathbb{K}(\mathbb{X}, \mathbb{Y})$ are such that $T \perp_B A$. Let $\mathcal{O}$ denote the collection of all s.i.p. on $\mathbb{Y}$. Then*

$$\begin{aligned}\|T\| &= \sup\{[Tx, y] : x \in S_{\mathbb{X}}, y \in S_{\mathbb{Y}}, [\ ,\ ] \in \mathcal{O}, [Ax, y] \geq 0\}\\ &= \sup\{[Tx, y] : x \in S_{\mathbb{X}}, y \in S_{\mathbb{Y}}, [\ ,\ ] \in \mathcal{O}, [Ax, y] \leq 0\}.\end{aligned}$$

A slight change in the arguments of [119, Theorem 2.5] enables us to generalize Theorem 4.8.3.

Theorem 4.8.4 *Let $\mathbb{X}$ be a reflexive Banach space and let $\mathbb{Y}$ be any Banach space. Suppose that $\mathbb{K}(\mathbb{X}, \mathbb{Y})$ is an M-ideal in $\mathbb{L}(\mathbb{X}, \mathbb{Y})$. Let $T, A \in \mathbb{L}(\mathbb{X}, \mathbb{Y})$ be such that $T \notin span\{A\}$ and $dist(T, \mathbb{K}(\mathbb{X}, \mathbb{Y})) < \|T\|$. Let $\mathcal{O}$ denote the collection of all s.i.p. on $\mathbb{Y}$. Then*
(i) $T \perp_B A$ and T is non-zero imply that

$$\begin{aligned} \|T\| &= sup\{[Tx, y] : x \in S_{\mathbb{X}}, y \in S_{\mathbb{Y}}, [\ ,\ ] \in \mathcal{O}, [Ax, y] \geq 0\} \\ &= sup\{[Tx, y] : x \in S_{\mathbb{X}}, y \in S_{\mathbb{Y}}, [\ ,\ ] \in \mathcal{O}, [Ax, y] \leq 0\}. \end{aligned}$$

(ii) Moreover, if $M_{T_0} = D \cup (-D)$, where D is a closed connected subset of $S_{\mathbb{X}}$, where $T_0 (= T - \lambda_0 A) \perp_B A$ then

$$\begin{aligned} dist(T, \text{span}\{A\}) &= \|T_0\| \\ &= \sup\{[T_0 x, y] : x \in S_{\mathbb{X}}, y \in S_{\mathbb{Y}}, [\ ,\ ] \in \mathcal{O}, [Ax, y] = 0\} \\ &= \sup\{[Tx, y] : x \in S_{\mathbb{X}}, y \in S_{\mathbb{Y}}, [\ ,\ ] \in \mathcal{O}, [Ax, y] = 0\}. \end{aligned}$$

Proof (i) Since $T \perp_B A$, from Theorem 4.4.5, (i) it follows that there exist $u, v \in M_T$ such that $Au \in (Tu)^+$ and $Av \in (Tv)^-$. Now, by Theorem 4.8.1, there exist s.i.p. $[,]_1$, $[,]_2 \in \mathcal{O}$ such that $[Au, Tu]_1 \geq 0$ and $[Av, Tv]_2 \leq 0$. Therefore,

$$\begin{aligned} \|T\| &\geq \sup\{[Tx, y] : x \in S_{\mathbb{X}}, y \in S_{\mathbb{Y}}, [\ ,\ ] \in \mathcal{O}, [Ax, y] \geq 0\} \\ &\geq \left[Tu, \frac{Tu}{\|Tu\|}\right]_1 \\ &= \|T\|. \end{aligned}$$

Thus, $\|T\| = \sup\{[Tx, y] : x \in S_{\mathbb{X}}, y \in S_{\mathbb{Y}}, [\ ,\ ] \in \mathcal{O}, [Ax, y] \geq 0\}$.
Similarly, using $[Av, Tv]_2 \leq 0$ for $v \in M_T$, it can be shown that $\|T\| = \sup\{[Tx, y] : x \in S_{\mathbb{X}}, y \in S_{\mathbb{Y}}, [\ ,\ ] \in \mathcal{O}, [Ax, y] \leq 0\}$.

(ii) In this case, from Theorem 4.4.5, (ii) it follows that there exists $u \in M_T$ such that $T_0 u \perp_B Au$. Now, [33, Proposition 5.3] ensures the existence of a s.i.p. $[\ ,\ ]_1$ on $\mathbb{Y}$ such that $[Au, T_0 u]_1 = 0$. Therefore,

$$\begin{aligned} dist(T, \text{span}\{A\}) &= \|T_0\| \\ &\geq \sup\{[T_0 x, y] : x \in S_{\mathbb{X}}, y \in S_{\mathbb{Y}}, [\ ,\ ] \in \mathcal{O}, [Ax, y] = 0\} \\ &= \sup\{[Tx, y] : x \in S_{\mathbb{X}}, y \in S_{\mathbb{Y}}, [\ ,\ ] \in \mathcal{O}, [Ax, y] = 0\} \\ &\geq \left[Tu, \frac{T_0 u}{\|T_0 u\|}\right]_1 \\ &= \left[T_0 u, \frac{T_0 u}{\|T_0 u\|}\right]_1 \\ &= \|T_0\|. \end{aligned}$$

This completes the proof. □

As an immediate consequence of Theorem 4.8.4 (ii), we get the following theorem.

Theorem 4.8.5 *Let $\mathbb{X}$ be a reflexive Banach space and let $\mathbb{Y}$ be any Banach space. Suppose $T, A \in \mathbb{K}(\mathbb{X}, \mathbb{Y})$ are such that $T \notin span\{A\}$. Let $\mathcal{O}$ denote the collection of all s.i.p. on $\mathbb{Y}$. Moreover, assume $M_{T_0} = D \cup (-D)$, where D is a closed connected subset of $S_{\mathbb{X}}$ and $T_0 (= T - \lambda_0 A) \perp_B A$. Then*

$$\begin{aligned} dist(T, span\{A\}) &= \|T_0\| \\ &= \sup\{[T_0 x, y] : x \in S_{\mathbb{X}}, y \in S_{\mathbb{Y}}, [\ ,\ ] \in \mathcal{O}, [Ax, y] = 0\} \\ &= \sup\{[Tx, y] : x \in S_{\mathbb{X}}, y \in S_{\mathbb{Y}}, [\ ,\ ] \in \mathcal{O}, [Ax, y] = 0\}. \end{aligned}$$

Clearly, Theorem 4.8.5 improves Theorem 4.8.2 by relaxing the conditions of reflexivity and strict convexity on $\mathbb{Y}$. In case of Hilbert spaces, where both these conditions are automatically satisfied, Theorem 4.8.5 further reduces to the following form.

Theorem 4.8.6 ([151, Theorem 5.2]) *Let $\mathbb{H}_1, \mathbb{H}_2$ be Hilbert spaces. Let $T, A \in \mathbb{K}(\mathbb{H}_1, \mathbb{H}_2)$ be such that $T \notin span\{A\}$. Then*

$$dist(T, span\{A\}) = \sup\{\langle Tx, y\rangle : x \in S_{\mathbb{H}_1}, y \in S_{\mathbb{H}_2}, \langle Tx, y\rangle = 0\}.$$

Proof For any scalar λ, $T - \lambda A$ is a bounded linear operator on $\mathbb{H}$ and so from Theorem 3.3.5, it follows that $M_{T-\lambda A}$ is the unit sphere of some subspace of $\mathbb{H}_1$, i.e., $M_{T-\lambda A}$ is of the form $D \cup (-D)$, where D is a closed connected subset of $S_{\mathbb{H}}$. Further, observe that the only s.i.p. on a Hilbert space is the inner product itself. Now, the proof of the theorem follows simply using Theorem 4.8.4 (ii). □

We can further generalize Theorem 4.8.4, by considering the following approximations:

Theorem 4.8.7 ([119, Theorem 2.6]) *Let $\mathbb{X}, \mathbb{Y}$ be any two Banach spaces. Suppose $T, A \in \mathbb{L}(\mathbb{X}, \mathbb{Y})$ are such that $T \perp_B A$. Let $\mathcal{O}$ denote the collection of all s.i.p. on $\mathbb{Y}$. Let $\epsilon > 0$ be arbitrary but fixed after choice. Then $\|T\| = max\ \{l_1(\epsilon), l_2(\epsilon)\} = max\ \{l_1(\epsilon), l_3(\epsilon)\}$, where,*
$l_1(\epsilon) = sup\{[Tx, y] : x \in S_{\mathbb{X}}, y \in S_{\mathbb{Y}}, [\ ,\ ] \in \mathcal{O}, |[Ax, y]| < \epsilon\}$,
$l_2(\epsilon) = sup\{[Tx, y] : x \in S_{\mathbb{X}}, y \in S_{\mathbb{Y}}, [\ ,\ ] \in \mathcal{O}, Ax \in (y)^{+\epsilon}\}$,
$l_3(\epsilon) = sup\{[Tx, y] : x \in S_{\mathbb{X}}, y \in S_{\mathbb{Y}}, [\ ,\ ] \in \mathcal{O}, Ax \in (y)^{-\epsilon}\}$.

Proof Since $T \perp_B A$, by Theorem 4.4.14, we have either (a) or (b) holds
(a) There exists a norming sequence $\{x_n\}$ for T such that $\|Ax_n\| \to 0$.
(b) There exist two norming sequences $\{x_n\}$, $\{y_n\}$ for T and $\{\epsilon_n\}$, $\{\delta_n\} \subseteq [0, 1)$ such that $\epsilon_n \to 0$, $\delta_n \to 0$, $Ax_n \in (Tx_n)^{+\epsilon_n}$ and $Ay_n \in (Ty_n)^{-\delta_n}$ for all $n \in \mathbb{N}$.
Suppose (a) holds. Then there exists $k \in \mathbb{N}$ such that $\|Ax_n\| < \epsilon$ for all $n \geq k$. Therefore, for any s.i.p. $[\ ,\ ] \in \mathcal{O}$, $|[Ax_n, \frac{Tx_n}{\|Tx_n\|}]| \leq \|Ax_n\| < \epsilon$ for all $n \geq k$. Now,

$$\begin{aligned}\|T\| \geq l_1(\epsilon) &= sup\{[Tx, y] : x \in S_{\mathbb{X}}, y \in S_{\mathbb{Y}}, [\ ,\ ] \in \mathcal{O}, |[Ax, y]| < \epsilon\}\\ &\geq sup\left\{\left[Tx_n, \frac{Tx_n}{\|Tx_n\|}\right] : [\ ,\ ] \in \mathcal{O}, n \geq k\right\}\\ &= sup\{\|Tx_n\| : n \geq k\}\\ &= \|T\|.\end{aligned}$$

Thus, $\|T\| = l_1(\epsilon)$. On the other hand, for $i = 1, 2$,

$$l_i(\epsilon) \leq \|T\| \Rightarrow max\{l_1(\epsilon), l_2(\epsilon)\} \leq \|T\| = l_1(\epsilon) \leq max\{l_1(\epsilon), l_2(\epsilon)\}.$$

Hence, $\|T\| = max\{l_1(\epsilon), l_2(\epsilon)\}$. Similarly, $\|T\| = max\{l_1(\epsilon), l_3(\epsilon)\}$.
Next, suppose that (b) holds. Then since $\epsilon_n \to 0$, there exists $m \in \mathbb{N}$ such that $\epsilon_n \leq \epsilon$ for all $n \geq m$. Clearly, $Ax_n \in (Tx_n)^{+\epsilon_n}$ for all $n \in \mathbb{N}$ implies that $Ax_n \in (Tx_n)^{+\epsilon}$ for all $n \geq m$. Thus, $Ax_n \in (\frac{Tx_n}{\|Tx_n\|})^{+\epsilon}$ for all $n \geq m$. Now,

$$\begin{aligned}\|T\| \geq l_2(\epsilon) &= sup\{[Tx, y] : x \in S_{\mathbb{X}}, y \in S_{\mathbb{Y}}, [\ ,\ ] \in \mathcal{O}, Ax \in (y)^{+\epsilon}\}\\ &\geq sup\left\{\left[Tx_n, \frac{Tx_n}{\|Tx_n\|}\right] : [\ ,\ ] \in \mathcal{O}, n \geq m\right\}\\ &= sup\{\|Tx_n\| : n \geq m\}\\ &= \|T\|.\end{aligned}$$

Thus, $\|T\| = l_2(\epsilon)$. Now, proceeding as before, we have $\|T\| = max\{l_1(\epsilon), l_2(\epsilon)\}$. Similarly, using $Ay_n \in (Ty_n)^{-\delta_n}$, $\delta_n \to 0$ and $\|Ty_n\| \to \|T\|$, we obtain $\|T\| = max\{l_1(\epsilon), l_3(\epsilon)\}$. This completes the proof of the theorem. □

Our next goal is to obtain analogous results for bounded linear functionals. The following two results serve our purpose quite well.

Theorem 4.8.8 ([119, Theorem 2.7]) *Let $\mathbb{X}$ be a Banach space such that $\mathbb{X}^*$ is strictly convex. Let $f, g \in \mathbb{X}^*$ be such that $f \perp_B g$. Then*

$$\begin{aligned}\|f\| &= sup\{f(x) : x \in S_{\mathbb{X}}, g(x) \geq 0\}\\ &= sup\{f(x) : x \in S_{\mathbb{X}}, g(x) \leq 0\}.\end{aligned}$$

Proof Since $f \perp_B g$ and $\mathbb{X}^*$ is strictly convex, by Theorem 4.4.16, there exist $\{x_n\}$, $\{y_n\}$ in $S_{\mathbb{X}}$ such that the following holds
(i) $|f(x_n)| \to \|f\|$ and $|f(y_n)| \to \|f\|$ as $n \to \infty$
(ii) $f(x_n).g(x_n) \geq 0$ and $f(y_n).g(y_n) \leq 0$ for all $n \in \mathbb{N}$.
Now,

$$\begin{aligned}\|f\| &\geq sup\{f(x).\alpha : x \in S_{\mathbb{X}}, |\alpha| = 1, g(x).\alpha \geq 0\}\\ &\geq sup\left\{f(x_n).\frac{f(x_n)}{|f(x_n)|} : n \in \mathbb{N}\right\}\\ &= sup\{|f(x_n)| : n \in \mathbb{N}\}\\ &= \|f\|.\end{aligned}$$

Thus, $\|f\| = sup\{f(x).\alpha : x \in S_{\mathbb{X}}, |\alpha| = 1, g(x).\alpha \geq 0\} = sup\{f(x) : x \in S_{\mathbb{X}}, g(x) \geq 0\}$. Similarly, using the property of the sequence $\{y_n\}$ we can show that $\|f\| = sup\{f(x) : x \in S_{\mathbb{X}}, g(x) \leq 0\}$. This completes the proof. □

Theorem 4.8.9 ([119, Theorem 2.8]) *Suppose $\mathbb{X}$ is a Banach space. Let $f, g \in \mathbb{X}^*$ be such that $f \perp_B g$. Let $\epsilon > 0$ be arbitrary but fixed after choice. Then $\|f\| = max\ \{l(\epsilon), k_1\} = max\ \{l(\epsilon), k_2)\}$, where,*
$l(\epsilon) = sup\{f(x) : x \in S_{\mathbb{X}}, |g(x)| < \epsilon\}$,
$k_1 = sup\{f(x) : x \in S_{\mathbb{X}}, g(x) \geq 0\}$,
$k_2 = sup\{f(x) : x \in S_{\mathbb{X}}, g(x) \leq 0\}$.

Proof Since $f \perp_B g$, by Theorem 4.4.17, we have either (a) or (b) holds
(a) there exists $\{x_n\}$ in $S_{\mathbb{X}}$ such that $|f(x_n)| \to \|f\|$ and $g(x_n) \to 0$.
(b) there exist $\{x_n\}, \{y_n\}$ in $S_{\mathbb{X}}$ such that
(i) $|f(x_n)| \to \|f\|$ and $|f(y_n)| \to \|f\|$ as $n \to \infty$
(ii) $f(x_n).g(x_n) \geq 0$ and $f(y_n).g(y_n) \leq 0$ for all $n \in \mathbb{N}$.
Suppose that (a) holds. Since $g(x_n) \to 0$, there exists $k \in \mathbb{N}$ such that $|g(x_n)| < \epsilon$ for all $n \geq k$. Now,

$$\begin{aligned}\|f\| &\geq sup\{f(x).\alpha : x \in S_{\mathbb{X}}, |\alpha| = 1, |g(x).\alpha| < \epsilon\}\\ &\geq sup\left\{f(x_n).\frac{f(x_n)}{|f(x_n)|} : n \geq k\right\}\\ &= sup\{|f(x_n)| : n \geq k\}\\ &= \|f\|.\end{aligned}$$

Thus, $\|f\| = sup\{f(x).\alpha : x \in S_{\mathbb{X}}, |\alpha| = 1, |g(x).\alpha| < \epsilon\} = sup\{f(x) : x \in S_{\mathbb{X}}, |g(x)| < \epsilon\} = l_1(\epsilon)$. Again, $k_1 \leq \|f\|$. Thus, $max\{l(\epsilon), k_1\} = l(\epsilon) = \|f\|$.
Now, suppose that (b) holds. Then using similar arguments as in Theorem 4.8.8, we obtain, $\|f\| = k_1 = k_2$. Again, $l(\epsilon) \leq \|f\|$. Thus, $max\{l(\epsilon), k_1\} = max\{l(\epsilon), k_2\} = \|f\|$. This completes the proof. □

Suppose $\mathcal{W}$ is a subset of $\mathbb{X}$ and $x_0 \in \mathbb{X}$. An element $w_0 \in \mathcal{W}$ is said to be a best approximation to x_0 out of $\mathcal{W}$, if $\|x_0 - w_0\| = \inf\{\|x_0 - w\| : w \in \mathcal{W}\} = dist(x_0, \mathcal{W})$. Observe that if $\mathcal{W}$ is a subspace of $\mathbb{X}$, then w_0 is a best approximation to x_0 out of $\mathcal{W}$ is equivalent to $x_0 - w_0 \perp_B \mathcal{W}$. Kolmogorov obtained the following characterization of best approximation in $C(Q)$, the space of all continuous functions on a compact topological space Q.

Theorem 4.8.10 ([141, Theorem 1.16, pp. 69]) *Suppose $\mathcal{W}$ is a subspace of $C(Q)$. Let $x \in C(Q) \setminus \overline{\mathcal{W}}$ and let $w_0 \in \mathcal{W}$. Then w_0 is a best approximation to x out of $\mathcal{W}$*

(i.e., $x - w_0 \perp_B \mathcal{W}$) if and only if for every $w \in \mathcal{W}$, there exists a $q = q_w \in Q$ such that the following hold:
(i) $\Re\{(x(q) - w_0(q))w(q)\} \geq 0$
(ii) $|x(q) - w_0(q)| = \max\{|x(t) - w_0(t)| : t \in Q\}$.

In [151, Theorem 5.3] the author presented a Kolmogorov type characterization of best approximation in the space of compact operators between Hilbert spaces. However, in [119] Sain et al. exhibited an example which shows that [151, Theorem 5.3] is not correct. They also provided a correct version of [151, Theorem 5.3] in the more general setting of Banach spaces in the same paper.

Theorem 4.8.11 ([119, Theorem 2.10]) *Let $\mathbb{X}$ be a reflexive Banach space and let $\mathbb{Y}$ be any Banach space. Suppose $\mathcal{Z}$ is a finite-dimensional subspace of $\mathbb{K}(\mathbb{X}, \mathbb{Y})$. Let $T \in \mathbb{K}(\mathbb{X}, \mathbb{Y}) \setminus \mathcal{Z}$. Further assume that, for any $\lambda \in \mathbb{R}$ and for any $A \in \mathcal{Z}$, $M_{T+\lambda A} = D_{\lambda,A} \cup (-D_{\lambda,A})$, where $D_{\lambda,A}$ is a non-empty connected subset of $S_{\mathbb{X}}$. Then there exists $A_0 \in \mathcal{Z}$ such that*

$$dist(T, \mathcal{Z}) = sup\{[Tx, y] : x \in S_{\mathbb{X}}, y \in S_{\mathbb{Y}}, [A_0x, y] = 0\}.$$

Moreover, A_0 is the best approximation to T in $\mathcal{Z}$.

Proof Since $\mathcal{Z}$ is a finite-dimensional subspace of $\mathbb{K}(\mathbb{X}, \mathbb{Y})$, there exists $A_0 \in \mathcal{Z}$ such that $dist(T, \mathcal{Z}) = dist(T, \text{span}\{A_0\}) = \|T - A_0\|$. Thus, by Theorem 4.8.5, we have

$$dist(T, \mathcal{Z}) = sup\{[Tx, y] : x \in S_{\mathbb{X}}, y \in S_{\mathbb{Y}}, [A_0x, y] = 0\}.$$

□

In Theorem 4.8.11, if we consider Hilbert spaces instead of Banach spaces, then the result assumes the following simpler form.

Theorem 4.8.12 ([119, Theorem 2.11]) *Let $\mathbb{H}_1, \mathbb{H}_2$ be Hilbert spaces and let $\mathcal{Z}$ be a finite-dimensional subspace of $\mathbb{K}(\mathbb{H}_1, \mathbb{H}_2)$. Let $T \in \mathbb{K}(\mathbb{H}_1, \mathbb{H}_2) \setminus \mathcal{Z}$. Then there exists $A_0 \in \mathcal{Z}$ such that*

$$dist(T, \mathcal{Z}) = sup\{[Tx, y] : x \in S_{\mathbb{X}}, y \in S_{\mathbb{Y}}, [A_0x, y] = 0\}.$$

Moreover, A_0 is the best approximation to T in $\mathcal{Z}$.

Proof From Theorem 3.3.5, it follows that $M_{T+\lambda A}$ is the unit sphere of some subspace of $\mathbb{H}_1$ for all $\lambda \in \mathbb{R}$, when $T, A \in \mathbb{K}(\mathbb{H}_1, \mathbb{H}_2)$. Therefore, for any $\lambda \in \mathbb{R}$ and for any $A \in \mathcal{Z}$, $M_{T+\lambda A} = D_{\lambda,A} \cup (-D_{\lambda,A})$, where $D_{\lambda,A}$ is a non-empty connected subset of $S_{\mathbb{H}_1}$. Thus, by Theorem 4.8.11, we obtain the desired result. □

Now, we obtain lower bounds of $dist(T, \text{span}\{A\})$, where T, A are $n \times n$ matrices over $\mathcal{K}$. In [18, Theorem 3.1], Bhatia and Sharma proved the following inequality for any positive unital linear map $\phi : M_n(\mathcal{K}) \to M_k(\mathcal{K})$ and $A \in M_n(\mathcal{K})$,

$$\phi(A^*A) - \phi(A)^*\phi(A) \leq dist(T, \text{span}\{I\})^2. \tag{4.8.5}$$

Considering different positive unital linear maps, they obtained lower bounds of $dist(T, \text{span}\{I\})^2$. For example, considering $T = (t_{ij})_{n\times n}$ and $\phi_j(T) = t_{jj}$ $(1 \leq j \leq n)$ yields the following inequality:

$$dist(T, \text{span}\{I\})^2 \geq \max_j \sum_{k\neq j} |t_{kj}|^2. \tag{4.8.6}$$

On the other hand, $\phi(T) = \frac{1}{n}tr(T)$ leads to the following inequality:

$$dist(T, \text{span}\{I\})^2 \geq \frac{1}{n}\|T\|_F^2 - \frac{1}{n^2}|tr(T)|^2, \tag{4.8.7}$$

where $\|T\|_F$ is the Frobenius norm of T defined by $\|T\|_F^2 = tr(T^*T)$.

In [84], the authors generalized the inequalities (4.8.6) and (4.8.7). For bounded linear operators T, A defined on arbitrary Hilbert space $\mathbb{H}$, assuming that $0 \notin \sigma_{app}(A)$, Paul introduced in [98] the following notation:

$$M_T(A) = \sup_{\|x\|=1} \left\{ \|Tx - \frac{\langle Tx, Ax\rangle}{\|Ax\|^2} Ax\| \right\}.$$

From [100, Theorem 4], it follows that $M_T(A) = dist(T, \text{span}\{A\})$. Furthermore, in [98, Theorem 4], the author proved that

$$M_T(A)^2 = \sup \left\{ g(T^*T) - \frac{|g(A^*T)|^2}{g(A^*A)} : g \text{ is a state and } g(A^*A) \neq 0 \right\},$$

where a state on $\mathbb{L}(\mathbb{H})$ is defined to be a norm one positive linear functional on $\mathbb{L}(\mathbb{H})$. Thus, for a state g with $g(A^*A) \neq 0$, we get

$$g(T^*T) - \frac{|g(A^*T)|^2}{g(A^*A)} \leq M_T(A)^2 = dist(T, \text{span}\{A\})^2. \tag{4.8.8}$$

Now, suppose $T = (t_{ij})_{n\times n}$ and $A = (a_{ij})_{n\times n}$. Assume that each row of A is non-zero. Then, for each $1 \leq j \leq n$, consider $g_j : M_n(\mathcal{K}) \to \mathcal{K}$ defined by $g_j(S) = s_{jj}$, where $S = (s_{ij})_{n\times n} \in M_n(\mathcal{K})$. Then

$$dist(T, \text{span}\{A\})^2 \geq \sum_{i=1}^{n} |t_{ij}|^2 - \frac{\sum_{i=1}^{n} |\overline{t_{ij}} a_{ij}|^2}{\sum_{i=1}^{n} |a_{ij}|^2}.$$

Therefore,

$$dist(T, \operatorname{span}\{A\})^2 \geq \max_{1\leq j\leq n} \left\{ \sum_{i=1}^{n} |t_{ij}|^2 - \frac{\sum_{i=1}^{n} |\overline{t_{ij}} a_{ij}|^2}{\sum_{i=1}^{n} |a_{ij}|^2} \right\}, \tag{4.8.9}$$

which generalizes (4.8.6). On the other hand, assume that $\|A\|_F \neq 0$ and consider $g : \mathbb{L}(\mathbb{H}) \to \mathcal{K}$ defined by $g(S) = \frac{1}{n} tr(S)$. Then we get

$$dist(T, \operatorname{span}\{A\})^2 \geq \frac{1}{n}\|T\|_F^2 - \frac{1}{n}\frac{|tr(T^*A)|^2}{\|A\|_F^2}, \tag{4.8.10}$$

which generalizes (4.8.7).

In this chapter, we have explored the B–J orthogonality of bounded linear operators to the point of greatest possible generalities in the setting of Banach spaces. The importance of norming points and of norming sequences corresponding to the concerned operator in the whole scheme of things is quite clear from our discussions. Moreover, as we will see in the next chapter, these notions also play central roles in describing approximate versions of B–J orthogonality of bounded linear operators. Our motivation behind investigating approximate B–J orthogonality of operators is twofold. First, it allows us to extract results on the exact orthogonality of operators as a particular case. Furthermore, perhaps more importantly, such an exploration allows us to understand the orthogonality of operators from the perspective of stability analysis and perturbation theory. In other words, we will try to understand the connection between approximate versions of orthogonality with the exact orthogonality condition, in the setting of bounded linear operators between Banach spaces. We will also notice that such a study has apparent relations with the geometric properties of the underlying spaces, which makes it all the more interesting and fruitful.

4.9 Further Readings

The notion of B–J orthogonality in C^*-algebra and Hilbert C^*-modules was also studied by many mathematicians. Arambašíc and Rajíc [9] proved that in a Hilbert C^*-module $\mathbb{V}$ over a C^*-algebra, for $x, y \in \mathbb{V}$, $x \perp_B y$ if and only if there exists a norm one positive linear functional ϕ such that $\phi(\langle x, x\rangle) = \|x\|^2$ and $\phi(\langle x, y\rangle) = 0$. As a consequence, the characterization of B–J orthogonality in a C^*-algebra is as follows: In a C^*-algebra $\mathcal{A}$, for $a, b \in \mathcal{A}$, $a \perp_B b$ if and only if there exists a Hilbert space $\mathbb{H}$, a representation $\pi : \mathcal{A} \to \mathbb{L}(\mathbb{H})$, and a unit vector $\xi \in \mathbb{H}$ such that $\pi(a)\xi = a$ and $\langle \pi(a)\xi, \pi(b)\xi\rangle = 0$. In [10] a strong version of B–J orthogonality was studied in a Hilbert C^*-module. Interested readers can see [11, 12, 152] for more results related to orthogonality in C^*-algebra and Hilbert C^*-module.

Chapter 5
Approximate B-J Orthogonality

5.1 Introduction

In order to describe the analytic structure and the geometry of an inner product space $(\mathbb{H}, \langle , \rangle)$, the usual orthogonality relation "$x \perp y \Leftrightarrow \langle x, y \rangle = 0$ for any $x, y \in \mathbb{H}$" plays the most fundamental role, without a shade of doubt. As we will see in this chapter, it is both useful and interesting to study an approximate version of orthogonality, in the more general setting of a Banach space $\mathbb{X}$. In an inner product space $\mathbb{H}$, a natural way to define approximate orthogonality is as follows:

$$x \perp^{\epsilon} y \Leftrightarrow |\langle x, y \rangle| \leq \epsilon \|x\| \|y\|, \text{ where } x, y \in \mathbb{H}, \epsilon \in [0, 1).$$

Due to the importance of B-J orthogonality in the study of geometry of operator spaces, the above notion has been generalized in Banach spaces. Dragomir [32] provided a suitable definition of approximate B-J orthogonality in a Banach space in the following way: Let $\epsilon \in [0, 1)$. Then for $x, y \in \mathbb{X}$, x is said to be approximate $\epsilon-$ B-J orthogonal to y if

$$\|x + \lambda y\| \geq (1 - \epsilon)\|x\| \ \forall \ \lambda \in \mathcal{K}.$$

Replacing $(1 - \epsilon)$ by $\sqrt{1 - \epsilon^2}$, Chmieliński [24] slightly modified the definition of approximate B-J orthogonality given by Dragomir in the following way: Let $\epsilon \in [0, 1)$. Then for $x, y \in \mathbb{X}$, x is said to be approximate B-J orthogonal to y if

$$\|x + \lambda y\| \geq \sqrt{1 - \epsilon^2}\|x\| \ \forall \ \lambda \in \mathcal{K}.$$

In this case, we write $x \perp_D^{\epsilon} y$.

In the same paper [24], Chmieliński introduced another notion of approximate B-J orthogonality, defined in the following way:

A. Mal et al., *Birkhoff–James Orthogonality and Geometry of Operator Spaces*, Infosys Science Foundation Series, https://doi.org/10.1007/978-981-99-7111-4_5

Let $\epsilon \in [0, 1)$. Then for $x, y \in \mathbb{X}$, x is said to be approximate B-J orthogonal to y if

$$\|x + \lambda y\|^2 \geq \|x\|^2 - 2\epsilon\|x\|\|\lambda y\| \ \forall \ \lambda \in \mathcal{K}.$$

In this case, we write $x \perp_B^{\epsilon} y$.

Note that both the notions $\perp_D^{\epsilon}$ and $\perp_B^{\epsilon}$ are homogeneous, i.e., $x \perp_D^{\epsilon} y$ implies that $\alpha x \perp_D^{\epsilon} \beta y$ and $x \perp_B^{\epsilon} y$ implies that $\alpha x \perp_B^{\epsilon} \beta y$ for all $\alpha, \beta \in \mathcal{K}$. Furthermore, in an inner product space $\mathbb{H}$, all the notions of approximate B-J orthogonality coincide, i.e.,

$$x \perp^{\epsilon} y \Leftrightarrow x \perp_D^{\epsilon} y \Leftrightarrow x \perp_B^{\epsilon} y.$$

However, in an arbitrary Banach space, the notions $\perp_D^{\epsilon}$ and $\perp_B^{\epsilon}$ are not equivalent in general. It is easy to find out Banach spaces (for example, $(\mathbb{R}^2, \|.\|_{\infty})$) where neither of these two notions imply the other. Moreover, an example was provided in [136, Example 2.1] to show that these two notions are not equivalent even in strictly convex, smooth Banach spaces. In view of this fact, it is logical to explore these two notions separately. In Sect. 5.2, we study the notion of $\perp_D^{\epsilon}$ on $\mathbb{X}$ from both analytic and geometric points of view. We show that this notion of approximate orthogonality is related to that of normal cones in a Banach space. We continue an analogous study in Sect. 5.3, for the notion of $\perp_B^{\epsilon}$ on $\mathbb{X}$. In Sect. 5.4, we explore the notion of $\perp_D^{\epsilon}$ for bounded linear operators on Banach spaces. Finally, in Sect. 5.5, we characterize $\perp_B^{\epsilon}$ for bounded linear operators between Banach spaces. Throughout the chapter, we will assume that the considered Banach (Hilbert) space is real, unless otherwise specified.

The study of approximate orthogonality in the setting of Banach spaces, and more specifically in operator spaces, serves primarily two purposes. On one hand, it helps us to visualize the quantitative version of approximate orthogonality in terms of normal cones. As it will be observed in this chapter, although $\perp_D^{\epsilon}$ and $\perp_B^{\epsilon}$ are different in a general Banach space, they do share similar geometric properties which further underlines the possibility of studying these two non-equivalent notions from a geometric perspective. On the other hand, particularizing our observations in the setting of operator spaces, we obtain a better idea about the role of the so-called norming sequences corresponding to an operator in studying B-J orthogonality of operators. The careful reader will remember that such norming sequences also played an important part in describing the exact B-J orthogonality of operators, especially in the infinite-dimensional setting.

5.2 $\perp_D^{\epsilon}$-Approximate Orthogonality on $\mathbb{X}$

We begin this section with a characterization of approximate B-J orthogonality ($\perp_D^{\epsilon}$) in terms of linear functionals. The idea of the proof is adapted from [59, Theorem 2.1]. However, for the sake of completeness, we present the detailed proof of the same.

Theorem 5.2.1 ([88, Theorem 3.3]) *Let $\mathbb{X}$ be a Banach space over $\mathcal{K}$. Let x be a non-zero element in $\mathbb{X}$ and H be a hyperspace in $\mathbb{X}$. Then for $\epsilon \in [0, 1)$, $x \perp_D^\epsilon H$ if and only if there exists $f \in S_{\mathbb{X}^*}$ such that $f(x) \geq \sqrt{1-\epsilon^2}\|x\|$, where $H = ker(f)$.*

Proof First suppose that there exists $f \in S_{\mathbb{X}^*}$ such that $f(x) \geq \sqrt{1-\epsilon^2}\|x\|$, where $H = \ker(f)$. Then for any $h \in H$ and for any scalar λ,

$$\sqrt{1-\epsilon^2}\|x\| \leq f(x) = f(x+\lambda h) \leq \|f\|\|x+\lambda h\| = \|x+\lambda h\|.$$

This implies that $\sqrt{1-\epsilon^2}\|x\| \leq \|x+\lambda h\|$ for any $h \in H$ and for any scalar λ. Thus, $x \perp_D^\epsilon H$.

Conversely, suppose that $x \perp_D^\epsilon H$. It is immediate that $\mathbb{X} = span\{x, H\}$. Define $g : \mathbb{X} \longrightarrow \mathcal{K}$ by $g(ax+h) = a$, where $a \in \mathcal{K}$ and $h \in H$. Clearly, g is linear and $\ker(g) = H$. Now, $|g(ax+h)| = |a| \leq \frac{\|ax+h\|}{\sqrt{1-\epsilon^2}\|x\|}$, since $x \perp_D^\epsilon h$. Thus, $\|g\| \leq \frac{1}{\sqrt{1-\epsilon^2}\|x\|} \Rightarrow \sqrt{1-\epsilon^2}\|g\|\|x\| \leq 1 = g(x)$. Let $f = \frac{1}{\|g\|}g$. Therefore, $f \in S_{\mathbb{X}^*}$, $f(x) \geq \sqrt{1-\epsilon^2}\|x\|$ and $H = \ker(g) = \ker(f)$. This establishes the theorem. □

The following corollary may be viewed as a generalization of the extremely useful result Theorem 2.1.14, characterizing B-J orthogonality in an arbitrary Banach space.

Corollary 5.2.2 *Let $\mathbb{X}$ be a Banach space over $\mathcal{K}$ and let $x, y \in \mathbb{X}$. Then for any $\epsilon \in [0, 1)$, $x \perp_D^\epsilon y$ if and only if there exists a linear functional $f \in S_{\mathbb{X}^*}$ such that $f(x) \geq \sqrt{1-\epsilon^2}\|x\|$ and $f(y) = 0$.*

Proof Let us first prove the sufficient part. Suppose there exists a linear functional $f \in S_{\mathbb{X}^*}$ such that $f(x) \geq \sqrt{1-\epsilon^2}\|x\|$ and $f(y) = 0$. Then for any scalar λ, $\sqrt{1-\epsilon^2}\|x\| \leq f(x) = f(x+\lambda y) \leq \|f\|\|x+\lambda y\| = \|x+\lambda y\|$. Therefore, $x \perp_D^\epsilon y$.

For the necessary part, let $x \perp_D^\epsilon y$. Assume that $Z = \text{span}\{x, y\}$. Then $\text{span}\{y\}$ is a hyperspace in Z. Now, from Theorem 5.2.1, it follows that there exists $g \in S_{Z^*}$ such that $g(z) \geq \sqrt{1-\epsilon^2}\|x\|$, where $\text{span}\{y\} = \ker(g)$. Consider a norm preserving extension f of g on $\mathbb{X}$. Then $f \in S_{\mathbb{X}^*}$ satisfies the desired conditions. □

Motivated by Theorem 4.4.4, we now characterize approximate B-J orthogonality ($\perp_D^\epsilon$) in terms of extreme linear functionals. To prove the desired theorem, we use the following two lemmas by Singer, which are also of independent interest.

Lemma 5.2.3 ([141, Lemma 1.1, p. 166]) *Let $\mathbb{X}$ be a k-dimensional Banach space over $\mathcal{K}$ and let $\phi \in S_{\mathbb{X}^*}$. Then there exist h extreme linear functionals $\phi_1, \phi_2, \ldots, \phi_h \in S_{\mathbb{X}^*}$, where $1 \leq h \leq k$ if the scalars are real and $1 \leq h \leq 2k-1$ if the scalars are complex, and h numbers $\lambda_1, \lambda_2, \ldots, \lambda_h > 0$ with $\sum_{i=1}^h \lambda_i = 1$, such that $\phi = \sum_{i=1}^h \lambda_i \phi_i$.*

Lemma 5.2.4 ([141, Lemma 1.2, p. 168]) *Let $\mathbb{X}$ be a Banach space over $\mathcal{K}$ and let G be a subspace of $\mathbb{X}$. Suppose ϕ is an extreme point of S_{G^*}. Then there exists an extension of ϕ to an extreme point of $S_{\mathbb{X}^*}$.*

The characterization of $\perp_D^\epsilon$ in terms of extreme linear functionals in a real Banach space was proved in [87, Theorem 2.2].

Theorem 5.2.5 *Let $\mathbb{X}$ be a Banach space and let $x, y \in \mathbb{X}$. Suppose $\epsilon \in [0, 1)$. Then $x \perp_D^\epsilon y$ if and only if there exist $\phi, \psi \in Ext(B_{\mathbb{X}^*})$, $u \in span\{x, y\} \cap S_{\mathbb{X}}$ and $t \in [0, 1]$ such that $\phi(u) = \psi(u) = 1$, $(1-t)\phi(x) + t\psi(x) \geq \sqrt{1-\epsilon^2}\|x\|$ and $(1-t)\phi(y) + t\psi(y) = 0$.*

For complex Banach space, the previous theorem assumes the following form.

Theorem 5.2.6 *Let $\mathbb{X}$ be a complex Banach space. Let $x, y \in \mathbb{X}$ and $\epsilon \in [0, 1)$. Then $x \perp_D^\epsilon y$ if and only if there exist $\rho, \phi, \psi \in Ext(B_{\mathbb{X}^*})$, $u \in span\{x, y\} \cap S_{\mathbb{X}}$ and $t_i \in [0, 1]$ for $i = 1, 2, 3$ such that the following hold:*

(a) $t_1 + t_2 + t_3 = 1$,
(b) $\rho(u) = \phi(u) = \psi(u) = 1$,
(c) $t_1\rho(x) + t_2\phi(x) + t_3\psi(x) \geq \sqrt{1-\epsilon^2}\|x\|$, *and*
(d) $t_1\rho(y) + t_2\phi(y) + t_3\psi(y) = 0$.

Proof First we prove the easier sufficient part of the theorem. Suppose that $f = t_1\rho + t_2\phi + t_3\psi$. Then $f(u) = 1$, $f(y) = 0$ and $\|f\| = 1$. Now, $f(x) = t_1\rho(x) + t_2\phi(x) + t_3\psi(x) \geq \sqrt{1-\epsilon^2}\|x\|$. Therefore from Corollary 5.2.2, it follows that $x \perp_D^\epsilon y$.

Now, we prove the necessary part of the theorem. Let $x \perp_D^\epsilon y$. Observe that x, y are linearly independent. Let $Z = span\{x, y\}$. Using Theorem 2.1.13, we get a scalar λ_0 such that $x - \lambda_0 y \perp_B y$. Suppose that $u = \frac{x-\lambda_0 y}{\|x-\lambda_0 y\|}$. Then $u \in Z \cap S_{\mathbb{X}}$ and $u \perp_B y$. Now, from Theorem 2.1.14, it follows that there exists $f \in S_{\mathbb{X}^*}$ such that $f(u) = 1$ and $f(y) = 0$. Thus,

$$f(u) = 1 \Rightarrow f(x - \lambda_0 y) = \|x - \lambda_0 y\| \Rightarrow f(x) = \|x - \lambda_0 y\| \geq \sqrt{1-\epsilon^2}\|x\|.$$

Here the last inequality follows from the assumption that $x \perp_D^\epsilon y$. Now, let $g = f|_Z$. Thus, $g \in S_{Z^*}$. Therefore, by Lemma 5.2.3, there exist $g_1, g_2, g_3 \in Ext(B_{Z^*})$ and $t_1, t_2, t_3 \in [0, 1]$ such that $t_1 + t_2 + t_3 = 1$ and $g = t_1 g_1 + t_2 g_2 + t_3 g_3$. Now, $g(u) = 1 \Rightarrow g_1(u) = g_2(u) = g_3(u) = 1$. Again, by Lemma 5.2.4, there exist $\rho, \phi, \psi \in Ext(B_{\mathbb{X}^*})$ such that $\rho|_Z = g_1$, $\phi|_Z = g_2$ and $\psi|_Z = g_3$. Clearly, $\rho(u) = g_1(u) = 1$, $\phi(u) = g_2(u) = 1$ and $\psi(u) = g_3(u) = 1$. Now,

$$\begin{aligned} t_1\rho(x) + t_2\phi(x) + t_3\psi(x) &= t_1 g_1(x) + t_2 g_2(x) + t_3 g_3(x) \\ &= g(x) \\ &= f(x) \\ &\geq \sqrt{1-\epsilon^2}\|x\| \text{ and} \\ t_1\rho(y) + t_2\phi(y) + t_3\psi(y) &= t_1 g_1(y) + t_2 g_2(y) + t_3 g_3(y) \\ &= g(y) = f(y) = 0. \end{aligned}$$

This completes the proof of the theorem. □

Now, we turn our attention towards exploring the notion of $\perp_D^{\epsilon}$ on a Banach space, from a geometric point of view. To serve our purpose, we introduce the following set:

Given $x \in \mathbb{X}$ and $\epsilon \in [0, 1)$, we define

$$F(x, \epsilon) = \{y \in \mathbb{X} : x \perp_D^{\epsilon} y\}.$$

Recall that a subset K of $\mathbb{X}$ is said to be a normal cone in $\mathbb{X}$ if

$$(i)\ K + K \subset K,\ (ii)\ \alpha K \subset K \ \text{ for all } \alpha \geq 0 \ \text{ and } (iii)\ K \cap (-K) = \{\theta\}.$$

One of the importance of normal cones in the study of geometry of Banach spaces is that it provides us with a natural partial ordering $\geq$, which is defined as $x \geq y$ if $x - y \in K$, for any $x, y \in \mathbb{X}$. Observe that in a two-dimensional Banach space $\mathbb{X}$, any normal cone K is completely determined by the intersection of K with the unit sphere $S_{\mathbb{X}}$. Keeping this in mind, when we say that K is a normal cone in a two-dimensional Banach space $\mathbb{X}$, determined by v_1, v_2, what we really mean is that $K \cap S_{\mathbb{X}} = \{\frac{(1-t)v_1 + tv_2}{\|(1-t)v_1 + tv_2\|} : t \in [0, 1]\}$. Of course, in this case $K = \{\alpha v_1 + \beta v_2 : \alpha, \beta \geq 0\}$. In the next theorem, we relate the set $F(x, \epsilon)$ with normal cones, in a two-dimensional Banach space.

Theorem 5.2.7 ([125, Theorem 2.1]) *Let $\mathbb{X}$ be a two-dimensional Banach space. Then for any $x \in S_{\mathbb{X}}$ and any $\epsilon \in [0, 1)$, there exists a normal cone K in $\mathbb{X}$ such that $F(x, \epsilon) = K \cup (-K)$.*

Proof Let $x \in S_{\mathbb{X}}$. Then there exists $y \in S_{\mathbb{X}}$ such that $x \perp_B y$. Consider

$$A = \{t \in [0, 1] : x \perp_D^{\epsilon} \{(1 - t)x + ty\}\},$$

$$B = \{t \in [0, 1] : x \perp_D^{\epsilon} \{-(1 - t)x + ty\}\}.$$

Clearly $1 \in A \cap B$ and $0 \notin A \cup B$, so that both A and B are non-empty proper subset of $[0, 1]$. We claim that both A and B are closed. Suppose $\{t_n\} \subseteq A$ is such that $t_n \to t$. Then $x \perp_D^{\epsilon} \{(1 - t_n)x + t_n y\} \Rightarrow \|x + \lambda\{(1 - t_n)x + t_n y\}\| \geq \sqrt{1 - \epsilon^2}$ for all $\lambda \in \mathbb{R}$. Letting $n \to \infty$ we get, $\|x + \lambda\{(1 - t)x + ty\}\| \geq \sqrt{1 - \epsilon^2}$ for all $\lambda \in \mathbb{R}$. Then $x \perp_D^{\epsilon} \{(1 - t)x + ty\}$ and so $t \in A$. This proves that A is closed. Similarly, we can prove that B is closed. This establishes our claim. Let $t_1 = \inf A,\ t_2 = \inf B$. Clearly $t_1 \neq 0,\ t_2 \neq 0$. Let $u_1 = (1 - t_1)x + t_1 y$, $u_2 = -(1 - t_2)x + t_2 y$ and $v_1 = \frac{u_1}{\|u_1\|}$, $v_2 = \frac{u_2}{\|u_2\|}$. Let K be the normal cone determined by v_1, v_2. We next show that $F(x, \epsilon) = K \cup (-K)$. If $z \in \mathbb{X}$ is such that $z = c\{(1 - t)x + ty\}$, where $c > 0$ and $t < t_1$ then from the definition of infimum, $x \not\perp_D^{\epsilon} z$. Let $w \in \mathbb{X}$ be such that $w = c\{(1 - t)x + ty\}$, where $c \geq 0$ and $1 \geq t \geq t_1$. We show that $x \perp_D^{\epsilon} w$. Let $\lambda > 0$. Choose $\eta_1 = 1 + \lambda c(1 - t) \geq 1$ and choose $\lambda_1 = \frac{\lambda ct}{\eta_1}$. Then $x + \lambda w = \eta_1(x + \lambda_1 y)$. Hence,

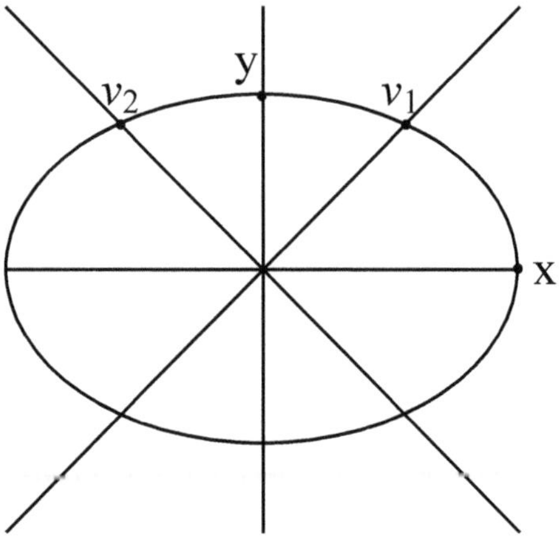

Fig. 5.1 Cone K

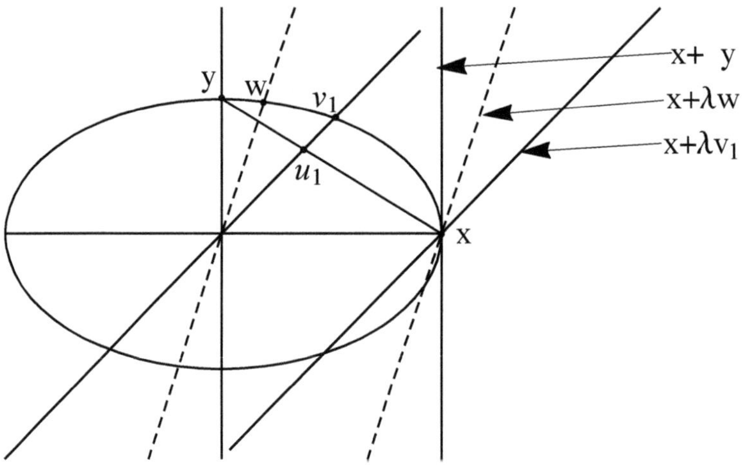

Fig. 5.2 $x \perp_D^{\epsilon} w$

$$\begin{aligned}\|x + \lambda w\| &= |\eta_1| \|x + \lambda_1 y\| \\ &\geq \|x + \lambda_1 y\| \\ &\geq 1 \\ &\geq \sqrt{1 - \epsilon^2}.\end{aligned}$$

Let $\lambda < 0$. Write $w = b\{(1 - s)x + s v_1\}$, where $b \geq 0$, $s \geq 1$. Choose $\eta_2 = 1 + \lambda b(1 - s) \geq 1$ and $\lambda_2 = \frac{\lambda b s}{\eta_2}$. Then $x + \lambda w = \eta_2(x + \lambda_2 v_1)$. Hence,

$$\begin{aligned}\|x + \lambda w\| &= |\eta_2|\|x + \lambda_2 v_1\| \\ &\geq \|x + \lambda_2 v_1\| \\ &\geq \sqrt{1-\epsilon^2}.\end{aligned}$$

Thus, for all $\lambda \in \mathbb{R}$, $\|x + \lambda w\| \geq \sqrt{1-\epsilon^2} \Rightarrow x \perp_D^\epsilon w$(Fig. 5.2). Similarly, if $z = c\{-(1-t)x + ty\}$, where $c > 0$ and $t < t_2$ then $x \not\perp_D^\epsilon z$ and if $w = d\{-(1-t)x + ty\}$, where $d \geq 0$ and $1 \geq t \geq t_2$ then $x \perp_D^\epsilon w$. Therefore, $F(x, \epsilon) = K \cup (-K)$(Fig. 5.1). □

Remark 5.2.8 ([125, Remark 2.2]) We further observe that if $\epsilon \in (0, 1)$ then $t_1 < 1$, for if $z \in S_{\mathbb{X}}$ and $\|z - y\| \leq \frac{1-\sqrt{1-\epsilon^2}}{1+\sqrt{1-\epsilon^2}}$ then $\|x + \lambda z\| \geq |\|\lambda z\| - \|x\|| \geq |\lambda| - 1 > \sqrt{1-\epsilon^2}$, whenever $|\lambda| > 1 + \sqrt{1-\epsilon^2}$. Also, if $|\lambda| \leq 1 + \sqrt{1-\epsilon^2}$ then

$$\begin{aligned}\|x + \lambda z\| &= \|x + \lambda y + \lambda(z - y)\| \\ &\geq \|x + \lambda y\| - |\lambda|\|z - y\| \\ &\geq 1 - (1 + \sqrt{1-\epsilon^2})\|z - y\| \\ &\geq \sqrt{1-\epsilon^2},\end{aligned}$$

so that $x \perp_D^\epsilon z$. This shows that there exists some t, $0 < t < 1$ such that $x \perp_D^\epsilon \{(1-t)x + ty\}$. Similarly, if $\epsilon \in (0, 1)$ then $t_2 < 1$.

Remark 5.2.9 ([125, Remark 2.3]) From the proof of Theorem 5.2.7, it is clear that if $\epsilon > 0$ then both v_1, v_2 cannot be simultaneously in x^+ (or x^-). In fact, for $\epsilon > 0$, exactly one of v_1, v_2 will be in x^+ and the other one will be in x^-. On the other hand, it is easy to observe that in two-dimensional smooth Banach spaces, $v_1 = v_2$, if $\epsilon = 0$.

Next, let us take $x \in S_{\mathbb{X}}$ and consider the subset $S(x, \epsilon) = \{z \in S_{\mathbb{X}} : \inf_{\lambda \in \mathbb{R}} \|x + \lambda z\| = \sqrt{1-\epsilon^2}\}$ of $F(x, \epsilon)$. In context of the previous theorem, it is possible to obtain a nice characterization of $S(x, \epsilon)$. We do the same in the following theorem:

Theorem 5.2.10 ([125, Theorem 2.4]) *Let $\mathbb{X}$ be a two-dimensional Banach space and let v_1, v_2 be as in Theorem 5.2.7. Then $S(x, \epsilon) = \{\pm v_1, \pm v_2\}$, if $\epsilon \in (0, 1)$ and $S(x, \epsilon) = \{y \in S_{\mathbb{X}} : x \perp_B y\}$ if $\epsilon = 0$.*

Proof Let $\epsilon \in (0, 1)$. Observe that each element in $S_{\mathbb{X}}$ can be written as some multiple of a convex combination of v_2, y or y, v_1 or v_1, x or $x, -v_2$. From Theorem 5.2.7, it follows that $\inf_\lambda \|x + \lambda z\| < \sqrt{1-\epsilon^2}$, if z is some multiple of a convex combination of v_1, x or $x, -v_2$. Now, suppose $z = \frac{(1-t)y+tv_1}{\|(1-t)y+tv_1\|}$ for some $0 \leq t < 1$. Let $\lambda < 0$. Consider $z = a\{(1-s)x + sv_1\}$, where $s > 1$ and $a = \frac{1}{\|(1-s)x+sv_1\|} > 0$. Let $\eta = 1 + a\lambda(1-s) > 1$ and $\lambda' = \frac{a\lambda s}{\eta}$. Then

$$x + \lambda z = \eta(x + \lambda' v_1) \Rightarrow \|x + \lambda z\| > \|x + \lambda' v_1\| \geq \sqrt{1-\epsilon^2}.$$

Now, let $\lambda \geq 0$. Write $z = b\{(1-s')x + s'y\}$, where $0 < s' \leq 1$ and $b = \frac{1}{\|(1-s')x+s'y\|}$. Let $\eta_1 = 1 + \lambda b(1-s') \geq 1$ and $\lambda_1 = \frac{\lambda b s'}{\eta_1}$. Then

$$x + \lambda z = \eta_1(x + \lambda_1 y) \Rightarrow \|x + \lambda z\| \geq \|x + \lambda_1 y\| \geq 1 > \sqrt{1-\epsilon^2}.$$

Therefore, $\inf_{\lambda \in \mathbb{R}} \|x + \lambda z\| > \sqrt{1-\epsilon^2}$. Similarly, if $z = \frac{(1-t)y+tv_2}{\|(1-t)y+tv_2\|}$ for some $0 \leq t < 1$, then we can show that $\inf_\lambda \|x + \lambda z\| > \sqrt{1-\epsilon^2}$. Thus, $z \in S(x, \epsilon)$ if and only if $z \in \{\pm v_1, \pm v_2\}$. This completes the proof of the first part. The second part follows easily. □

Suppose $P_{x,y}(\epsilon)$ denotes the restriction of $F(x, \epsilon)$ to the subspace spanned by x and y. Equipped with the characterization of $F(x, \epsilon)$ in two-dimensional Banach spaces, it is now possible to describe the set $F(x, \epsilon)$ in an arbitrary Banach space. The following theorem, the proof of which is immediate, illustrates our claim.

Theorem 5.2.11 ([125, Theorem 2.5]) *Let $\mathbb{X}$ be a Banach space and $x \in \mathbb{X}$. Suppose $\epsilon \in [0, 1)$. Then $F(x, \epsilon) = \bigcup_{y \in \mathbb{X}} P_{x,y}(\epsilon)$. In particular, $F(x, \epsilon)$ is a union of two-dimensional normal cones.*

Now, the natural question that arises is whether the converse of Theorem 5.2.7 holds. The following example was exhibited in [125] to show that in general the converse of the Theorem 5.2.7 does not hold.

Example 5.2.12 ([125, Example 2.7]) Let $\mathbb{X} = \ell_\infty^2$. Let K be the normal cone determined by $(-\frac{1}{2}, 1)$ and $(-1, 1)$. If possible, suppose that there exist $x \in S_{\mathbb{X}}$ and $\epsilon \in [0, 1)$ such that $F(x, \epsilon) = K \cup (-K)$. Let $x \perp_B z$ and $z \in S_{\mathbb{X}}$. Then $z \in K \cup (-K)$. Without loss of generality, assume that $z \in K$. Then $z = (1-t)(-\frac{1}{2}, 1) + t(-1, 1)$ for some $t \in [0, 1]$. Then clearly $x = \pm(1, 1)$. Since $x \perp_B (-1, 0)$, we must have, $(-1, 0) \in F(x, \epsilon)$. But $(-1, 0) \notin K \cup (-K)$. This contradiction proves that there does not exist any $x \in S_{\mathbb{X}}$ and $\epsilon \in [0, 1)$ such that $F(x, \epsilon) = K \cup (-K)$.

Observe that in the previous example, we have considered a non-smooth Banach space. The following theorem shows that, in a smooth Banach space, the converse of Theorem 5.2.7 indeed holds true.

Theorem 5.2.13 ([125, Theorem 2.6]) *Let $\mathbb{X}$ be a two-dimensional smooth Banach space. Let K be a normal cone in $\mathbb{X}$. Then there exist $x \in S_{\mathbb{X}}$ and $\epsilon \in [0, 1)$ such that $F(x, \epsilon) = K \cup (-K)$.*

Proof Let the cone K be determined by $v_1, v_2 \in S_{\mathbb{X}}$. Observe that $v_1 \neq -v_2$, since K is a normal cone. First, let us assume that $v_1 \neq v_2$. Consider the two sets

$$W_1 = \{x \in S_{\mathbb{X}} : \inf_\lambda \|x + \lambda v_1\| > \inf_\lambda \|x + \lambda v_2\|\},$$

$$W_2 = \{x \in S_{\mathbb{X}} : \inf_\lambda \|x + \lambda v_2\| > \inf_\lambda \|x + \lambda v_1\|\}.$$

Clearly, $W_1 \neq \emptyset$ and $W_2 \neq \emptyset$, since $v_2 \in W_1$ and $v_1 \in W_2$. We show that W_1 is open. Let $z \in W_1$. Let $l_1 = \inf_\lambda \|z + \lambda v_1\| > \inf_\lambda \|z + \lambda v_2\| = l_2$. Choose $\epsilon' > 0$ such that $\epsilon' < \frac{l_1 - l_2}{2}$. We claim that $B(z, \epsilon') \cap S_{\mathbb{X}} \subseteq W_1$. By a standard compactness argument, there exists λ_0 such that $\|z + \lambda_0 v_2\| = l_2$. Now, suppose that $w \in B(z, \epsilon') \cap S_{\mathbb{X}}$. Then for any $\lambda \in \mathbb{R}$, we have

$$\begin{aligned}\|w + \lambda v_1\| &= \|(z + \lambda v_1) - (z - w)\| \\ &\geq \|z + \lambda v_1\| - \|z - w\| \\ &> l_1 - \epsilon'.\end{aligned}$$

On the other hand,

$$\begin{aligned}\|w + \lambda_0 v_2\| &= \|(z + \lambda_0 v_2) + (w - z)\| \\ &\leq \|z + \lambda_0 v_2\| + \|w - z\| \\ &< l_2 + \epsilon'.\end{aligned}$$

Therefore, $\inf_\lambda \|w + \lambda v_1\| > l_1 - \epsilon' > l_2 + \epsilon' > \inf_\lambda \|w + \lambda v_2\|$. This implies that $w \in W_1$. Hence, $B(z, \epsilon') \cap S_{\mathbb{X}} \subseteq W_1$. Therefore, W_1 is open in $S_{\mathbb{X}}$. Similarly, it can be shown that W_2 is open. Again, $W_1 \cap W_2 = \emptyset$. Since $S_{\mathbb{X}}$ is connected, $W_1 \cup W_2 \neq S_{\mathbb{X}}$. Choose $x \in S_{\mathbb{X}} \setminus (W_1 \cup W_2)$. Then clearly $\inf_\lambda \|x + \lambda v_1\| = \inf_\lambda \|x + \lambda v_2\|$. Observe that since $\mathbb{X}$ is smooth, and $v_1 \neq \pm v_2$, x cannot be B-J orthogonal to both v_1 and v_2. Since $\inf_\lambda \|x + \lambda v_1\| = \inf_\lambda \|x + \lambda v_2\|$, it is now easy to conclude that x is not B-J orthogonal to either of v_1, v_2. Therefore, there exists $\epsilon \in (0, 1)$ such that $\inf_\lambda \|x + \lambda v_1\| = \inf_\lambda \|x + \lambda v_2\| = \sqrt{1 - \epsilon^2}$. Now, by Theorems 5.2.7 and 5.2.10, it is clear that $F(x, \epsilon) = K \cup (-K)$. Next, we consider the case $v_1 = v_2$. Clearly, K is simply a half-line in this case, given by $K = \{\lambda v_1 : \lambda \geq 0\}$. By Theorem 2.1.13, there exists $x \in S_{\mathbb{X}}$ such that $x \perp_B v_1 (= v_2)$. Furthermore, since $\mathbb{X}$ is smooth, if $y \in S_{\mathbb{X}}$ is such that $x \perp_B y$, then $y = \pm v_1$. Therefore, by choosing $\epsilon = 0$, it is now immediate that $F(x, \epsilon) = K \cup (-K)$. This establishes the theorem. □

The set $F(x, \epsilon)$ satisfies a certain kind of uniqueness property. In the next theorem, we explore this uniqueness property for a two dimensional Banach space.

Theorem 5.2.14 ([125, Theorem 2.8]) *Let $\mathbb{X}$ be a two-dimensional Banach space. Let $x_1, x_2 \in S_{\mathbb{X}}$ and $\epsilon_1, \epsilon_2 \in (0, 1)$. If $F(x_1, \epsilon_1) = F(x_2, \epsilon_2)$ then $x_1 = \pm x_2$ and $\epsilon_1 = \epsilon_2$.*

Proof We prove the theorem in the following two steps.
Step 1: If $F(x, \epsilon) = F(y, \epsilon)$ for some $x, y \in S_{\mathbb{X}}$, then $x = \pm y$.

Let $F(x, \epsilon) = F(y, \epsilon) = K \cup (-K)$, where K is the normal cone determined by $v_1, v_2 \in S_{\mathbb{X}}$. From the proof of Theorem 5.2.7 it follows that x is either in the cone determined by $v_1, -v_2$ or in the cone determined by $-v_1, v_2$. Moreover, the same is true for y also. Let x be an element in the cone determined by $v_1, -v_2$ so that $x = a\{(1 - t_1)v_1 - t_1 v_2\}$ for some $t_1 \in (0, 1)$ and $a > 0$. Suppose y is in the cone determined by $v_1, -v_2$ and $y = b\{(1 - t)x - t v_2\}$ where $b > 0$ and $0 \leq t \leq$

1. Clearly $t \neq 1$. If possible, suppose that $t \neq 0$. Let $(x + \beta v_2) \perp_B v_2$ and $(y + \mu v_1) \perp_B v_1$. Then from Theorem 5.2.10, it follows that $\|x + \beta v_2\| = \|y + \mu v_1\| = \sqrt{1 - \epsilon^2}$. Clearly, $x + \beta v_2 = k(y + \lambda v_2)$ for some scalars k, λ. Thus,

$$x + \beta v_2 = kb(1 - t)x - kbtv_2 + k\lambda v_2 \Rightarrow k = \frac{1}{b(1 - t)}.$$

If $k > 1$ then $\sqrt{1 - \epsilon^2} = \|x + \beta v_2\| = |k|\|y + \lambda v_2\| > \|y + \lambda v_2\| \geq \sqrt{1 - \epsilon^2}$, a contradiction. Therefore, $k = \frac{1}{b(1-t)} \leq 1$. Again, $y + \mu v_1 = k_1(x + \lambda_1 v_1)$ for some scalar k_1, λ_1. Now, it follows that

$$b(1 - t)x - bt[\frac{1 - t_1}{t_1}v_1 - \frac{1}{at_1}x] + \mu v_1 = k_1 x + k_1 \lambda_1 v_1 \Rightarrow k_1 = b(1 - t) + \frac{bt}{at_1} \Rightarrow k_1 > 1.$$

Therefore, $\sqrt{1 - \epsilon^2} = \|y + \mu v_1\| = |k_1|\|x + \lambda_1 v_1\| > \|x + \lambda_1 v_1\| \geq \sqrt{1 - \epsilon^2}$, a contradiction. Hence we must have $t = 0$. However, since $x, y \in S_{\mathbb{X}}$, this implies that $x = y$. Similarly, $y = b\{(1 - t)x + tv_1\}$ for some $b > 0$ implies that $t = 0$ and once again we have, $y = x$. On the other hand, if y is in the cone determined by $-v_1, v_2$ then it can be shown using similar arguments that $x = -y$.

Step 2: Claim $F(x_1, \epsilon_1) = F(x_2, \epsilon_2) \Rightarrow \epsilon_1 = \epsilon_2$.

Let $F(x_1, \epsilon_1) = F(x_2, \epsilon_2) = K \cup (-K)$, where K is the normal cone determined by $v_1, v_2 \in S_{\mathbb{X}}$. Then from Theorem 5.2.10, it follows that $\inf_{\lambda \in \mathbb{R}} \|x_2 + \lambda v_1\| = \inf_{\lambda \in \mathbb{R}} \|x_2 + \lambda v_2\| = \sqrt{1 - \epsilon_2{}^2}$ and $\inf_{\lambda \in \mathbb{R}} \|x_1 + \lambda v_1\| = \inf_{\lambda \in \mathbb{R}} \|x_1 + \lambda v_2\| = \sqrt{1 - \epsilon_1{}^2}$. Therefore, if $x_1 = \pm x_2$ then $\epsilon_1 = \epsilon_2$ and we are done. Hence assume that $x_1 \neq \pm x_2$. Clearly $x_1, x_2 \notin K \cup (-K)$. From Remark 5.2.9, we have, either $v_1 \in x_1{}^+, v_2 \in x_1{}^-$ or $v_1 \in x_1{}^-, v_2 \in x_1{}^+$. Without loss of generality, assume that $v_1 \in x_1{}^+, v_2 \in x_1{}^-$. First suppose that $x_2 = \frac{(1-t)x_1 + tv_1}{\|(1-t)x_1 + tv_1\|}$ for some $0 < t < 1$. Then it is easy to show that $v_1 \in x_2{}^+, v_2 \in x_2{}^-$. Note that,

$$1 = \|\frac{(1 - t)x_1 + tv_1}{\|(1 - t)x_1 + tv_1\|}\| = \frac{1 - t}{\|(1 - t)x_1 + tv_1\|}\|x_1 + \frac{t}{1 - t}v_1\| \geq \frac{1 - t}{\|(1 - t)x_1 + tv_1\|},$$

since $v_1 \in x_1{}^+$. Now, for any $\lambda \in \mathbb{R}$,

$$\begin{aligned}
\|x_2 + \lambda v_1\| &= \left\| \frac{(1 - t)x_1 + tv_1}{\|(1 - t)x_1 + tv_1\|} + \lambda v_1 \right\| \\
&= \frac{1 - t}{\|(1 - t)x_1 + tv_1\|}\|x_1 + \left(\frac{t}{1 - t} + \frac{\lambda\|(1 - t)x_1 + tv_1\|}{1 - t}\right)v_1\| \\
\Rightarrow \inf_{\lambda \in \mathbb{R}} \|x_2 + \lambda v_1\| &= \frac{1 - t}{\|(1 - t)x_1 + tv_1\|} \inf_{\lambda \in \mathbb{R}} \|x_1 + \lambda v_1\| \\
&\leq \inf_{\lambda \in \mathbb{R}} \|x_1 + \lambda v_1\| \\
\Rightarrow \sqrt{1 - \epsilon_2{}^2} &\leq \sqrt{1 - \epsilon_1{}^2}.
\end{aligned}$$

We can write $x_2 = \frac{(1-t)x_1+tv_2}{\|(1-t)x_1+tv_2\|}$ for some $0 < t < 1$. This implies that

$$\begin{aligned} x_1 &= \frac{\|(1-t)x_1 + tv_2\|}{1-t} x_2 - \frac{t}{1-t} v_2 \\ \Rightarrow 1 &= \frac{\|(1-t)x_1 + tv_2\|}{1-t} \left\| x_2 - \frac{t}{\|(1-t)x_1 + tv_2\|} v_2 \right\| \\ &\geq \frac{\|(1-t)x_1 + tv_2\|}{1-t}, \text{ since } v_2 \in x_2^{-}. \end{aligned}$$

Therefore, for all $\lambda \in \mathbb{R}$,

$$\begin{aligned} x_1 + \lambda v_2 &= \frac{\|(1-t)x_1 + tv_2\|}{1-t} x_2 - \frac{t}{1-t} v_2 + \lambda v_2 \\ \Rightarrow \inf_{\lambda \in \mathbb{R}} \|x_1 + \lambda v_2\| &= \frac{\|(1-t)x_1 + tv_2\|}{1-t} \inf_{\lambda \in \mathbb{R}} \|x_2 + \lambda v_2\| \\ &\leq \inf_{\lambda \in \mathbb{R}} \|x_2 + \lambda v_2\| \\ \Rightarrow \sqrt{1 - {\epsilon_1}^2} &\leq \sqrt{1 - {\epsilon_2}^2}. \end{aligned}$$

Thus, $\epsilon_1 = \epsilon_2$. Similarly, if we assume that $x_2 = \frac{(1-t)x_1-tv_2}{\|(1-t)x_1-tv_2\|}$, for some $0 < t < 1$, then we can apply similar arguments to prove that $\epsilon_1 = \epsilon_2$. This completes the proof. □

Remark 5.2.15 ([125, Remark 2.9]) For $\epsilon = 0$, $F(x_1, 0) = F(x_2, 0) \Rightarrow x_1 = \pm x_2$ holds if the space is strictly convex. Suppose $F(x_1, 0) = F(x_2, 0)$. Then there exists z such that $x_1 \perp_B z$ and $x_2 \perp_B z$. Being strictly convex, B-J orthogonality is left unique and so $x_1 = \pm x_2$.

Remark 5.2.16 ([125, Remark 2.10]) It follows from Theorems 5.2.13 and 5.2.14 that given any normal cone K in a two-dimensional smooth Banach space $\mathbb{X}$, there exist a unique (up to multiplication by ± 1) $x \in S_{\mathbb{X}}$ and a unique $\epsilon \in [0, 1)$ such that $F(x, \epsilon) = K \cup (-K)$. This observation strengthens Theorem 5.2.13 to a considerable extent.

Observe that Theorems 5.2.11 and 5.2.14 prove that the uniqueness property of the set $F(x, \epsilon)$ also holds in an arbitrary Banach space.

Theorem 5.2.17 ([125, Theorem 2.11]) *Let $\mathbb{X}$ be a Banach space. Let $x_1, x_2 \in S_{\mathbb{X}}$ and $\epsilon_1, \epsilon_2 \in (0, 1)$. If $F(x_1, \epsilon_1) = F(x_2, \epsilon_2)$ then $x_1 = \pm x_2$ and $\epsilon_1 = \epsilon_2$.*

Slightly moving from the geometric structure of $F(x, \epsilon)$, we obtain a connection among $F(x, \epsilon)$, normal cones and norm one linear functionals.

Theorem 5.2.18 ([136, Theorem 2.19]) *Let $\mathbb{X}$ be a two-dimensional smooth Banach Space and $x \in S_{\mathbb{X}}$. Suppose $\epsilon \in [0, 1)$. Then the following conditions are equivalent.*

(i) *$F(x, \epsilon) = K \cup (-K)$, where K is the normal cone determined by $u_1, u_2 \in S_{\mathbb{X}}$ and $u_1 \neq \pm u_2$.*

(ii) *There exist $\phi_i \in S_{\mathbb{X}^*}$ such that $\phi_i(x) = \sqrt{1 - \epsilon^2}$, $\phi_i(u_i) = 0$ for $i = 1, 2$ and $\phi_1(u_2)\phi_2(u_1) < 0$.*

Proof (i) $\Rightarrow$ (ii). Since the normal cone K is determined by $u_1, u_2 \in S_{\mathbb{X}}$, by Theorem 5.2.10, we have $\inf_{\lambda \in \mathbb{R}} \| x + \lambda u_i \| = \sqrt{1-\epsilon^2}$, for $i = 1, 2$. Now, as $x \perp_D^{\epsilon} u_i$, by Corollary 5.2.2, for $i = 1, 2$, there exist $\phi_i \in S_{\mathbb{X}^*}$ such that $\phi_i(u_i) = 0$ and $\phi_i(x) \geq \sqrt{1-\epsilon^2}$. Again,

$$\phi_i(x) = \inf_{\lambda \in \mathbb{R}} \phi_i(x + \lambda u_i) \leq \inf_{\lambda \in \mathbb{R}} \| x + \lambda u_i \| = \sqrt{1-\epsilon^2}.$$

Thus, $\phi_i(x) = \sqrt{1-\epsilon^2}$, for $i = 1, 2$. Now, we show that $\phi_1(u_2)\phi_2(u_1) < 0$. Clearly, $x \notin F(x, \epsilon)$. Without loss of generality, we may assume that $x = \frac{(1-t)u_1 - tu_2}{\|(1-t)u_1 - tu_2\|}$ for some $t \in (0, 1)$. Now,

$$(1-\epsilon^2) = \phi_1(x)\phi_2(x) = -\frac{t(1-t)}{\|(1-t)u_1 - tu_2\|^2}\phi_1(u_2)\phi_2(u_1).$$

Thus, $\phi_1(u_2)\phi_2(u_1) = -\frac{(1-\epsilon^2)\|(1-t)u_1 - tu_2\|^2}{t(1-t)} < 0$.

(ii) $\Rightarrow$ (i). Since $\phi_1(u_2)\phi_2(u_1) < 0$ and $\phi_i(u_i) = 0$ for $i = 1, 2$, $u_1 \neq \pm u_2$. Let K be the normal cone determined by u_1 and $u_2 \in S_X$. We first show that $K \cup (-K) \subseteq F(x, \epsilon)$. Let $w \in K \cup (-K)$. If $w = 0$, then clearly $w \in F(x, \epsilon)$. So let $w = a\{tu_1 + (1-t)u_2\}$, where $a \in \mathbb{R} \setminus \{0\}$ and $t \in [0, 1]$. If $t = 0$, then $w = au_2$ and hence by (ii), $x \perp_D^{\epsilon} w$. Thus, $w \in F(x, \epsilon)$. Similarly, $t = 1$ implies that $w \in F(x, \epsilon)$. So let $t \in (0, 1)$. Now, $\phi_1(w)\phi_2(w) = a^2 t(1-t)\phi_1(u_2)\phi_2(u_1) < 0$. Thus there exists $s \in [0, 1]$ such that $(s\phi_1 + (1-s)\phi_2)(w) = 0$. Let $\psi = \frac{s\phi_1 + (1-s)\phi_2}{\|s\phi_1 + (1-s)\phi_2\|}$. Then $\psi \in S_{\mathbb{X}^*}$. Clearly, $\psi(w) = 0$. Now, $\psi(x) = \frac{\sqrt{1-\epsilon^2}}{\|s\phi_1 + (1-s)\phi_2\|} \geq \sqrt{1-\epsilon^2}$. Therefore, from Corollary 5.2.2, it follows that $x \perp_D^{\epsilon} w$. Hence, $w \in F(x, \epsilon)$. Thus, we have $K \cup (-K) \subseteq F(x, \epsilon)$.

Next, we show that $F(x, \epsilon) \subseteq K \cup (-K)$. Let $y \notin K \cup (-K)$. If possible, let $y \in F(x, \epsilon)$. Then by Corollary 5.2.2, there exists $\phi \in S_{\mathbb{X}^*}$ such that $\phi(x) \geq \sqrt{1-\epsilon^2}$ and $\phi(y) = 0$. Without loss of generality we assume that $y = b\{(1-t)u_1 - tu_2\}$ where $b > 0$ and $t \in (0, 1)$. Now, $-\phi_1(y)\phi_2(y) = b^2 t(1-t)\phi_1(u_2)\phi_2(u_1) < 0$, as $\phi_1(u_2)\phi_2(u_1) < 0$. Therefore, there exists $s \in (0, 1)$ such that $(s\phi_2 - (1-s)\phi_1)(y) = 0$. Since $\mathbb{X}$ is two-dimensional, there exists unique $\phi \in S_{\mathbb{X}^*}$ (up to multiplication by ± 1) such that $\phi(y) = 0$. Thus without loss of generality we assume that $\phi = \frac{s\phi_2 - (1-s)\phi_1}{\|s\phi_2 - (1-s)\phi_1\|}$. Since $\mathbb{X}$ is smooth, $\mathbb{X}^*$ is strictly convex. Therefore, $\|s\phi_2 - (1-s)\phi_1\| > ||s| - |(1-s)|| = |2s-1|$. Hence, $|\phi(x)| < \frac{|2s-1|\sqrt{1-\epsilon^2}}{|2s-1|} = \sqrt{1-\epsilon^2}$, a contradiction. Thus, $y \notin K \cup (-K)$ implies that $y \notin F(x, \epsilon)$. Therefore, $F(x, \epsilon) \subseteq K \cup (-K)$. This completes the proof. $\square$

5.3 $\perp_B^\epsilon$-Approximate Orthogonality on $\mathbb{X}$

Observe that in a Hilbert space $\mathbb{H}$, if $x \perp^\epsilon y$ holds, i.e., $|\langle x, y\rangle| \leq \epsilon\|x\|\|y\|$ for some $x(\neq 0)$, $y \in \mathbb{H}$ and $\epsilon \in [0, 1)$, then taking $z = -\frac{\langle x,y\rangle}{\|x\|^2} + y$, we get that $x \perp z$ and $\|z - y\| \leq \epsilon\|y\|$. If $x = 0$ and $x \perp^\epsilon y$ holds, then we can simply consider $z = y$ to get $x \perp z$ and $\|z - y\| \leq \epsilon\|y\|$. Moreover, in such cases, $z \in \text{span}\{x, y\}$. On the other hand, if $x \perp z$ and if $\|z - y\| \leq \epsilon\|y\|$ holds, then we get

$$|\langle x, y\rangle| = |\langle x, y - z\rangle| \leq \|x\|\|y - z\| \leq \epsilon\|x\|\|y\|.$$

Thus, in a Hilbert space, the following characterization of approximate orthogonality holds:

$$x \perp^\epsilon y \Leftrightarrow \exists\, z \in \text{ span}\{x, y\} \text{ such that } x \perp z \text{ and } \|z - y\| \leq \epsilon\|y\|. \tag{5.3.1}$$

This fact motivated Chmieliński et al. [25] to search for an analogous characterization of $\perp_B^\epsilon$ in a Banach space. For this characterization, we need to introduce the notion of norm derivatives. The functions $\rho_+^{'}, \rho_-^{'} : \mathbb{X} \times \mathbb{X} \to \mathbb{R}$ are defined as

$$\rho_\pm^{'}(x, y) := \lim_{t \to 0\pm} \frac{\|x + ty\|^2 - \|x\|^2}{2t}, \text{ where } x, y \in \mathbb{X}$$

are called norm derivatives. For a detailed study on norm derivatives and its properties, we refer the readers to [6]. The notion of norm derivatives is related to the notion of s.i.p. in the following way:

$$\rho_+^{'}(x, y) = \sup\{[y, x] : [.,.] \in \mathcal{O}\} \text{ and} \tag{5.3.2}$$

$$\rho_-^{'}(x, y) = \inf\{[y, x] : [.,.] \in \mathcal{O}\}, \tag{5.3.3}$$

where $\mathcal{O}$ is the collection of all s.i.p. on $\mathbb{X}$. Chmieliński and Wójcik proved a characterization of $\perp_B^\epsilon$, in terms of norm derivatives.

Theorem 5.3.1 ([26, Theorem 3.1]) *Let $\mathbb{X}$ be a Banach space and let $\epsilon \in [0, 1)$. Then for any $x, y \in \mathbb{X}$ and any $\alpha \in \mathbb{R}$, we have*

$$x \perp_B^\epsilon (y - \alpha x) \Leftrightarrow \rho_-^{'}(x, y) - \epsilon\|x\|\|y - \alpha x\| \leq \alpha\|x\|^2 \leq \rho_+^{'}(x, y) + \epsilon\|x\|\|y - \alpha x\|.$$

Proof Let $x \perp_B^\epsilon (y - \alpha x)$. Then for $\lambda > 0$, we have

$$-\epsilon\|x\|\|y - \alpha x\| \leq \frac{\|x + \lambda(y - \alpha x)\|^2 - \|x\|^2}{2\lambda}.$$

Letting $\lambda \to 0^+$, we get $-\epsilon\|x\|\|y - \alpha x\| \leq \rho_+^{'}(x, y - \alpha x) = \rho_+^{'}(x, y) - \alpha\|x\|^2$. Again considering $\lambda < 0$, we get $\rho_-^{'}(x, y) - \alpha\|x\|^2 \leq \epsilon\|x\|\|y - \alpha x\|$. On the other hand, the result follows easily if $x = 0$ or $y = 0$. Assume $x \neq 0$, $y \neq 0$. It follows from $\rho_-'(x, y) - \epsilon\|x\|\|y - \alpha x\| \leq \alpha\|x\|^2$ that $\rho_-'(x, y - \alpha x) \leq \epsilon\|x\|\|y - \alpha x\|$. Also

$$2 \lim_{t \to 0^-} \frac{\|x + t(y - \alpha x)\|^2 - \|x\|^2}{2t} \leq 2\epsilon\|x\|\|y - \alpha x\|.$$

Choose $\gamma \in (0, 1)$ so that

$$\lim_{t \to 0^-} \frac{\|x + t(y - \alpha x)\|^2 - \|x\|^2}{t} \leq 2(\epsilon + \gamma)\|x\|\|y - \alpha x\|.$$

Then we can find $\delta_1 < 0$ such that

$$\forall\, t \in [\delta_1, 0), \quad \frac{\|x + t(y - \alpha x)\|^2 - \|x\|^2}{t} \leq 2(\epsilon + \gamma)\|x\|\|y - \alpha x\|,$$

and so

$$\forall\, t \in [\delta_1, 0), \ \|x\|^2 < \|x + t(y - \alpha x)\|^2 + 2(\epsilon + \gamma)\|x\|\|t(y - \alpha x)\|. \quad (5.3.4)$$

Proceeding similarly from the inequality $-\epsilon\|x\|\|y - \alpha x\| \leq \rho_+'(x, y) - \alpha\|x\|^2 = \rho_+'(x, y - \alpha x)$ we can find $\delta_2 > 0$ such that

$$\forall\, t \in (0, \delta_2], \ \|x\|^2 < \|x + t(y - \alpha x)\|^2 + 2(\epsilon + \gamma)\|x\|\|t(y - \alpha x)\|. \quad (5.3.5)$$

Consider the function $\phi : \mathbb{R} \to \mathbb{R}$ defined by

$$\phi(t) = \|x + t(y - \alpha x)\|^2 + 2(\epsilon + \gamma)\|x\|\|t(y - \alpha x)\|.$$

It can be shown that the function ϕ is convex and from (5.3.4) and (5.3.5) it follows that $\phi(0) = \min\{\phi(t) : t \in [\delta_1, \delta_2]\}$. By the convexity of ϕ, we get $\phi(0) = \min\{\phi(t) : t \in \mathbb{R}\}$. Thus

$$\forall\, t \in \mathbb{R} \setminus \{0\}, \ \|x\|^2 < \|x + t(y - \alpha x)\|^2 + 2(\epsilon + \gamma)\|x\|\|t(y - \alpha x)\|.$$

Choose $\lambda \neq 0$. Then we have

$$\|x\|^2 < \|x + \lambda(y - \alpha x)\|^2 + 2(\epsilon + \gamma)\|x\|\|t(y - \alpha x)\|.$$

Since $\gamma \in (0, 1)$ was arbitrary, letting $\gamma \to 0^+$, we get

$$\|x\|^2 \leq \|x + \lambda(y - \alpha x)\|^2 + 2\epsilon\|x\|\|t(y - \alpha x)\|.$$

Clearly this holds for $\lambda = 0$ and so we get $x \perp_B^{\epsilon} (y - \alpha x)$. □

To prove the desired characterization of $\perp_B^{\epsilon}$, we also require the following result by Chmieliński et al.

Proposition 5.3.2 ([25, Proposition 2.1]) *Let $\mathbb{X}$ be a Banach space and let $x, y \in \mathbb{X}$ be such that $x \perp_B^{\epsilon} y$ for some $\epsilon \in [0, 1)$. Then for each $n \in \mathbb{N}$, there exists a s.i.p. $[.,.]_n$ in $\mathbb{X}$ such that*

$$|[y, x]_n| \leq \left(\epsilon + \frac{1}{n}\right)\|x\|\|y\|. \tag{5.3.6}$$

Proof Consider any integer n such that $\frac{1}{n} \in (0, 1-\epsilon)$. Applying (5.3.2), (5.3.3) and Theorem 5.3.1, we can choose s.i.p. $[.,.]_{n1}$ and $[.,.]_{n2}$ such that the following hold:

$$[y, x]_{n1} < \left(\epsilon + \frac{1}{n}\right)\|x\|\|y\| \text{ and } -\left(\epsilon + \frac{1}{n}\right)\|x\|\|y\| < [y, x]_{n2}.$$

It follows that for some $\lambda_n \in [0, 1]$, we have

$$-\left(\epsilon + \frac{1}{n}\right)\|x\|\|y\| \leq \lambda_n[y, x]_{n1} + (1-\lambda_n)[y, x]_{n2} \leq \left(\epsilon + \frac{1}{n}\right)\|x\|\|y\|.$$

Now, considering the s.i.p. $[.,.]_n = \lambda_n[.,.]_{n1} + (1-\lambda_n)[.,.]_{n2}$, and using the above inequalities, we get the desired result. □

Now we are ready to present a characterization of $\perp_B^{\epsilon}$ analogous to (5.3.1).

Theorem 5.3.3 ([25, Theorem 2.2]) *Let $\mathbb{X}$ be a Banach space. Suppose that $x, y \in \mathbb{X}$ and $\epsilon \in [0, 1)$. Then*

$$x \perp_B^{\epsilon} y \text{ if and only if } \exists\, z \in \text{span}\{x, y\} \text{ such that } x \perp_B z \text{ and } \|z - y\| \leq \epsilon\|y\|.$$

Proof Let $x \perp_B^{\epsilon} y$. If $x = 0$, then the result holds trivially. Therefore, assume that $x \neq 0$. From Proposition 5.3.2 it follows that for each $n \in \mathbb{N}$, there exists a s.i.p. $[.,.]_n$ in $\mathbb{X}$ such that

$$|[y, x]_n| \leq \left(\epsilon + \frac{1}{n}\right)\|x\|\|y\|.$$

Define $z_n = -\frac{[y,x]_n}{\|x\|^2}x + y$. Then $z_n \in \text{span}\{x, y\}$ and $[z_n, x]_n = 0$. Now, for each scalar λ, we have

$$\|x + \lambda z_n\|\|x\| \geq |[x + \lambda z_n, x]_n| = \|x\|^2 \Rightarrow \|x + \lambda z_n\| \geq \|x\|.$$

Thus, $x \perp_B z_n$. On the other hand, applying (5.3.6), we get $\|z_n - y\| = \frac{|[y,x]_n|}{\|x\|} \leq \left(\epsilon + \frac{1}{n}\right)\|y\|$. Since $\{z_n\}$ is a bounded sequence in the two-dimensional space span$\{x, y\}$, it has a convergent subsequence say, $\{z_{n_k}\}$. Suppose that $z_{n_k} \to z$. Then $z \in \text{span}\{x, y\}$. Therefore, from $x \perp_B z_n$ and $\|z_n - y\| \leq \left(\epsilon + \frac{1}{n}\right)\|y\|$, it follows that $x \perp_B z$ and $\|z - y\| \leq \epsilon\|y\|$.

Conversely, suppose that there exists $z \in \text{span}\{x, y\}$ such that $x \perp_B z$ and $\|z - y\| \leq \epsilon\|y\|$. Applying Theorem 2.1.14, we get $f \in S_{\mathbb{X}^*}$ such that $f(x) = \|x\|$ and $f(z) = 0$. Now, for any scalar λ,

$$|f(\lambda y)| = |f(\lambda y - \lambda z)| \leq \|\lambda y - \lambda z\| \leq \epsilon\|\lambda y\|.$$

Therefore, for any scalar λ, we have

$$\begin{aligned}
\|x + \lambda y\|^2 &\geq |f(x + \lambda y)|^2 \\
&\geq ||f(x)| - |f(\lambda y)||^2 \\
&= |f(x)|^2 - 2|f(x)||f(\lambda y)| + |f(\lambda y)|^2 \\
&\geq \|x\|^2 - 2\epsilon\|x\|\|\lambda y\|,
\end{aligned}$$

which means that $x \perp_B^\epsilon y$. This completes the proof. □

Imitating the proof of the last part of the previous theorem, $\perp_B^\epsilon$ can be characterized in terms of linear functionals. To avoid repetition, we skip the details of the proof here.

Theorem 5.3.4 ([25, Theorem 2.3]) *Let $\mathbb{X}$ be a Banach space. Suppose that $x(\neq 0), y \in \mathbb{X}$ and $\epsilon \in [0, 1)$. Then $x \perp_B^\epsilon y$ if and only if there exists $f \in J(x)$ such that $|f(y)| \leq \epsilon\|y\|$.*

Remark 5.3.5 Observe that from the proof of Theorem 5.3.3, it follows that the sufficient part also holds for complex Banach space. Recently, in [147, Theorem 2.2], Theorems 5.3.3 and 5.3.4 have been proved for complex Banach spaces.

We next provide a characterization of $\perp_B^\epsilon$ in terms of extreme linear functionals, which generalizes Theorem 4.4.4.

Theorem 5.3.6 ([25, Theorem 2.4]) *Let $\mathbb{X}$ be a Banach space. Suppose that $x(\neq 0), y \in \mathbb{X}$ and $\epsilon \in [0, 1)$. Then $x \perp_B^\epsilon y$ if and only if there exist $\phi, \psi \in Ext(B_{\mathbb{X}^*})$ and $\alpha \in [0, 1]$ such that $\phi(x) = \psi(x) = \|x\|$ and $|\alpha\phi(y) + (1 - \alpha)\psi(y)| \leq \epsilon\|y\|$. Moreover, the sufficient condition for $x \perp_B^\epsilon y$ is true also for complex Banach space.*

Proof First, suppose that $x \perp_B^\epsilon y$. Applying Theorem 5.3.3, we get $z \in \text{span}\{x, y\}$ such that $x \perp_B z$ and $\|z - y\| \leq \epsilon\|y\|$. From Theorem 4.4.4, it follows that there exist $\phi, \psi \in Ext(B_{\mathbb{X}^*})$ and $\alpha \in [0, 1]$ such that $\phi(x) = \psi(x) = \|x\|$ and $\alpha\phi(z) + (1 - \alpha)\psi(z) = 0$. Therefore,

$$\begin{aligned}
|\alpha\phi(y) + (1 - \alpha)\psi(y)| &= |\alpha\phi(z - y) + (1 - \alpha)\psi(z - y)| \\
&\leq \alpha\|z - y\| + (1 - \alpha)\|z - y\| \\
&= \|z - y\| \leq \epsilon\|y\|.
\end{aligned}$$

Conversely, suppose that $\mathbb{X}$ is a real or complex Banach space and there exist $\phi, \psi \in Ext(B_{\mathbb{X}^*})$ and $\alpha \in [0, 1]$ such that $\phi(x) = \psi(x) = \|x\|$ and $|\alpha\phi(y) + (1 -$

$\alpha)\psi(y)| \leq \epsilon\|y\|$. Let $f = \alpha\phi + (1-\alpha\psi)$. Then clearly, $|f(y)| \leq \epsilon\|y\|$. Furthermore, $f(x) = \|x\|$, i.e., $\|f\| = 1$ and $f \in J(x)$. Hence, by Theorem 5.3.4, we conclude that $x \perp_B^\epsilon y$. □

Applying Theorems 5.3.4 and 5.3.6, Chmieliński et al. characterized approximate B-J orthogonality ($\perp_B^\epsilon$) of continuous functions, in a particular setting. Suppose that Q is a locally compact topological space. Suppose $C(Q)$ is the space of all real valued continuous functions defined on Q, endowed with the supremum norm. Consider the subspace $C_0(Q)$ of $C(Q)$ defined as follows:

$$C_0(Q) = \Big\{ f \in C(Q) : \forall\, \epsilon > 0, \text{ the set } \{t \in Q : |f(t)| \geq \epsilon\} \text{ is compact} \Big\}.$$

Note that, for $f \in C_0(Q)$, the set $M_f = \{t \in Q : |f(t)| = \|f\|\}$ is always non-empty and compact.

Theorem 5.3.7 ([25, Theorem 3.6]) *Let $f, g \in C_0(Q)$, $f \neq 0 \neq g$. Assume that M_f is connected. Then the following conditions are equivalent:*
(a) $f \perp_B^\epsilon g$.
(b) *There exists $t_0 \in M_f$ such that $|g(t_0)| \leq \epsilon\|g\|$.*

Proof (a) $\Rightarrow$ (b): Let $f \perp_B^\epsilon g$. Then it follows from Theorem 5.3.6 that there exist $\phi, \psi \in Ext(B_{C_0(Q)^*})$ and $\alpha \in [0,1]$ such that $\phi(f) = \psi(f) = \|f\|$ and

$$|\alpha\phi(g) + (1-\alpha)\psi(g)| \leq \epsilon\|g\|. \tag{5.3.7}$$

From [141, Chaps. 1, 1.10, 1.11] it is known that the extreme points ϕ, ψ of $C_0(Q)^*$ have the form $\pm\rho_u, \pm\rho_v$, for some $u, v \in M_f$, where $\rho_t : C_0(Q) \to \mathbb{R}$ is defined as $\rho_t(h) = h(t)$. Since M_f is connected, f takes values of the same sign on M_f. Without loss of generality we assume that $f(t) = \|f\|$ for all $t \in M_f$. Then ϕ, ψ take the form ρ_u, ρ_v. Thus, $\phi(g) = \rho_u(g) = g(u)$ and $\psi(g) = \rho_v(g) = g(v)$. Hence, from the inequality (5.3.7), we get

$$|\alpha g(u) + (1-\alpha)g(v)| \leq \epsilon\|g\|. \tag{5.3.8}$$

If $|g(u)| \leq \epsilon\|g\|$ or $|g(v)| \leq \epsilon\|g\|$, then we take $t_0 = u$ or $t_0 = v$, respectively. If $g(u) < -\epsilon\|g\|$ and $g(v) > \epsilon\|g\|$ (or $g(v) < -\epsilon\|g\|$ and $g(u) > \epsilon\|g\|$) we use the fact that M_f is connected and the Darboux property for g to get $-\epsilon\|g\| \leq g(t_0) \leq \epsilon\|g\|$ for some $t_0 \in M_f$, as required. The remaining cases $g(u), g(v) > \epsilon\|g\|$ or $g(u), g(v) < \epsilon\|g\|$ are ruled out as they would contradict (5.3.8).

(b)$\Rightarrow$ (a): Since $t_0 \in M_f$, for the functional $\phi = \rho_{t_0}$ (or $\phi = -\rho_{t_0}$), we have $\|\phi\| = 1$ and $\phi(f) = |f(t_0)| = \|f\|$, and thus $\phi \in J(f)$. Moreover, $|\phi(g)| = |\rho_{t_0}(g)| = |g(t_0)| \leq \epsilon\|g\|$. Therefore, (a) follows from Theorem 5.3.4. □

Now, we explore $\perp_B^\epsilon$ from a geometric point of view. We relate this notion of approximate B-J orthogonality with that normal cones. For our purpose, we define the following set:

$$G(x, \epsilon) = \{y \in \mathbb{X} : x \perp_B^{\epsilon} y\}.$$

It was first proved in [125, Theorem 2.14] that for a smooth point x in a two-dimensional Banach space $\mathbb{X}$ and $\epsilon \in [0, 1)$, $G(x, \epsilon) = K \cup (-K)$ for some normal cone K in $\mathbb{X}$. The result was further generalized in [136] for possibly non-smooth points in a two-dimensional Banach space. This requires the following easy propositions.

Proposition 5.3.8 ([136, Proposition 2.2]) *Let $\mathbb{X}$ be a two-dimensional Banach space. Suppose $x \in S_{\mathbb{X}}$ is a smooth point and $\epsilon \in [0, 1)$. Let $G(x, \epsilon) = K \cup (-K)$, where K is the normal cone determined by $v_1, v_2 \in S_{\mathbb{X}}$. Then for $\phi \in J(x)$, $|\phi(v_1)| = |\phi(v_2)| = \epsilon$. Moreover, if $\epsilon > 0$, then $\phi(v_1)\phi(v_2) < 0$.*

Proof Clearly $v_1, v_2 \in K \subseteq G(x, \epsilon)$ and so $x \perp_B^{\epsilon} v_1, x \perp_B^{\epsilon} v_2$. If $\epsilon = 0$, then from Theorem 2.1.15 it follows that $\phi(v_1) = \phi(v_2) = 0$. Let $\epsilon > 0$. Then by Theorem 5.3.4, we have $|\phi(v_1)| \leq \epsilon$ for $\phi \in J(x)$. We claim that $|\phi(v_1)| = \epsilon$. If possible, let $|\phi(v_1)| < \epsilon$. Since ϕ is continuous, there exists a neighborhood $B(v_1, \delta)$ of v_1 such that $|\phi(y)| < \epsilon$ for all $y \in B(v_1, \delta) \cap S_{\mathbb{X}}$ for some $\delta > 0$. Hence, $y \in G(x, \epsilon)$ for all $y \in B(v_1, \delta) \cap S_{\mathbb{X}}$. Since the cone K is determined by v_1, v_2, $B(v_1, \delta) \cap S_{\mathbb{X}}$ contains a vector y which is not in $K \cup (-K)$. This shows that $G(x, \epsilon) \nsubseteq K \cup (-K)$, a contradiction. So $|\phi(v_1)| = \epsilon$. Proceeding similarly we can show that $|\phi(v_2)| = \epsilon$.

Next, we show that $\phi(v_1)\phi(v_2) < 0$. Let $z \in S_{\mathbb{X}}$ and $\phi(z) = 0$. Clearly, $x \perp_B^{\epsilon} z$. Hence, $z \in G(x, \epsilon) = K \cup (-K)$. Without loss of generality we assume that $z \in K$. Then $z = a\{(1-t)v_1 + tv_2\}$ for some $t \in [0, 1]$ and $a > 0$. If $t = 1$, then $z = av_2 \Rightarrow \phi(z) = a\phi(v_2)$, a contradiction, since $\phi(z) = 0$ and $|\phi(v_2)| = \epsilon > 0$. Hence, $t < 1$. Similarly, $t > 0$. Now,

$$\phi(z) = 0 \Rightarrow a(1-t)\phi(v_1) + at\phi(v_2) = 0 \Rightarrow \phi(v_1)\phi(v_2) < 0,$$

since $a > 0$ and $t \in (0, 1)$. This completes the proof. □

We now prove the converse of Proposition 5.3.8.

Proposition 5.3.9 ([136, Proposition 2.3]) *Let $\mathbb{X}$ be a two-dimensional Banach space. Suppose $x \in S_{\mathbb{X}}$ is a smooth point and $\epsilon \in (0, 1)$ be such that for $\phi \in J(x)$, $|\phi(v_1)| = |\phi(v_2)| = \epsilon$ for some $v_1, v_2 \in S_{\mathbb{X}}$ with $v_1 \neq \pm v_2$ and $\phi(v_1)\phi(v_2) < 0$. Then $G(x, \epsilon) = K \cup (-K)$, where K is the normal cone determined by v_1 and v_2.*

Proof Without loss of generality we assume that $\phi(v_1) = \epsilon$ and $\phi(v_2) = -\epsilon$. We now show that $G(x, \epsilon) = K \cup (-K)$. Let $w \in K$. Then $w = a((1-t)v_1 + tv_2)$ for some $a \geq 0$ and $t \in [0, 1]$. Now,

$$\phi(w) = a((1-t)\phi(v_1) + t\phi(v_2)) = a(1-t)\epsilon + at(-\epsilon) = a(1-2t)\epsilon.$$

Therefore, $|\phi(w)| = a|(1-2t)|\epsilon$. On the other hand,

$$\|w\| = a\|(1-t)v_1 + tv_2\| \geq a|\|(1-t)v_1\| - t\|v_2\|| = a|1-2t|.$$

So $|\phi(w)| = a|(1-2t)|\epsilon \le \epsilon\|w\|$. So from Theorem 5.3.4, we get $x \perp_B^\epsilon w$. Therefore, $w \in G(x, \epsilon)$. Similarly, $w \in (-K) \Rightarrow w \in G(x, \epsilon)$. Thus, $K \cup (-K) \subseteq G(x, \epsilon)$.

Next, we show that $G(x, \epsilon) \subseteq K \cup (-K)$. Clearly, $x \notin K \cup (-K)$, since $x \notin G(x, \epsilon)$. Without loss of generality we assume that $x = \frac{(1-s)v_1 - sv_2}{\|(1-s)v_1 - sv_2\|}$ for some $s \in (0, 1)$. Let $w \notin K \cup (-K)$. Then either $w = a((1-t)x - tv_2)$ where $a \neq 0$ and $t \in (0, 1)$ or $w = a((1-t)v_1 + tx)$ where $a \neq 0$ and $t \in (0, 1)$. For the first case, $\phi(w) = a(1-t) + at\epsilon$. So $|\phi(w)| = |a|\ |1 - t + t\epsilon| > |a|\epsilon$ as $0 < \epsilon < 1$. Moreover,

$$\|w\| = |a|\|(1-t)x - tv_2\| \le |a|(1 - t + t) = |a|.$$

So $|\phi(w)| > |a|\epsilon \ge \epsilon\ \|w\|$, which by Theorem 5.3.4, shows that $w \notin G(x, \epsilon)$. Now, for the second case, $\phi(w) = a(1-t)\epsilon + at$. So

$$|\phi(w)| = |a||(1-t)\epsilon + t| > |a|\epsilon$$

and

$$\|w\| = |a|\|(1-t)v_1 + tx\| \le |a|(1 - t + t) = |a|.$$

Therefore, $|\phi(w)| > |a|\epsilon \ge \epsilon\ \|w\|$, which again by Theorem 5.3.4, shows that $w \notin G(x, \epsilon)$. So $w \in G(x, \epsilon)$ implies $w \in K \cup (-K)$. Therefore, $G(x, \epsilon) \subseteq K \cup (-K)$. Thus, $G(x, \epsilon) = K \cup (-K)$. This completes the proof. □

Now, we are ready to describe the set $G(x, \epsilon)$ in terms of normal cones in a two-dimensional Banach space.

Theorem 5.3.10 ([136, Theorem 2.4]) *Let $\mathbb{X}$ be a two-dimensional Banach space. Then for any $x \in S_{\mathbb{X}}$ and $\epsilon \in [0, 1)$, there exists a normal cone K in $\mathbb{X}$ such that $G(x, \epsilon) = K \cup (-K)$.*

Proof Let $\epsilon \in [0, 1)$. From Theorem 5.2.7 we infer that there exists a normal cone K in $\mathbb{X}$ such that $F(x, \epsilon) = K \cup (-K)$. Therefore, $F(x, 0) = K_0 \cup (-K_0)$, where K_0 is a normal cone in $\mathbb{X}$. Let the cone K_0 be determined by $u_1, u_2 \in S_{\mathbb{X}}$. Then $x \perp_B u_i$, for $i = 1, 2$. Thus, by Theorem 2.1.15, there exist $\phi_i \in J(x)$ such that $\phi_i(u_i) = 0$, for $i = 1, 2$. Without loss of generality we assume that $\phi_1(u_2) \le 0$. Clearly, there exist $t_1, t_2 \in (0, 1)$ such that

$$v_1 = \frac{(1-t_1)x + t_1u_1}{\|(1-t_1)x + t_1u_1\|},\quad v_2 = \frac{-(1-t_2)x + t_2u_2}{\| -(1-t_2)x + t_2u_2\|}$$

and $\phi_1(v_1) = \epsilon$, $\phi_2(v_2) = -\epsilon$. Let K be the normal cone determined by $v_1, v_2 \in S_{\mathbb{X}}$. Clearly, $K_0 \subseteq K$ and $x \notin K \cup (-K)$. We show that $G(x, \epsilon) = K \cup (-K)$. Let $v \in K \cap S_{\mathbb{X}}$. If $v \in K_0$, then clearly $v \in G(x, \epsilon)$. Let $v = \frac{(1-t)v_1 + tu_1}{\|(1-t)v_1 + tu_1\|}$ for some $t \in [0, 1]$. Then $\phi_1(v) = \frac{(1-t)\epsilon}{\|(1-t)v_1 + tu_1\|}$. If possible, let $\|(1-t)v_1 + tu_1\| < (1-t)$. Then $u_1 \notin v_1^+ \Rightarrow u_1 \in v_1^-$. Now, there exists $s \in (0, 1)$ such that $x = \frac{(1-s)v_1 - su_1}{\|(1-s)v_1 - su_1\|}$. Therefore,

$$1 = \phi_1(x) = \frac{(1-s)\epsilon}{\|(1-s)v_1 - su_1\|} \leq \frac{(1-s)\epsilon}{1-s} = \epsilon,$$

a contradiction. Thus, $\|(1-t)v_1 + tu_1\| \geq (1-t)$. Now, it follows that

$$\phi_1(v) = \frac{(1-t)\epsilon}{\|(1-t)v_1 + tu_1\|} \leq \frac{(1-t)\epsilon}{(1-t)} = \epsilon.$$

Thus, $v \in G(x, \epsilon)$. Similarly, $v = \frac{(1-t)v_2+tu_2}{\|(1-t)v_2+tu_2\|}$ for some $t \in [0, 1]$ implies that $|\phi_2(v)| \leq \epsilon$. Thus, $v \in G(x, \epsilon)$. Hence, by homogeneity of approximate B-J orthogonality, we have, $K \cup (-K) \subseteq G(x, \epsilon)$.

Next, let $v \in S_{\mathbb{X}} \setminus (K \cup (-K))$. Then either $v = \pm\frac{(1-t)x+tv_1}{\|(1-t)x+tv_1\|}$ or $v = \pm\frac{(1-t)x-tv_2}{\|(1-t)x-tv_2\|}$ for some $t \in [0, 1)$. First let $v = \pm\frac{(1-t)x+tv_1}{\|(1-t)x+tv_1\|}$ for some $t \in [0, 1)$. If possible, let $v \in G(x, \epsilon)$. Then there exists $\phi \in J(x)$ such that $|\phi(v)| \leq \epsilon$. Thus, $|1 - t + t\phi(v_1)| \leq |\frac{(1-t)+t\phi(v_1)}{\|(1-t)x+tv_1\|}| \leq \epsilon$. This gives that

$$\phi(v_1) < \epsilon \Rightarrow \frac{(1-t_1) + t_1\phi(u_1)}{\|(1-t_1)x + t_1u_1\|} < \epsilon = \phi_1(v_1) = \frac{(1-t_1)}{\|(1-t_1)x + t_1u_1\|}.$$

Thus, $\phi(u_1) < 0$. Therefore, there exists $r \in (0, 1)$ such that $\phi(\frac{(1-r)x+ru_1}{\|(1-r)x+ru_1\|}) = 0$. Hence, $x \perp_B (1-r)x + ru_1$, i.e., $(1-r)x + ru_1 \in F(x, 0)$ but $(1-r)x + ru_1 \notin K_0 \cup (-K_0)$, a contradiction. Therefore, $v \notin G(x, \epsilon)$. Similarly, $v = \pm\frac{(1-t)x-tv_2}{\|(1-t)x-tv_2\|}$ for some $t \in [0, 1)$ implies that $v \notin G(x, \epsilon)$. Thus, $G(x, \epsilon) = K \cup (-K)$. □

Suppose we denote the restriction of $G(x, \epsilon)$ to the subspace span$\{x, y\}$ by $Q_{x,y}(\epsilon)$. Now, as an immediate consequence of the previous theorem, we get the following description of $G(x, \epsilon)$ in an arbitrary Banach space.

Theorem 5.3.11 ([136, Theorem 2.6]) *Let $\mathbb{X}$ be a Banach space and $x \in \mathbb{X}$. Suppose $\epsilon \in [0, 1)$. Then $G(x, \epsilon) = \bigcup_{y \in X} Q_{x,y}(\epsilon)$, i.e., $G(x, \epsilon)$ is a union of two-dimensional normal cones.*

The natural question that arises from the study so far is whether the converse of Theorem 5.3.10 is true. More precisely, we can ask whether for each normal cone K in a two-dimensional Banach space $\mathbb{X}$, there exist $x \in S_{\mathbb{X}}$ and $\epsilon \in [0, 1)$ such that $G(x, \epsilon) = K \cup (-K)$. Easy examples can be provided to show that it is not true in general. Such an example was exhibited on the space ℓ_∞^2 in [136, Example 2.7]. The following theorem shows that even in a two-dimensional smooth Banach space the converse of Theorem 5.3.10 is not true, if the space is not strictly convex.

Theorem 5.3.12 ([136, Theorem 2.8]) *Let $\mathbb{X}$ be a two-dimensional Banach space which is not strictly convex. Then there exists a normal cone K in $\mathbb{X}$ such that for no $x \in S_{\mathbb{X}}$ and $\epsilon \in [0, 1)$, $G(x, \epsilon) = K \cup (-K)$.*

Proof Let $\mathbb{X}$ be not strictly convex, then there exist v_1 and $v_2 \in S_{\mathbb{X}}$ such that $v_1 \neq \pm v_2$ and $((1-t)v_1 + tv_2) \in S_{\mathbb{X}}$ for all $t \in [0, 1]$. Let K be the normal cone determined by v_1 and $-v_2$. If possible, suppose $x \in S_{\mathbb{X}}$ and $\epsilon \in [0, 1)$ with $G(x, \epsilon) = K \cup (-K)$.

Since $x \notin G(x, \epsilon)$, without loss of generality we assume that $x = ((1-t)v_1 + tv_2)$ for some $t \in (0, 1)$. Then x is a smooth point and so $J(x) = \{\phi\}$, for some $\phi \in S_{\mathbb{X}^*}$. We first claim that $\epsilon \neq 0$. If $\epsilon = 0$, then by Proposition 5.3.8, $\phi(v_1) = \phi(-v_2) = 0$. Now, $\phi(x)$=1 implies that $(1-t)\phi(v_1) + t\phi(v_2) = 1$. But then $0 = 1$, which is a contradiction. Now, if $\epsilon \in (0, 1)$, then by Proposition 5.3.8, $|\phi(v_1)| = |\phi(-v_2)| = \epsilon$ and $\phi(v_1)\,\phi(-v_2) < 0$. Let $\phi(v_1) = \epsilon$ and $\phi(-v_2) = -\epsilon$, i.e., $\phi(v_2) = \epsilon$. Then $\phi(x) = 1$ implies that $(1-t)\phi(v_1) + t\phi(v_2) = 1$. So we get $\epsilon = 1$, a contradiction. Similarly, $\phi(v_1) = -\epsilon$ and $\phi(-v_2) = \epsilon$ gives a contradiction. Hence, there does not exist any $x \in S_{\mathbb{X}}$ and $\epsilon \in [0, 1)$ such that $G(x, \epsilon) = K \cup (-K)$. This completes the proof. □

However, if the two-dimensional space is considered to be smooth and strictly convex then the converse of Theorem 5.3.10 is indeed true.

Theorem 5.3.13 ([136, Theorem 2.9]) *Let $\mathbb{X}$ be a two-dimensional strictly convex and smooth Banach space. Let K be a normal cone determined by $v_1, v_2 \in S_{\mathbb{X}}$. Then there exist $x \in S_{\mathbb{X}}$ and $\epsilon \in [0, 1)$ such that $G(x, \epsilon) = K \cup (-K)$.*

Proof Clearly, $v_1 \neq -v_2$, as K is a normal cone. Otherwise, $-v_2 \in K \cap (-K)$, a contradiction. If $v_1 = v_2$, then choose $x \in S_{\mathbb{X}}$ such that $x \perp_B v_1$. Then for $\epsilon = 0$, $G(x, \epsilon) = G(x, 0) = \{y \in \mathbb{X} : x \perp_B y\} = \{\alpha v_1 : \alpha \in \mathbb{R}\} = K \cup (-K)$. Let $v_1 \neq \pm v_2$. Let $w = v_1 + v_2$. Then following Theorem 2.1.13 there exists $x \in S_{\mathbb{X}}$ such that $x \perp_B w$. Therefore, for $\phi \in J(x)$, $\phi(w) = 0$ which shows $\phi(v_1) = -\phi(v_2)$. Let $|\phi(v_1)| = |\phi(v_2)| = \epsilon$. Then $0 \le \epsilon \le 1$. If $\epsilon = 1$, then $|\phi((1-t)v_1 + tv_2)| = 1$ for all $t \in [0, 1]$. Thus, $(1-t)v_1 + tv_2 \in S_{\mathbb{X}}$ for all $t \in [0, 1]$, contradicting that $\mathbb{X}$ is strictly convex. Therefore, $\epsilon < 1$. If $\epsilon = 0$, then $\phi(v_1) = \phi(v_2) = 0$. So $x \perp_B v_1$ and $x \perp_B v_2$. Thus, B-J orthogonality is not right unique. Thus, from Theorem 2.3.3, it follows that $\mathbb{X}$ is not smooth, a contradiction. Therefore, $\epsilon \in (0, 1)$. Clearly, $\phi(v_1)\phi(v_2) < 0$. Hence, by Proposition 5.3.9, $G(x, \epsilon) = K \cup (-K)$. This completes the proof □

Now, we turn our attention to some uniqueness property of the set $G(x, \epsilon)$, where $x \in S_{\mathbb{X}}$ and $\epsilon \in [0, 1)$. We explore whether $G(x_1, \epsilon_1) = G(x_2, \epsilon_2)$ for $x_1, x_2 \in S_{\mathbb{X}}$ and $\epsilon_1, \epsilon_2 \in [0, 1)$ imply that $x_1 = \pm x_2$ and $\epsilon_1 = \epsilon_2$. The following easy example shows that this is not true in general.

Example 5.3.14 ([136, Example 2.10]) Consider the space $\mathbb{X} = \ell_\infty^2$. Let $x_1 = (1, \frac{1}{2})$ and $x_2 = (1, -\frac{1}{2})$. Choose $\epsilon_1 = \epsilon_2 = \frac{1}{2}$. Then it is straightforward to check that

$$\begin{aligned} G(x_1, \epsilon_1) \cap S_{\mathbb{X}} &= \left\{ t\left(\frac{1}{2}, 1\right) + (1-t)\left(-\frac{1}{2}, 1\right) : t \in [0, 1] \right\} \\ &\quad \cup \left\{ t\left(-\frac{1}{2}, -1\right) + (1-t)\left(\frac{1}{2}, -1\right) : t \in [0, 1] \right\} \\ &= G(x_2, \epsilon_2) \cap S_{\mathbb{X}}. \end{aligned}$$

However, $x_1 \neq \pm\, x_2$.

In the following theorem, we show that some restriction on the space gives the desired uniqueness of the set $G(x, \epsilon)$. We should reiterate here that such results exhibit the strength of nice geometric properties such as smoothness and strict convexity.

Theorem 5.3.15 ([136, Theorem 2.11]) *Let $\mathbb{X}$ be a two-dimensional strictly convex and smooth Banach space. Suppose there exist $x_1, x_2 \in S_{\mathbb{X}}$ and $\epsilon_1, \epsilon_2 \in [0, 1)$ such that $G(x_1, \epsilon_1) = G(x_2, \epsilon_2)$. Then $x_1 = \pm\, x_2$ and $\epsilon_1 = \epsilon_2$.*

Proof Let $G(x_1, \epsilon_1) = G(x_2, \epsilon_2) = K \cup (-K)$, where K is the normal cone determined by $v_1, v_2 \in S_{\mathbb{X}}$. Let $w = v_1 + v_2$. If $\epsilon_1 = 0$, then $x_1 \perp_B v_1$ and $x_1 \perp_B v_2$. Since $\mathbb{X}$ is smooth, by Theorem 2.3.2, $x_1 \perp_B (v_1 + v_2)$, i.e., $x_1 \perp_B w$. If $\epsilon_1 \in (0, 1)$, then by Proposition 5.3.8, for $\phi \in J(x_1)$, $|\phi(v_1)| = |\phi(v_2)| = \epsilon_1$ and $\phi(v_1)\phi(v_2) < 0$. Then clearly, $\phi(w) = 0$. So $x_1 \perp_B w$. Similarly, we can show that $x_2 \perp_B w$. Since $\mathbb{X}$ is strictly convex, by Theorem 2.2.1, B-J orthogonality is left unique. Therefore, $x_2 = a\, x_1$, for some scalar a. Again, $x_1, x_2 \in S_{\mathbb{X}}$ implies that $x_2 = \pm\, x_1$. If $x_1 = x_2$, then $\phi \in J(x_2)$. Hence, by Proposition 5.3.8, $|\phi(v_1)| = |\phi(v_2)| = \epsilon_2$. Thus, $\epsilon_1 = \epsilon_2$. Similarly, $x_1 = -x_2$ implies that $\epsilon_1 = \epsilon_2$. This completes the proof. □

As an immediate consequence of Theorems 5.3.11 and 5.3.15, we have the following uniqueness theorem for $G(x, \epsilon)$ in a strictly convex and smooth Banach space.

Theorem 5.3.16 ([136, Theorem 2.12]) *Let $\mathbb{X}$ be an arbitrary strictly convex and smooth Banach space. Suppose there exist $x_1, x_2 \in S_{\mathbb{X}}$ and $\epsilon_1, \epsilon_2 \in [0, 1)$ such that $G(x_1, \epsilon_1) = G(x_2, \epsilon_2)$. Then $x_1 = \pm\, x_2$ and $\epsilon_1 = \epsilon_2$.*

If we only assume that the two-dimensional Banach space $\mathbb{X}$ is smooth, then we obtain the following uniqueness theorem for $G(x, \epsilon)$.

Theorem 5.3.17 ([136, Theorem 2.13]) *Let $\mathbb{X}$ be a two-dimensional smooth Banach space. Suppose $x \in S_{\mathbb{X}}$ and $\epsilon_1, \epsilon_2 \in [0, 1)$ such that $G(x, \epsilon_1) = G(x, \epsilon_2)$. Then $\epsilon_1 = \epsilon_2$.*

Proof Let $G(x, \epsilon_1) = G(x, \epsilon_2) = K \cup (-K)$, where K is the normal cone determined by $v_1, v_2 \in S_{\mathbb{X}}$. Then by Proposition 5.3.8, for $\phi \in J(x)$, $|\phi(v_1)| = \epsilon_1 = \epsilon_2$, hence the theorem. □

Combining Theorems 5.3.10, 5.3.13, and 5.3.15, we obtain the following characterization of normal cones in a two-dimensional strictly convex and smooth Banach space.

Theorem 5.3.18 ([136, Theorem 2.14]) *Let $\mathbb{X}$ be a two-dimensional strictly convex and smooth Banach space. Then K is a normal cone in $\mathbb{X}$ if and only if there exist a unique $x \in S_{\mathbb{X}}$ (up to multiplication by ± 1) and a unique $\epsilon \in [0, 1)$ such that $G(x, \epsilon) = K \cup (-K)$.*

We end this section with a relation between the sets $T(G(x, \epsilon))$ and $G(Tx, \delta)$ for a norm one linear operator T defined on a two-dimensional strictly convex and smooth Banach space, where $x \in M_T$.

Theorem 5.3.19 ([136, Theorem 2.18]) *Let $\mathbb{X}$ be a two-dimensional strictly convex and smooth Banach space. Suppose $T \in \mathbb{L}(\mathbb{X})$ is such that T is onto, $\|T\| = 1$ and $x \in M_T$. Let $\epsilon \in (0, 1)$. Suppose that $G(x, \epsilon) = K \cup (-K)$, where K is the normal cone determined by $v_1, v_2 \in S_{\mathbb{X}}$. Then $T(G(x, \epsilon)) = G(Tx, \delta)$ for some $\delta \in [0, 1)$ if and only if $\|Tv_1\| = \|Tv_2\| = k$ and $\delta = \frac{\epsilon}{k}$.*

Proof First suppose that $T(G(x, \epsilon)) = G(Tx, \delta)$ for some $\delta \in [0, 1)$. Now, $T(G(x, \epsilon)) = T(K \cup (-K)) = T(K) \cup (-T(K))$. Clearly $T(K)$ is a normal cone in $\mathbb{X}$ determined by $\frac{Tv_1}{\|Tv_1\|}$ and $\frac{Tv_2}{\|Tv_2\|}$. For $\psi \in J(Tx)$, we have $|\psi(\frac{Tv_1}{\|Tv_1\|})| = |\psi(\frac{Tv_1}{\|Tv_1\|})| = \delta$ (see Proposition 5.3.8). Hence, $|\psi \circ T(v_1)| = \delta\|Tv_1\|$ and $|\psi \circ T(v_2)| = \delta\|Tv_2\|$. Again $\psi \in J(Tx) \Rightarrow \psi(Tx) = \|Tx\| = 1$. Thus, $\psi \circ T \in J(x)$. Since $G(x, \epsilon) = K \cup (-K)$, using Proposition 5.3.8, we have $|\psi \circ T(v_1)| = |\psi \circ T(v_2)| = \epsilon$, for $\psi \circ T \in J(x)$. Therefore, $\epsilon = \delta\|Tv_1\| = \delta\|Tv_2\|$. This implies that $\|Tv_1\| = \|Tv_2\| = k$ (say) and $\delta = \frac{\epsilon}{k}$.

Conversely, suppose that $\|Tv_1\| = \|Tv_2\| = k$. Since $G(x, \epsilon) = K \cup (-K)$, by Proposition 5.3.8, for $\phi \in J(x)$, we have $|\phi(v_1)| = |\phi(v_2)| = \epsilon$ and $\phi(v_1)\phi(v_2) < 0$. Clearly, $\phi(v_1 + v_2) = 0$. Thus, $x \perp_B (v_1 + v_2)$. Hence, by Proposition 3.2.6, $Tx \perp_B (Tv_1 + Tv_2)$. Now, if $\psi \in J(Tx)$, then $\psi(Tv_1 + Tv_2) = 0 \Rightarrow \psi(\frac{Tv_1+Tv_2}{k}) = 0$. Let $|\psi(\frac{Tv_1}{k})| = |\psi(\frac{Tv_2}{k})| = \delta$. Proceeding similarly as in Theorem 5.3.13, we can show that $\delta \in (0, 1)$ and $G(Tx, \delta) = T(K) \cup (-T(K)) = T(G(x, \epsilon))$. Now, $\psi \in J(Tx) \Rightarrow \psi(Tx) = \|Tx\| = 1$. This implies that $\psi \circ T \in J(x)$. Since x is smooth, $\psi \circ T = \phi$. Thus, $\epsilon = |\phi(v_1)| = |\psi \circ T(v_1)| = \delta k$. Hence, $\delta = \frac{\epsilon}{k}$. This completes the proof. □

5.4 $\perp_D^{\epsilon}$-Approximate Orthogonality on $\mathbb{L}(\mathbb{X}, \mathbb{Y})$

The purpose of this section is to characterize $\perp_D^{\epsilon}$ in the space of bounded linear operators. First we explore the problem in the setting of Hilbert spaces.

Theorem 5.4.1 *Let $\mathbb{H}_1, \mathbb{H}_2$ be Hilbert spaces over $\mathbb{K}$. Suppose that $T, A \in \mathbb{K}(\mathbb{H}_1, \mathbb{H}_2)$ and $\epsilon \in [0, 1)$. Then $T \perp_D^{\epsilon} A$ if and only if there exist a scalar λ_0 and $x \in M_{T-\lambda_0 A}$ such that $\langle Tx, Ax\rangle = \lambda_0\|Ax\|^2$ and $\|T - \lambda_0 A\| \geq \sqrt{1-\epsilon^2}\|T\|$.*

Proof First, suppose that $T \perp_D^{\epsilon} A$. Then T, A are linearly independent. By Theorem 2.1.13, there exists a scalar λ_0 such that $T - \lambda_0 A \perp_B A$. Observe that $M_{T-\lambda_0 A} \neq \emptyset$ and from Theorem 3.3.5 it follows that $M_{T-\lambda_0 A}$ is the unit sphere of some subspace of $\mathbb{H}_1$. Therefore, from Theorem 4.2.1, we get $x \in M_{T-\lambda_0 A}$ such that $Tx - \lambda_0 Ax \perp Ax$. Thus, $\langle Tx - \lambda_0 Ax, Ax\rangle = 0$, i.e., $\langle Tx, Ax\rangle = \lambda_0\|Ax\|^2$. Moreover, from the definition of $T \perp_D^{\epsilon} A$, we get $\|T - \lambda_0 A\| \geq \sqrt{1-\epsilon^2}\|T\|$.

Conversely, suppose that there exist a scalar λ_0 and $x \in M_{T-\lambda_0 A}$ such that $\langle Tx, Ax\rangle = \lambda_0\|Ax\|^2$ and $\|T - \lambda_0 A\| \geq \sqrt{1-\epsilon^2}\|T\|$. Therefore, $\langle Tx - \lambda_0 Ax, Ax\rangle = 0$, i.e., $Tx - \lambda_0 Ax \perp Ax$. Now, for any scalar λ, we have

$$\begin{aligned}
\|T + \lambda A\|^2 &= \|T - \lambda_0 A + (\lambda + \lambda_0)A\|^2 \\
&\geq \|(T - \lambda_0 A)x + (\lambda + \lambda_0)Ax\|^2 \\
&= \|(T - \lambda_0 A)x\|^2 + \|(\lambda + \lambda_0)Ax\|^2 \\
&\geq \|(T - \lambda_0 A)x\|^2 = \|T - \lambda_0 A\|^2 \\
\Rightarrow \|T + \lambda A\| &\geq \|T - \lambda_0 A\| \geq \sqrt{1-\epsilon^2}\|T\|.
\end{aligned}$$

Thus, $T \perp_D^{\epsilon} A$. This establishes the theorem. □

Instead of considering compact operators on a Hilbert space, if we deal with bounded linear operators, we get the following result.

Theorem 5.4.2 *Let $\mathbb{H}$ be a Hilbert space over $\mathbb{K}$. Suppose that $T, A \in \mathbb{L}(\mathbb{H})$ and $\epsilon \in [0, 1)$. Then $T \perp_D^\epsilon A$ if and only if there exist a scalar λ_0 and a sequence $\{x_n\} \subset S_{\mathbb{H}}$ such that $\|Tx_n - \lambda_0 Ax_n\| \to \|T - \lambda_0 A\|$, $\langle Tx_n - \lambda_0 Ax_n, Ax_n\rangle \to 0$ and $\|T - \lambda_0 A\| \geq \sqrt{1-\epsilon^2}\|T\|$.*

Proof First suppose that $T \perp_D^\epsilon A$. Choose a scalar λ_0 such that $T - \lambda_0 A \perp_B A$. Applying Theorem 4.2.1, we get a sequence $\{x_n\} \subset S_{\mathbb{H}}$ such that

$$\|Tx_n - \lambda_0 Ax_n\| \to \|T - \lambda_0 A\| \text{ and } \langle Tx_n - \lambda_0 Ax_n, Ax_n\rangle \to 0.$$

Moreover, from the definition of $T \perp_D^\epsilon A$, we get $\|T - \lambda_0 A\| \geq \sqrt{1-\epsilon^2}\|T\|$.

Conversely, suppose that there exist a scalar λ_0 and a sequence $\{x_n\} \subset S_{\mathbb{H}}$ such that $\|Tx_n - \lambda_0 Ax_n\| \to \|T - \lambda_0 A\|$, $\langle Tx_n - \lambda_0 Ax_n, Ax_n\rangle \to 0$ and $\|T - \lambda_0 A\| \geq \sqrt{1-\epsilon^2}\|T\|$. Therefore, from Theorem 4.2.1, it follows that $T - \lambda_0 A \perp_B A$. Now, for any scalar λ, we have

$$\|T + \lambda A\| = \|T - \lambda_0 A + (\lambda + \lambda_0)A\| \geq \|T - \lambda_0 A\| \geq \sqrt{1-\epsilon^2}\|T\|.$$

Thus, $T \perp_D^\epsilon A$. This establishes the theorem. □

Next, we explore $\perp_D^\epsilon$ of compact (bounded) linear operators defined on Banach spaces. A characterization of $T \perp_D^\epsilon A$, where T, A are compact operators on a reflexive Banach space, was first obtained in [106, Theorem 2.1]. Following similar ideas as in the last two theorems, we obtain another characterization of the same.

Theorem 5.4.3 *Let $\mathbb{X}, \mathbb{Y}$ be real Banach spaces such that $\mathbb{X}$ is reflexive. Suppose that $T, A \in \mathbb{K}(\mathbb{X}, \mathbb{Y})$ and $\epsilon \in [0, 1)$. Then $T \perp_D^\epsilon A$ if and only if there exist a scalar λ_0 and $x, y \in M_{T-\lambda_0 A}$ such that the following hold:*

$$Ax \in (Tx - \lambda_0 Ax)^+, \; Ay \in (Ty - \lambda_0 Ay)^- \text{ and } \|T - \lambda_0 A\| \geq \sqrt{1-\epsilon^2}\|T\|.$$

Moreover, if the scalars are considered to be complex, then $T \perp_D^\epsilon A$ if and only if there exists a scalar λ_0 and for each $\alpha \in U$ there exist $x = x(\alpha), y = y(\alpha) \in M_{T-\lambda_0 A}$ such that the following hold:

$$Ax \in (Tx - \lambda_0 Ax)_\alpha^+, \; Ay \in (Ty - \lambda_0 A)_\alpha^- \text{ and } \|T - \lambda_0 A\| \geq \sqrt{1-\epsilon^2}\|T\|.$$

Proof First we assume that the scalars are real. Suppose that $T \perp_D^\epsilon A$. Choose a scalar λ_0 such that $T - \lambda_0 A \perp_B A$. Applying Theorem 4.3.7, we get $x, y \in M_{T-\lambda_0 A}$ such that $Ax \in (Tx - \lambda_0 Ax)^+$ and $Ay \in (Ty - \lambda_0 Ay)^-$. Moreover, from the definition of $T \perp_D^\epsilon A$, we get $\|T - \lambda_0 A\| \geq \sqrt{1-\epsilon^2}\|T\|$.

Conversely, from the existence of $x, y \in M_{T-\lambda_0 A}$ with $Ax \in (Tx - \lambda_0 Ax)^+$ and $Ay \in (Ty - \lambda_0 Ay)^-$ and applying Theorem 4.3.7, we get $T - \lambda_0 A \perp_B A$. Now, for any scalar λ, we have

$$\|T + \lambda A\| = \|T - \lambda_0 A + (\lambda + \lambda_0)A\| \geq \|T - \lambda_0 A\| \geq \sqrt{1-\epsilon^2}\|T\|.$$

Thus, $T \perp_D^\epsilon A$.

For complex scalers, the theorem can be proved similarly using Theorem 4.5.2 instead of using Theorem 4.3.7. □

Now, we study $T \perp_D^\epsilon A$, where the domain of the compact operators T, A is not necessarily reflexive.

Theorem 5.4.4 ([87, Theorem 2.4]) *Let $\mathbb{X}, \mathbb{Y}$ be Banach spaces. Suppose that $T, A \in \mathbb{K}(\mathbb{X}, \mathbb{Y})$ and $\epsilon \in [0, 1)$. Then $T \perp_D^\epsilon A$ if and only if there exist $S \in span\{T, A\}$, $y_1^*, y_2^* \in Ext(B_{\mathbb{Y}^*})$, $x_1^{**}, x_2^{**} \in Ext(B_{\mathbb{X}^{**}})$ and $t \in [0, 1]$ such that the following hold:*

(a) $\|S\| = 1$,
(b) $x_i^{**}(S^* y_i^*) = 1$ *for* $i = 1, 2$,
(c) $(1-t)x_1^{**}(T^* y_1^*) + t x_2^{**}(T^* y_2^*) \geq \sqrt{1-\epsilon^2}\|T\|$, *and*
(d) $(1-t)x_1^{**}(A^* y_1^*) + t x_2^{**}(A^* y_2^*) = 0$.

Proof We first prove the necessary part of the theorem. Let $T \perp_D^\epsilon A$. Then by Theorem 5.2.5, there exist $S \in \text{span}\{T, A\}$, $\phi, \psi \in Ext(B_{\mathbb{K}(\mathbb{X},\mathbb{Y})^*})$ and $t \in [0, 1]$ such that $\|S\| = 1$, $\phi(S) = \psi(S) = 1$, $(1-t)\phi(T) + t\psi(T) \geq \sqrt{1-\epsilon^2}\|T\|$ and $(1-t)\phi(A) + t\psi(A) = 0$. By Theorem 4.4.1, we obtain $\phi = x_1^{**} \otimes y_1^*$ and $\psi = x_2^{**} \otimes y_2^*$, where $x_i^{**} \in Ext(B_{\mathbb{X}^{**}})$, $y_i^* \in Ext(B_{\mathbb{Y}^*})$, for $i = 1, 2$. Therefore,

$$(1-t)x_1^{**}(T^* y_1^*) + t x_2^{**}(T^* y_2^*) = (1-t)(x_1^{**} \otimes y_1^*)(T) + t(x_2^{**} \otimes y_2^*)(T) \geq \sqrt{1-\epsilon^2}\|T\|$$

and

$$(1-t)x_1^{**}(A^* y_1^*) + t x_2^{**}(A^* y_2^*) = (1-t)(x_1^{**} \otimes y_1^*)(A) + t(x_2^{**} \otimes y_2^*)(A) = 0.$$

Now, $\phi(S) = \psi(S) = 1 \Rightarrow (x_i^{**} \otimes y_i^*)(S) = 1$ for $i = 1, 2$, i.e., $x_i^{**}(S^* y_i^*) = 1$.

For the sufficient part, suppose that there exist $S \in \text{span}\{T, A\}$, $y_1^*, y_2^* \in Ext(B_{\mathbb{Y}^*})$, $x_1^{**}, x_2^{**} \in Ext(B_{\mathbb{X}^{**}})$ and $t \in [0, 1]$ such that conditions (a) to (d) hold. Let $\phi_i = x_i^{**} \otimes y_i^*$ for $i = 1, 2$. Then by Theorem 4.4.1, $\phi_i \in Ext(B_{\mathbb{K}(\mathbb{X},\mathbb{Y})^*})$, for $i = 1, 2$. Now, $\phi_i(S) = (x_i^{**} \otimes y_i^*)(S) = 1$ for $i = 1, 2$. Moreover,

$$(1-t)\phi_1(T) + t\phi_2(T) = (1-t)(x_1^{**} \otimes y_1^*)(T) + t(x_2^{**} \otimes y_2^*)(T) \geq \sqrt{1-\epsilon^2}\|T\|$$

and

$$(1-t)\phi_1(A) + t\phi_2(A) = (1-t)(x_1^{**} \otimes y_1^*)(A) + t(x_2^{**} \otimes y_2^*)(A) = 0.$$

Thus, by Theorem 5.2.5, $T \perp_D^\epsilon A$. □

Now, following the same method as in Theorem 5.4.4 and using Theorem 4.4.2, we get the following result. The proof is omitted, as it is rather straightforward in view of the above discussions.

Theorem 5.4.5 ([87, Theorem 2.5]) *Let $\mathbb{X}$ be a reflexive Banach space and let $\mathbb{Y}$ be a Banach space. Suppose that $T, A \in \mathbb{K}(\mathbb{X}, \mathbb{Y})$ and $\epsilon \in [0, 1)$. Then $T \perp_D^\epsilon A$ if and only if there exist $S \in span\{T, A\}$, $y_1^*, y_2^* \in Ext(B_{\mathbb{Y}^*})$, $x_1, x_2 \in Ext(B_{\mathbb{X}})$ and $t \in [0, 1]$ such that the following hold:*

(a) $\|S\| = 1$,

(b) $y_i^*(Sx_i) = 1$ *for* $i = 1, 2$,

(c) $(1-t)y_1^*(Tx_1) + ty_2^*(Tx_2) \geq \sqrt{1-\epsilon^2}\|T\|$, *and*

(d) $(1-t)y_1^*(Ax_1) + ty_2^*(Ax_2) = 0$.

In the next theorem, we characterize $T \perp_D^\epsilon A$ for bounded linear operators defined on any real Banach spaces. Recall that a sequence $\{x_n\}$ of unit vectors is said to be a norming sequence for a bounded linear operator T, if $\|Tx_n\| \to \|T\|$.

Theorem 5.4.6 *Let $\mathbb{X}$, $\mathbb{Y}$ be real Banach spaces. Suppose that $T, A \in \mathbb{L}(\mathbb{X}, \mathbb{Y})$ and $\epsilon \in [0, 1)$. Then $T \perp_D^\epsilon A$ if and only if there exists a scalar λ_0 such that $\|T - \lambda_0 A\| \geq \sqrt{1-\epsilon^2}$ and either*

(a) *there exists a norming sequence $\{x_n\}$ for $T - \lambda_0 A$ such that $\|Ax_n\| \to 0$ or*

(b) *there exist two norming sequences $\{x_n\}, \{y_n\}$ for $T - \lambda_0 A$ and sequences of positive scalars $\{\epsilon_n\}, \{\delta_n\}$ such that $\epsilon_n \to 0$, $\delta_n \to 0$, $Ax_n \in (Tx_n - \lambda_0 Ax_n)^{+\epsilon_n}$ and $Ay_n \in (Ty_n - \lambda_0 Ay_n)^{-\delta_n}$.*

Proof First suppose that $T \perp_D^\epsilon A$. Now, there exists a scalar λ_0 such that $T - \lambda_0 A \perp_B A$. Clearly, $\|T - \lambda_0 A\| \geq \sqrt{1-\epsilon^2}$. Moreover, applying Theorem 4.4.14, we get either (a) or (b).

Conversely, suppose that there exists a scalar λ_0 such that $\|T - \lambda_0 A\| \geq \sqrt{1-\epsilon^2}$. Now, from (a) and (b) and using Theorem 4.4.14, we get that $T - \lambda_0 A \perp_B A$. It follows that for any scalar λ,

$$\|T + \lambda A\| = \|T - \lambda_0 A + (\lambda + \lambda_0)A\| \geq \|T - \lambda_0 A\| \geq \sqrt{1-\epsilon^2}\|T\|.$$

Therefore, $T \perp_D^\epsilon A$. □

We would like to remark that analogous to the previous theorem, applying Theorem 4.5.1, we can also characterize $T \perp_D^\epsilon A$ for bounded linear operators defined on a complex Banach spaces. We end this section with an easy proposition, which generalizes Proposition 3.2.3.

Proposition 5.4.7 ([88, Proposition 3.1]) *Let $\mathbb{X}$, $\mathbb{Y}$ be Banach spaces over $\mathcal{K}$. Suppose that $T \in \mathbb{L}(\mathbb{X}, \mathbb{Y})$ and $x \in M_T$. Then for any $\epsilon \in [0, 1)$ and $y \in \mathbb{X}$, $Tx \perp_D^\epsilon Ty \Rightarrow x \perp_D^\epsilon y$.*

Proof Let $Tx \perp_D^\epsilon Ty$. Then for any scalar λ, $\|Tx + \lambda Ty\| \geq \sqrt{1-\epsilon^2}\|Tx\|$. Therefore, for any scalar λ, $\|T\|\|x + \lambda y\| \geq \|Tx + \lambda Ty\| \geq \sqrt{1-\epsilon^2}\|Tx\| = \sqrt{1-\epsilon^2}\|T\|\|x\|$. Thus, from $\|x + \lambda y\| \geq \sqrt{1-\epsilon^2}\|x\|$, it follows that $x \perp_D^\epsilon y$. □

5.5 $\perp_B^\epsilon$-Approximate Orthogonality on $\mathbb{L}(\mathbb{X}, \mathbb{Y})$

The purpose of this section is to explore $\perp_B^\epsilon$ in the space of bounded linear operators. We begin this section with an easy proposition, which is a generalization of Proposition 3.2.3. The following proposition was proved in [136, Proposition 2.15] assuming that $\mathbb{X} = \mathbb{Y}$ and the scalars are real.

Proposition 5.5.1 *Let $\mathbb{X}, \mathbb{Y}$ be Banach spaces over $\mathcal{K}$. Suppose that $T \in \mathbb{L}(\mathbb{X}, \mathbb{Y})$ and $x \in M_T$. Then for $\epsilon \in [0, 1)$ and $y \in \mathbb{X}$, $Tx \perp_B^\epsilon Ty \Rightarrow x \perp_B^\epsilon y$.*

Proof Since $Tx \perp_B^\epsilon Ty$, for any scalar λ, $\|Tx + \lambda Ty\|^2 \geq \|Tx\|^2 - 2\epsilon\|Tx\| \|\lambda Ty\|$. It follows that

$$\begin{aligned}\|T\|^2\|x + \lambda y\|^2 &\geq \|Tx + \lambda Ty\|^2 \\ &\geq \|Tx\|^2 - 2\epsilon\|Tx\|\|\lambda Ty\| \\ &\geq \|T\|^2\|x\|^2 - 2\epsilon\|T\|\|x\||\lambda|\|T\|\|y\|.\end{aligned}$$

Thus, $\|x + \lambda y\|^2 \geq \|x\|^2 - 2\epsilon\|x\|\|\lambda y\|$ for any scalar λ. Therefore, $x \perp_B^\epsilon y$. This completes the proof of the proposition. □

Let us also record the following easy corollary:

Corollary 5.5.2 ([136, Corollary 2.16]) *Let $\mathbb{X}, \mathbb{Y}$ be a Banach space. Let $T \in \mathbb{L}(\mathbb{X}, \mathbb{Y})$ be such that T is onto, $\|T\| = 1$ and $x \in M_T$. Then for $\epsilon \in [0, 1)$, $G(Tx, \epsilon) \subseteq T(G(x, \epsilon))$.*

Proof Let $z \in G(Tx, \epsilon)$. Then $Tx \perp_B^\epsilon z$. Since T is onto, there exists $y \in \mathbb{X}$ such that $Ty = z$. Thus, $Tx \perp_B^\epsilon Ty$. Hence, by Proposition 5.5.1, $x \perp_B^\epsilon y$. So $y \in G(x, \epsilon) \Rightarrow z = Ty \in T(G(x, \epsilon))$. Thus, $G(Tx, \epsilon) \subseteq T(G(x, \epsilon))$. □

The following example shows that the converse of Proposition 5.5.1 is not true in general.

Example 5.5.3 ([136, Example 2.17]) Let $\mathbb{X} = \ell_2^2$. Define $T \in \mathbb{L}(\ell_2^2)$ by $T(a, b) = (a, 0)$ for all $(a, b) \in \ell_2^2$. Then clearly, $(1, 0) \in M_T$. Let $\epsilon = \frac{\sqrt{3}}{2}$, $x = (1, 0)$ and $y = (-\frac{\sqrt{3}}{2}, 1)$. Then

$$|\langle x, y\rangle| = \left|\left\langle(1, 0), \left(-\frac{\sqrt{3}}{2}, 1\right)\right\rangle\right| = \frac{\sqrt{3}}{2} < \epsilon\|x\|\|y\|.$$

Thus, $x \perp^\epsilon y$. But

$$|\langle Tx, Ty\rangle| = \left|\left\langle(1, 0), \left(-\frac{\sqrt{3}}{2}, 0\right)\right\rangle\right| = \|Tx\|\|Ty\| > \epsilon\|Tx\|\|Ty\|.$$

Thus, $Tx \not\perp^\epsilon Ty$.

Chmieliński et al. initiated the problem of characterizing approximate B-J orthogonality ($\perp_B^\epsilon$) of operators in the setting of Hilbert spaces. In particular, they deduced the following extension of Theorem 4.2.1.

Theorem 5.5.4 ([25, Theorem 3.2]) *Let $\mathbb{H}$ be a Hilbert space. Suppose that $T, A \in \mathbb{L}(\mathbb{H})$ and $\epsilon \in [0, 1)$. Then $T \perp_B^\epsilon A$ if and only if there exists a sequence $\{x_n\} \subset S_{\mathbb{H}}$ such that $\|Tx_n\| \to \|T\|$ and $\lim_{n\to\infty} |\langle Tx_n, Ax_n\rangle| \leq \epsilon\|T\|\|A\|$.*

Proof Let $T \perp_B^\epsilon A$. From Theorem 5.3.3, it follows that there exists some $S \in \mathbb{L}(\mathbb{H})$, such that $T \perp_B S$ and $\|S - A\| \leq \epsilon\|A\|$. Now, $T \perp_B S$ implies that there exists a sequence $\{x_n\} \subset S_{\mathbb{H}}$ such that $\|Tx_n\| \to \|T\|$ and $\langle Tx_n, Sx_n\rangle \to 0$ as $n \to \infty$ (see Theorem 4.2.1). Therefore,

$$\begin{aligned}|\langle Tx_n, Ax_n\rangle| &\leq |\langle Tx_n, Sx_n\rangle| + |\langle Tx_n, Ax_n - Sx_n\rangle|\\ &\leq |\langle Tx_n, Sx_n\rangle| + \|T\|\|A - S\|\\ &\leq |\langle Tx_n, Sx_n\rangle| + \epsilon\|T\|\|A\|.\end{aligned}$$

Without loss of generality assume that $\{|\langle Tx_n, Ax_n\rangle|\}$ is convergent. Thus, we get $\lim_{n\to\infty} |\langle Tx_n, Ax_n\rangle| \leq \epsilon\|T\|\|A\|$.

Conversely, suppose that there exists a sequence $\{x_n\} \subset S_{\mathbb{H}}$ such that $\|Tx_n\| \to \|T\|$ and $\lim_{n\to\infty} |\langle Tx_n, Ax_n\rangle| \leq \epsilon\|T\|\|A\|$. Now, for each scalar λ and each $n \in \mathbb{N}$, we have

$$\begin{aligned}\|T + \lambda A\|^2 \geq \|Tx_n + \lambda Ax_n\|^2 &\geq \|Tx_n\|^2 + 2\lambda\langle Tx_n, Ax_n\rangle\\ &\geq \|Tx_n\|^2 - 2|\lambda||\langle Tx_n, Ax_n\rangle|.\end{aligned}$$

Letting $n \to \infty$, we get $\|T + \lambda A\|^2 \geq \|T\|^2 - 2\epsilon\|T\|\|\lambda A\|$. Therefore, $T \perp_B^\epsilon A$. □

Next, we extend Theorem 4.2.4 for approximate B-J orthogonality ($\perp_B^\epsilon$).

Theorem 5.5.5 ([106, Theorem 3.1]) (i) *Let $\mathbb{H}$ be a Hilbert space and $T \in \mathbb{L}(\mathbb{H})$. Suppose $\epsilon \in [0, 1)$. Then for any $A \in \mathbb{L}(\mathbb{H})$, $T \perp_B^\epsilon A \Leftrightarrow |\langle Tx, Ax\rangle| \leq \epsilon\|T\|\|A\|$ for some $x \in M_T$ if and only if $M_T = S_{H_0}$ for some finite-dimensional subspace H_0 of $\mathbb{H}$ and $\|T\|_{H_0^\perp} < \|T\|$.*

(ii) *Moreover, if $M_T \subseteq M_A$ then $T \perp_B^\epsilon A \Leftrightarrow Tx \perp^\epsilon Ax$ for some $x \in M_T$ if and only if $M_T = S_{H_0}$ for some finite-dimensional subspace H_0 of $\mathbb{H}$ and $\|T\|_{H_0^\perp} < \|T\|$.*

Proof Without loss of generality we may and do assume that $\|T\| = 1$.

(i) We first prove the necessary part. By the hypothesis $M_T \neq \emptyset$. So, from Theorem 3.3.5, it follows that M_T is the unit sphere of some subspace H_0, say, of $\mathbb{H}$. We claim that H_0 is finite-dimensional. Suppose on the contrary H_0 is infinite-dimensional. Then we can find a set $\{e_n : n \in \mathbb{N}\}$ of orthonormal vectors in H_0. Extend the set to a complete orthonormal basis $\mathcal{B} = \{e_\alpha : \alpha \in \Lambda \supset \mathbb{N}\}$ of $\mathbb{H}$. For each $e_\alpha \in H_0 \cap \mathcal{B}$, we have

$$\|T^*T\| = \|T\|^2 = \|Te_\alpha\|^2 = \langle T^*Te_\alpha, e_\alpha\rangle \leq \|T^*Te_\alpha\|\|e_\alpha\| \leq \|T^*T\|.$$

Therefore, we get $T^*Te_\alpha = \lambda_\alpha e_\alpha$ for some scalar λ_α. Thus, $\{Te_\alpha : e_\alpha \in H_0 \cap \mathcal{B}\}$ is a set of orthonormal vectors in $\mathbb{H}$. For $\epsilon \in [0, 1)$ choose a sequence $\{c_n\}$ such that $\epsilon < c_n < 1$ and $c_n \to \epsilon$. Define $A : \mathcal{B} \longrightarrow \mathbb{H}$ as follows:

$$\begin{aligned} A(e_n) &= c_n Te_n, n \in \mathbb{N} \\ A(e_\alpha) &= Te_\alpha,\ e_\alpha \in H_0 \cap \mathcal{B} \setminus \{e_n : n \in \mathbb{N}\} \\ A(e_\alpha) &= 0,\ e_\alpha \in \mathcal{B} - H_0 \cap \mathcal{B}. \end{aligned}$$

As $\{Te_\alpha : e_\alpha \in H_0 \cap \mathcal{B}\}$ is a set of orthonormal vectors in $\mathbb{H}$, A can be extended to a bounded linear operator on $\mathbb{H}$. Observe that $\|A\| = 1$. Now, for any scalar λ,

$$\|T + \lambda A\|^2 \geq \|(T + \lambda A)e_n\|^2 = \|(1 + \lambda c_n)Te_n\|^2 = 1 + \lambda^2 {c_n}^2 + 2\lambda c_n \to 1 + \lambda^2\epsilon^2 + 2\lambda\epsilon.$$

It is easy to see that $\|T + \lambda A\|^2 \geq \|T\|^2 - 2\epsilon\|T\|\|\lambda A\|$ for all λ and hence $T \perp_B^{\epsilon} A$.

We next show that there exists no $x \in M_T$ such that $|\ \langle Tx, Ax\rangle\ | \leq \epsilon\|T\|\|A\|$. Let $x = \sum_\alpha \langle x, e_\alpha\rangle e_\alpha\ \in M_T$. Then

$$\langle Tx, Ax\rangle = \sum_n c_n\ |\ \langle x, e_n\rangle\ |^2\ \|Te_n\|^2 + \sum_{\alpha \notin \mathbb{N}} |\ \langle x, e_\alpha\rangle\ |^2\ \|Te_\alpha\|^2 > \epsilon\|x\|^2$$

and so $|\ \langle Tx, Ax\rangle\ | > \epsilon\|T\|\|A\|$ for each $x \in M_T$. Thus, $T \perp_B^{\epsilon} A$ but there exists no $x \in M_T$ such that $|\ \langle Tx, Ax\rangle\ | \leq \epsilon\|T\|\|A\|$. This contradiction establishes our claim that H_0 must be finite dimensional.

Next, we claim that $\|T\|_{{H_0}^\perp} < \|T\|$. Suppose that $\|T\|_{{H_0}^\perp} = \|T\|$. Since T does not attain norm on $H_0^\perp$, there exists a sequence $\{e_n\}$ in $H_0^\perp$ such that $\|Te_n\| \to \|T\|$. We have $\mathbb{H} = H_0 \oplus {H_0}^\perp$. Define $A : \mathbb{H} \longrightarrow \mathbb{H}$ as follows:

$$Az = Tx,\ \text{where } z = x + y, x \in H_0, y \in {H_0}^\perp$$

Then it is easy to check that A is bounded on $\mathbb{H}$ and $\|A\| = 1$. Also for any scalar λ, $\|T + \lambda A\|^2 \geq \|(T + \lambda A)e_n\|^2 = \|Te_n\|^2$ holds for each $n \in \mathbb{N}$. Thus,

$$\|T + \lambda A\|^2 \geq \|T\|^2 \geq \|T\|^2 - 2\epsilon\|T\|\|\lambda A\|$$

for all λ. Hence $T \perp_B^{\epsilon} A$. But for all $x \in M_T$,

$$|\langle Tx, Ax\rangle| \geq \langle Tx, Ax\rangle = \langle Tx, Tx\rangle = \|Tx\|^2 = 1 > \epsilon\|T\|\|A\|.$$

This contradiction completes the proof of the necessary part of the theorem.

We next prove the sufficient part. Let $T \perp_B^{\epsilon} A$. Then by Theorem 5.3.3 there exists $S \in \mathbb{L}(\mathbb{H})$ such that $T \perp_B S$ and $\|A - S\| \leq \epsilon\|A\|$. Since $M_T = S_{H_0}$, where H_0 is a finite-dimensional subspace of $\mathbb{H}$ and $\|T\|_{H_0^\perp} < \|T\|$, by Theorem 4.2.4, there exists $x \in M_T$ such that $Tx \perp Sx$. Clearly, $\|Sx - Ax\| \leq \|S - A\| \leq \epsilon\|A\|$. Now,

$$\begin{aligned}\langle Tx, Ax\rangle &= \langle Tx, Sx\rangle + \langle Tx, Ax - Sx\rangle \\ \Rightarrow |\langle Tx, Ax\rangle| &\leq |\langle Tx, Sx\rangle| + |\langle Tx, Ax - Sx\rangle| \\ &\leq \|Tx\|\|Ax - Sx\| \leq \epsilon\|T\|\|A\|.\end{aligned}$$

Conversely, let $|\langle Tx, Ax\rangle| \leq \epsilon\|T\|\|A\|$ for some $x \in M_T$. Then for any $\lambda \in \mathbb{R}$,

$$\begin{aligned}\|T + \lambda A\|^2 &\geq \|(T + \lambda A)x\|^2 \\ &= \|Tx\|^2 + 2\lambda\langle Tx, Ax\rangle + |\lambda|^2\|Ax\|^2 \\ &\geq \|Tx\|^2 + 2\lambda\langle Tx, Ax\rangle \\ &\geq \|Tx\|^2 - 2|\lambda||\langle Tx, Ax\rangle| \\ &\geq \|T\|^2 - 2\epsilon|\lambda|\|T\|\|A\|.\end{aligned}$$

Hence $T \perp_B^\epsilon A$. This completes the proof of (i).

(ii) Since $M_T \subseteq M_A$, so $x \in M_T$ implies $\|Tx\| = \|T\|$, $\|Ax\| = \|A\|$ and hence (ii) follows from (i). □

The following two corollaries are simple consequences of the previous theorem.

Corollary 5.5.6 ([25, Theorem 3.2], [106, Corollary 3.1.1], [115, Theorem 4.6]) *Let $\mathbb{H}$ be a finite-dimensional Hilbert space and let $T \in \mathbb{L}(\mathbb{H})$. Suppose $\epsilon \in [0, 1)$. Then for any $A \in \mathbb{L}(\mathbb{H})$, $T \perp_B^\epsilon A \Leftrightarrow |\langle Tx, Ax\rangle| \leq \epsilon\|T\|\|A\|$ for some $x \in M_T$.*

Moreover, if $M_T \subseteq M_A$ then $T \perp_B^\epsilon A \Leftrightarrow Tx \perp^\epsilon Ax$ for some $x \in M_T$.

Corollary 5.5.7 ([106, Corollary 3.1.2]) *Let $\mathbb{H}$ be a Hilbert space and $T \in \mathbb{K}(\mathbb{H})$. Suppose $\epsilon \in [0, 1)$. If $A \in \mathbb{L}(\mathbb{H})$ and $M_T \subseteq M_A$ then $T \perp_B^\epsilon A$ if and only if there exists $x \in M_T$ such that $Tx \perp^\epsilon Ax$.*

Proof Observe that $M_T \neq \emptyset$, since T is a compact operator on a Hilbert space. Then from Theorem 3.3.5, it follows that $M_T = S_{H_0}$ for some subspace H_0 of $\mathbb{H}$. Now, it is easy to observe that for each $x \in H_0$, $T^*Tx = \|T\|^2x$. Since T^*T is also a compact operator, so H_0 must be a finite-dimensional subspace of $\mathbb{H}$. Since $H_0^\perp$ is reflexive and $T \in \mathbb{K}(\mathbb{H})$, clearly $\|T\|_{H_0^\perp} < \|T\|$. Hence, by applying part (ii) of Theorem 5.5.5, we have $T \perp_B^\epsilon A$ if and only if there exists $x \in M_T$ such that $Tx \perp^\epsilon Ax$. □

We would like to remark that an alternative proof of Corollary 5.5.7 can be found in [25, Theorem 3.4]. Our next objective is to generalize the results of this section in the setting of Banach spaces.

Theorem 5.5.8 ([106, Theorem 3.3]) *Let $\mathbb{X}$ be a reflexive Banach space and let $\mathbb{Y}$ be a Banach space. Suppose $T, A \in \mathbb{K}(\mathbb{X}, \mathbb{Y})$ and $\epsilon \in [0, 1)$. Then $T \perp_B^\epsilon A$ if and only if there exist $x, y \in M_T$ such that $\|Tx + \lambda Ax\|^2 \geq \|T\|^2 - 2\epsilon\|T\|\|\lambda A\|$ for all $\lambda \geq 0$ and $\|Ty + \lambda Ay\|^2 \geq \|T\|^2 - 2\epsilon\|T\|\|\lambda A\|$ for all $\lambda \leq 0$.*

Proof We first prove the necessary part. Since $T\perp_B^\epsilon A$, by Theorem 5.3.3, there exists a linear operator $S \in \text{span}\{T, A\}$ such that $T\perp_B S$ and $\|S - A\| \leq \epsilon\|A\|$. Since T, A are compact and $S \in \text{span}\{T, A\}$, S is also compact. Therefore, by Theorem 4.3.7, there exist $x, y \in M_T$ such that $Sx \in (Tx)^+$ and $Sy \in (Ty)^-$. Clearly, $\|Sx - Ax\| \leq \|S - A\| \leq \epsilon\|A\|$. Let $\lambda \geq 0$. If $\|Tx\| - 2\epsilon\|\lambda A\| < 0$, then $\|Tx\|^2 - 2\epsilon\|Tx\|\|\lambda A\| < 0 \leq \|Tx + \lambda Ax\|^2$, i.e., $\|T\|^2 - 2\epsilon\|T\|\|\lambda A\| < \|Tx + \lambda Ax\|^2$. Now, suppose that $\|Tx\| - 2\epsilon\|\lambda A\| \geq 0$. Then

$$\begin{aligned}\|Tx + \lambda Ax\|^2 &= \|(Tx + \lambda Sx) - \lambda(Sx - Ax)\|^2 \\ &\geq \|Tx + \lambda Sx\|^2 - 2|\lambda|\|Tx + \lambda Sx\|\|Sx - Ax\| \\ &= \|Tx + \lambda Sx\|(\|Tx + \lambda Sx\| - 2|\lambda|\|Sx - Ax\|) \\ &\geq \|Tx\|(\|Tx\| - 2\epsilon\|\lambda A\|) \\ &\geq \|Tx\|^2 - 2\epsilon\|Tx\|\|\lambda A\| \\ &= \|T\|^2 - 2\epsilon\|T\|\|\lambda A\|.\end{aligned}$$

Similarly, using $Sy \in (Ty)^-$ and $\|S - A\| \leq \epsilon\|A\|$, we can show that $\|Ty + \lambda Ay\|^2 \geq \|T\|^2 - 2\epsilon\|T\|\|\lambda A\|$ for all $\lambda \leq 0$.

We next prove the sufficient part. Suppose there exist $x, y \in M_T$ such that $\|Tx + \lambda Ax\|^2 \geq \|T\|^2 - 2\epsilon\|T\|\|\lambda A\|$ for all $\lambda \geq 0$ and $\|Ty + \lambda Ay\|^2 \geq \|T\|^2 - 2\epsilon\|T\|\|\lambda A\|$ for all $\lambda < 0$. Then for any $\lambda \geq 0$, $\|T + \lambda A\|^2 \geq \|Tx + \lambda Ax\|^2 \geq \|T\|^2 - 2\epsilon\|T\|\|\lambda A\|$. Similarly, for any $\lambda \leq 0$, $\|T + \lambda A\|^2 \geq \|Ty + \lambda Ay\|^2 \geq \|T\|^2 - 2\epsilon\|T\|\|\lambda A\|$. Therefore, $T\perp_B^\epsilon A$. □

The previous theorem can be generalized in the following way, by assuming that $\mathbb{K}(\mathbb{X}, \mathbb{Y})$ is an M-ideal in $\mathbb{L}(\mathbb{X}, \mathbb{Y})$.

Theorem 5.5.9 ([87, Theorem 3.3]) (i) *Let $\mathbb{X}$ be a reflexive Banach space and let $\mathbb{Y}$ be a Banach space. Suppose that $\mathbb{K}(\mathbb{X}, \mathbb{Y})$ is an M-ideal in $\mathbb{L}(\mathbb{X}, \mathbb{Y})$. Let $T \in \mathbb{L}(\mathbb{X}, \mathbb{Y})$ be such that $\|T\| = 1$ and $dist(T, \mathbb{K}(\mathbb{X}, \mathbb{Y})) < 1$. Then for any $A \in \mathbb{L}(\mathbb{X}, \mathbb{Y})$ and $\epsilon \in [0, 1)$, $T \perp_B^\epsilon A$ if and only if there exist $x_1, x_2 \in M_T \cap Ext(B_{\mathbb{X}})$ such that*

$$\|Tx_1 + \lambda Ax_1\|^2 \geq \|T\|^2 - 2\epsilon\|T\|\|\lambda A\| \ \forall \lambda \geq 0 \text{ and}$$

$$\|Tx_2 + \lambda Ax_2\|^2 \geq \|T\|^2 - 2\epsilon\|T\|\|\lambda A\| \ \forall \lambda \leq 0.$$

(ii) *In addition, if $M_T = D \cup (-D)$, where D is a closed connected subset of $S_{\mathbb{X}}$, then $T \perp_B^\epsilon A$ if and only if there exists $x \in M_T$ such that $\|Tx + \lambda Ax\|^2 \geq \|T\|^2 - 2\epsilon\|T\|\|\lambda A\|$ for all $\lambda \in \mathbb{R}$.*

Proof (i) The proof of the sufficient part follows easily. We only prove the necessary part. Let $T \perp_B^\epsilon A$. Then by Theorem 5.3.3, there exists $S \in \text{span}\{T, A\}$ such that $T \perp_B S$ and $\|S - A\| \leq \epsilon\|A\|$. Since $T \perp_B S$, by Theorem 4.4.5, there exist $x_1, x_2 \in M_T \cap Ext(B_{\mathbb{X}})$ such that $Sx_1 \in (Tx_1)^+$ and $Sx_2 \in (Tx_2)^-$. Now, for $i = 1, 2$, $\|Sx_i - Ax_i\| \leq \|S - A\| \leq \epsilon\|A\|$. Suppose that $\lambda \geq 0$. If $\|Tx_1\| - 2\epsilon\|\lambda A\| <$

0, then $\|T\|^2 - 2\epsilon\|T\|\|\lambda A\| = \|Tx_1\|^2 - 2\epsilon\|Tx_1\|\|\lambda A\| < 0 \leq \|Tx_1 + \lambda Ax_1\|^2$. If $\|Tx_1\| - 2\epsilon\|\lambda A\| \geq 0$, then

$$\begin{aligned}
\|Tx_1 + \lambda Ax_1\|^2 &= \|Tx_1 + \lambda Sx_1 - \lambda Sx_1 + \lambda Ax_1\|^2 \\
&\geq (\|Tx_1 + \lambda Sx_1\| - |\lambda|\|Sx_1 - Ax_1\|)^2 \\
&\geq \|Tx_1 + \lambda Sx_1\|^2 - 2\|Tx_1 + \lambda Sx_1\||\lambda|\|Sx_1 - Ax_1\| \\
&\geq \|Tx_1 + \lambda Sx_1\|^2 - 2\|Tx_1 + \lambda Sx_1\||\lambda|\epsilon\|A\| \\
&\geq \|Tx_1 + \lambda Sx_1\|(\|Tx_1 + \lambda Sx_1\| - 2\epsilon\|\lambda A\|) \\
&\geq \|Tx_1\|(\|Tx_1\| - 2\epsilon\|\lambda A\|) \; [Since\ Sx_1 \in (Tx_1)^+] \\
&= \|T\|^2 - 2\epsilon\|T\|\|\lambda A\|.
\end{aligned}$$

Thus, we get $\|Tx_1 + \lambda Ax_1\|^2 \geq \|T\|^2 - 2\epsilon\|T\|\|\lambda A\|$ for all $\lambda \geq 0$. Similarly, using $Sx_2 \in (Tx_2)^-$, we can show that $\|Tx_2 + \lambda Ax_2\|^2 \geq \|T\|^2 - 2\epsilon\|T\|\|\lambda A\|$ for all $\lambda \leq 0$.

(ii) The proof follows easily from Theorem 4.4.5 and the technique used in part (i) of this theorem. □

The next theorem follows easily from part (ii) of Theorem 5.5.9. For a direct proof of the theorem, see [106, Theorem 3.2].

Theorem 5.5.10 *Let $\mathbb{X}$ be a reflexive Banach space and let $\mathbb{Y}$ be a Banach space. Suppose $\epsilon \in [0, 1)$. Let $T, A \in \mathbb{K}(\mathbb{X}, \mathbb{Y})$ and let $M_T = D \cup (-D)$, where D is a non-empty compact connected subset of $S_{\mathbb{X}}$. Then $T \perp_B^\epsilon A$ if and only if there exists $x \in M_T$ such that $\|Tx + \lambda Ax\|^2 \geq \|T\|^2 - 2\epsilon\|T\|\|\lambda A\|$.*

Moreover, if $M_T \subseteq M_A$, then $T \perp_B^\epsilon A$ if and only if $Tx \perp_B^\epsilon Ax$.

Since an arbitrary bounded linear operator may not attain its norm, to characterize approximate B-J orthogonality for bounded linear operators, we need to consider norming sequence for a bounded linear operator instead of norm attaining vectors.

Theorem 5.5.11 ([106, Theorem 3.4]) *Let $\mathbb{X}, \mathbb{Y}$ be Banach spaces. Suppose $T \in \mathbb{L}(\mathbb{X}, \mathbb{Y})$ is a non-zero operator and $\epsilon \in [0, 1)$. Then for any $A \in \mathbb{L}(\mathbb{X}, \mathbb{Y})$, $T \perp_B^\epsilon A$ if and only if either (a) or (b) holds.*

(a) *There exists a norming sequence $\{x_n\}$ for T such that* $\lim_{n\to\infty} \|Ax_n\| \leq \epsilon\|A\|$.

(b) *There exist two norming sequences $\{x_n\}, \{y_n\}$ for T and two sequences of positive real numbers $\{\epsilon_n\}, \{\delta_n\}$ such that*

(i) $\epsilon_n \to 0,\ \delta_n \to 0$, *as* $n \to \infty$.

(ii) $\|Tx_n + \lambda Ax_n\|^2 \geq (1 - \epsilon_n{}^2)\|Tx_n\|^2 - 2\epsilon\sqrt{1 - \epsilon_n{}^2}\|Tx_n\|\|\lambda A\|$ *for all* $\lambda \geq 0$.

(iii) $\|Ty_n + \lambda Ay_n\|^2 \geq (1 - \delta_n{}^2)\|Ty_n\|^2 - 2\epsilon\sqrt{1 - \delta_n{}^2}\|Ty_n\|\|\lambda A\|$ *for all* $\lambda \leq 0$.

Proof We first prove the necessary part. Since $T \perp_B^\epsilon A$, so following Theorem 5.3.3, we can find $S \in span\{T, A\}$ such that $T \perp_B S$ and $\|S - A\| \leq \epsilon\|A\|$. Since $T \perp_B S$ so by Theorem 4.4.14, either (a) or (b) holds.

(a) There exists a norming sequence $\{x_n\}$ for T such that $\|Sx_n\| \to 0$.

(b) There exist two norming sequences $\{x_n\}, \{y_n\}$ for T and two sequences of positive real numbers $\{\epsilon_n\}, \{\delta_n\}$ such that $\epsilon_n \to 0,\ \delta_n \to 0$, as $n \to \infty$ and $Sx_n \in (Tx_n)^{(+\epsilon_n)}$, $Sy_n \in (Ty_n)^{(-\delta_n)}$ for all $n \in \mathbb{N}$.

First suppose that (a) holds. Then $\|Sx_n - Ax_n\| \le \|S - A\| \le \epsilon\|A\|$. This implies that $\lim_{n\to\infty} \|Ax_n\| \le \epsilon\|A\|$.

Now, suppose that (b) holds. Then for all $\lambda \ge 0$, we have

$$\begin{aligned}\|Tx_n + \lambda Ax_n\|^2 &= \|(Tx_n + \lambda Sx_n) + \lambda(Ax_n - Sx_n)\|^2\\ &\ge |\|Tx_n + \lambda Sx_n\| - |\lambda|\|Ax_n - Sx_n\||^2\\ &\ge \|Tx_n + \lambda Sx_n\|^2 - 2\|Tx_n + \lambda Sx_n\||\lambda|\|Sx_n - Ax_n\|\\ &\ge (1 - {\epsilon_n}^2)\|Tx_n\|^2 - 2\epsilon\sqrt{1 - {\epsilon_n}^2}\|Tx_n\|\|\lambda A\|.\end{aligned}$$

Similarly, using the norming sequence $\{y_n\}$ we get condition (iii). This proves the necessary part.

Next, we prove the sufficient part. Suppose that (a) holds. Then for any $\lambda \in \mathbb{R}$,

$$\|T + \lambda A\|^2 \ge \|(T + \lambda A)x_n\|^2 \ge |\|Tx_n\| - |\lambda|\|Ax_n\||^2 \ge \|Tx_n\|^2 - 2\|Tx_n\||\lambda|\|Ax_n\|.$$

Letting $n \to \infty$, we get $\|T + \lambda A\|^2 \ge \|T\|^2 - 2\epsilon\|T\|\|\lambda A\|$. Hence $T \perp_B^\epsilon A$. Now, suppose (b) holds. Then for any $\lambda \ge 0$, we have

$$\|T + \lambda A\|^2 \ge \|(T + \lambda A)x_n\|^2 \ge (1 - {\epsilon_n}^2)\|Tx_n\|^2 - 2\epsilon\sqrt{1 - {\epsilon_n}^2}\|Tx_n\|\|\lambda A\|.$$

Letting $n \to \infty$, we have $\|T + \lambda A\|^2 \ge \|T\|^2 - 2\epsilon\|T\|\|\lambda A\|$. Similarly, for any $\lambda \le 0$, we have

$$\|T + \lambda A\|^2 \ge \|(T + \lambda A)y_n\|^2 \ge (1 - {\delta_n}^2)\|Ty_n\|^2 - 2\epsilon\sqrt{1 - {\delta_n}^2}\|Ty_n\|\|\lambda A\|.$$

Letting $n \to \infty$, we have $\|T + \lambda A\|^2 \ge \|T\|^2 - 2\epsilon\|T\|\|\lambda A\|$. Hence $T \perp_B^\epsilon A$. $\square$

The previous theorem can be improved if we assume that the operator T satisfies some additional conditions.

Theorem 5.5.12 ([87, Theorem 3.1]) *Let $\mathbb{X}, \mathbb{Y}$ be Banach spaces. Suppose $T \in \mathbb{L}(\mathbb{X}, \mathbb{Y})$ is such that $M_T = \{\pm x_0\}$, for some $x_0 \in S_{\mathbb{X}}$. Let span$\{x_0\}$ be an L-ideal of $\mathbb{X}$ and P be the L-projection with range$(P) =$ span$\{x_0\}$. Suppose that $\|T\|_{ker(P)} < \|T\|$. Then for any $A \in \mathbb{L}(\mathbb{X}, \mathbb{Y})$ and any $\epsilon \in [0, 1)$, $T \perp_B^\epsilon A$ if and only if $\|Tx_0 + \lambda Ax_0\|^2 \ge \|Tx_0\|^2 - 2\epsilon\|Tx_0\|\|\lambda A\|$ for all $\lambda \in \mathbb{R}$.*

Moreover, if $x_0 \in M_A$, then $T \perp_B^\epsilon A$ if and only if $Tx_0 \perp_B^\epsilon Ax_0$.

Proof The proof of the sufficient part is trivial. We only prove the necessary part. Let $T \perp_B^\epsilon A$ and $\|T\|_{\text{ker}(P)} = m < \|T\|$. We claim that if $\{x_n\}$ is any norming sequence for T, then $\{x_n\}$ has a subsequence converging to ax_0, where $|a| = 1$. Clearly, for each $n \in \mathbb{N}$, $x_n = a_n x_0 + y_n$, where $a_n \in \mathbb{R}$, $y_n \in \ker(P)$ and $1 = \|x_n\| = |a_n| + \|y_n\|$. Thus $\{a_n\}$ is bounded and so there exists a subsequence $\{a_{n_k}\}$ converging a, say. Clearly $| a |\leq 1$. Now,

$$\|T(\frac{y_n}{\|y_n\|})\| \leq \|T\|_{\ker(P)} = m \Rightarrow \|Ty_n\| \leq m\|y_n\|.$$

Therefore,

$$\begin{aligned}
\|Tx_{n_k}\| &= \|a_{n_k} Tx_0 + Ty_{n_k}\| \\
&\leq |a_{n_k}|\|Tx_0\| + \|Ty_{n_k}\| \\
&\leq |a_{n_k}|\|T\| + m\|y_{n_k}\| \\
&\leq |a_{n_k}|\|T\| + m(1 - |a_{n_k}|) \\
\Rightarrow \|Tx_{n_k}\| - m &\leq |a_{n_k}|(\|T\| - m) \\
\Rightarrow \|T\| - m &\leq |a|(\|T\| - m), \text{ [taking limit } k \to \infty] \\
\Rightarrow 1 &\leq |a|.
\end{aligned}$$

Thus, $a_{n_k} \to a$, where $|a| = 1$. Hence, $y_{n_k} \to 0$ and $x_{n_k} \to ax_0$, where $|a| = 1$. This proves our claim. Since $T \perp_B^{\epsilon} A$, from Theorem 5.5.11, we have that either (a) or (b) holds:

(a) There exists a norming sequence $\{x_n\}$ for T such that $\lim_{n\to\infty} \|Ax_n\| \leq \epsilon\|A\|$.

(b) There exist two norming sequences $\{x_n\}, \{y_n\}$ for T and two sequences of positive real numbers $\{\epsilon_n\}, \{\delta_n\}$ such that

(i) $\epsilon_n \to 0,\ \delta_n \to 0$.

(ii) $\|Tx_n + \lambda Ax_n\|^2 \geq (1 - \epsilon_n^2)\|Tx_n\|^2 - 2\epsilon\sqrt{1 - \epsilon_n^2}\|Tx_n\|\|\lambda A\|$ for all $\lambda \geq 0$.

(iii) $\|Ty_n + \lambda Ay_n\|^2 \geq (1 - \delta_n^2)\|Ty_n\|^2 - 2\epsilon\sqrt{1 - \delta_n^2}\|Ty_n\|\|\lambda A\|$ for all $\lambda \leq 0$.

First suppose that (a) holds. Then there exists a subsequence $\{x_{n_k}\}$ of $\{x_n\}$ converging to ax_0, where $|a| = 1$. Therefore, $\|Ax_0\| = \lim_{k\to\infty} \|Ax_{n_k}\| \leq \epsilon\|A\|$. Thus, for all $\lambda \in \mathbb{R}$, $\|Tx_0 + \lambda Ax_0\|^2 \geq (\|Tx_0\| - |\lambda|\|Ax_0\|)^2 \geq \|Tx_0\|^2 - 2\|Tx_0\||\lambda|\|Ax_0\| \geq \|Tx_0\|^2 - 2\epsilon\|Tx_0\|\|\lambda A\|$.

Now, suppose that (b) holds. Then there exist subsequences $\{x_{n_k}\}$ and $\{y_{n_k}\}$ of $\{x_n\}$ and $\{y_n\}$ respectively, such that $x_{n_k} \to ax_0$ and $y_{n_k} \to bx_0$, where $|a| = |b| = 1$. Now, in (ii) and (iii) of (b), taking limit we obtain, $\|Tx_0 + \lambda Ax_0\|^2 \geq \|Tx_0\|^2 - 2\epsilon\|Tx_0\|\|\lambda A\|$ for all $\lambda \in \mathbb{R}$.

Clearly, if $x_0 \in M_A$, then $T \perp_B^{\epsilon} A$ if and only if $Tx_0 \perp_B^{\epsilon} Ax_0$. This completes the proof of the theorem. □

Remark 5.5.13 ([87, Remark 3.1]) Note that if $\mathbb{X}$ is an L^1-predual space, then for every $x^* \in Ext(B_{\mathbb{X}^*})$, span$\{x^*\}$ is an L-ideal of $\mathbb{X}^*$. Now, if $T \in \mathbb{L}(\mathbb{X})$ is such that $M_{T^*} = \{\pm x^*\}$ and $\|T^*\|_{\ker(P)} < \|T^*\|$, where P is the L-projection with range$(P) =$ span$\{x^*\}$, then T^* satisfies the hypothesis of Theorem 5.5.12.

The following consequence of Theorem 5.5.12 is also worth mentioning. The proof is omitted, as it is very easy.

Corollary 5.5.14 ([87, Corollary 3.1.1]) *Let $\mathbb{X}$, $\mathbb{Y}$ be Banach spaces. Suppose $T \in \mathbb{L}(\mathbb{X}, \mathbb{Y})$ is such that $M_T = \{\pm x_0\}$, for some $x_0 \in S_{\mathbb{X}}$. Let span$\{x_0\}$ be an L-ideal of $\mathbb{X}$ and P be the L-projection with range$(P) =$ span$\{x_0\}$. Suppose that $\|T\|_{ker(P)} < \|T\|$. Then for any $A \in \mathbb{L}(\mathbb{X}, \mathbb{Y})$, $T \perp_B A$ if and only if $Tx_0 \perp_B Ax_0$.*

Having studied the approximate and exact orthogonality of bounded linear operators in considerable detail, it is now time for us to reap the benefits by applying the results obtained in the study of orthogonality towards successfully understanding well-established geometric concepts such as that of smoothness, in general, Banach spaces and in spaces of operators between them. While the study of smoothness in Banach spaces is a classical one, the reader will observe in the next chapter that in order to describe the same in the setting of linear operators, B-J orthogonality techniques developed in the earlier chapters play pivotal roles. Indeed, the program of completely classifying the smooth points in spaces of operators between Banach spaces remained incomplete in its complete generality, until very recently. One of the focal points of the next chapter (and this monograph) is to complete this classical program in light of B-J orthogonality techniques. We also discuss the concept of k-smoothness, a natural extension of the concept of smoothness, in the space of bounded linear operators, that further extends the scope of our exploration.

Chapter 6
Smoothness and k–Smoothness of Operators

6.1 Introduction

The concept of smoothness is one of the fundamental aspects of the theory of Banach spaces, both from geometric and analytic points of view. Indeed, a cursory look at the contents of any classical or modern textbook on the subject matter would suffice to argue in favor of our claim. The reason behind this can perhaps be best explained by illustrating the importance of smoothness in ensuring certain niceties in the framework of Banach spaces, which may very well be missing otherwise. Let us recall that an element $x \in S_{\mathbb{X}}$ is said to be smooth if there exists a unique supporting hyperplane to the unit ball $B_{\mathbb{X}}$ at x. In case of a non-zero vector $x \in \mathbb{X}$, we say that x is smooth if the corresponding unit vector $\frac{x}{\|x\|}$ is smooth. The specialty of smooth points, especially from a geometric perspective, should be clear from the following two diagrams:

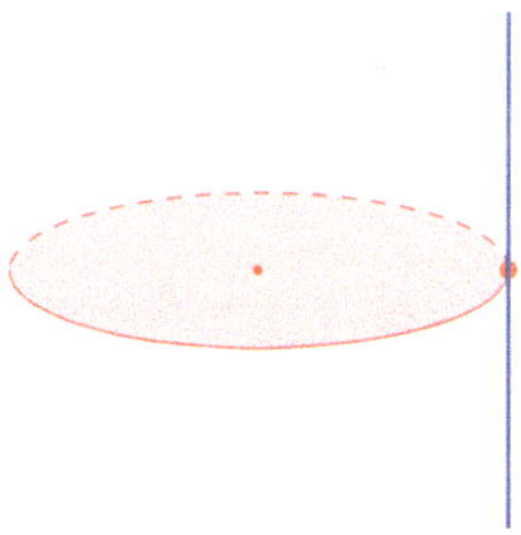

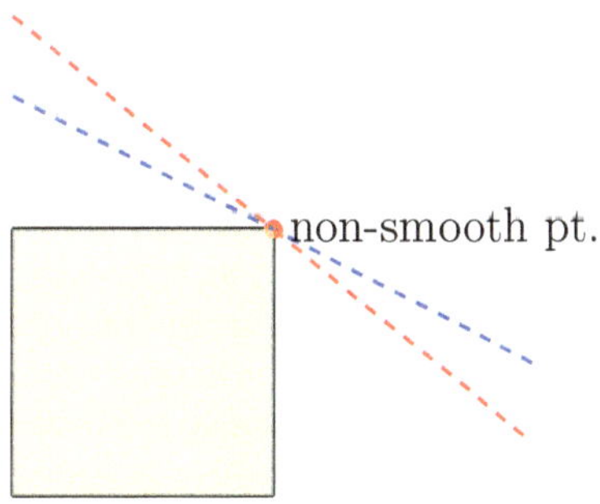

In an equivalent analytic language, an element $x \in S_{\mathbb{X}}$ is said to be smooth if there exists a unique norm one linear functional f on $\mathbb{X}$ such that $f(x) = 1$. Corresponding to any $x \in S_{\mathbb{X}}$, consider the set $J(x) = \{f \in S_{\mathbb{X}^*} : f(x) = 1\}$. Then it is easy to observe that $J(x)$ is a weak* compact closed convex subset of $\mathbb{X}^*$. The element x is

A. Mal et al., *Birkhoff–James Orthogonality and Geometry of Operator Spaces*,
Infosys Science Foundation Series, https://doi.org/10.1007/978-981-99-7111-4_6

smooth if and only if the set $J(x)$ is a singleton. Using the Hahn–Banach theorem, one can easily prove that x is a smooth point if and only if the set $H_x = \{y \in \mathbb{X} : x \perp_B y\}$ is a subspace of $\mathbb{X}$, which is in turn equivalent to the condition that B-J orthogonality is right additive at x. It should be noted that B-J orthogonality is said to be right additive at x if for all $y, z \in \mathbb{X}$, $x \perp_B y$ and $x \perp_B z$ together imply that $x \perp_B (y + z)$. In order to investigate the smoothness of a bounded linear operator T we will heavily rely on this particular characterization involving operator B-J orthogonality. To be more precise, we will verify whether $T \perp_B A_1$ and $T \perp_B A_2$ imply that $T \perp_B (A_1 + A_2)$, where A_1 and A_2 are arbitrary bounded linear operators between the corresponding spaces. Identifying the smooth points in the Banach space of all bounded linear operators, endowed with the usual supremum norm, is a classical problem in the geometric theory of linear operators. We would like to illustrate that the concept of B-J orthogonality can play a central role in answering the above query. Indeed, using some of B-J orthogonality techniques developed earlier, we will obtain a complete solution to this intriguing problem, in terms of some related concepts, such as that of operator norm attainment and norming sequences. However, the road ahead is quite long and we will be consciously dealing with the problem using a step-by-step approach, starting with the simplest case of bounded linear operators on a Hilbert space. As we will see in this chapter, the methods employed in the setting of Hilbert spaces will allow us to efficiently look for suitable modifications of the developed theory, to make it work successfully in the framework of Banach spaces.

It is clear that smoothness in Banach spaces is a local concept which can also be considered in the global context in the following natural way: A Banach space $\mathbb{X}$ is said to be smooth if every non-zero element x in $\mathbb{X}$ is smooth. Of course, the local study of smoothness is more potent than the corresponding global study, primarily for two reasons. Firstly, it is clear that in case of any Banach space, the global information on smoothness follows directly from the local one. But perhaps more importantly, we will see that a local study of smoothness and $k-$smoothness allows us to have much better idea about the finer geometric structure of the corresponding Banach space. As mentioned above, our approach towards understanding the local property of smoothness in Banach spaces, and more specifically, in operator spaces, involves some of the concepts and results related to B-J orthogonality developed in the earlier chapters. Starting with the easier case of operators on a Hilbert space, we gradually proceed towards studying the smoothness of (compact) operators between Banach spaces. The reader will once again observe the importance of the operator norm attainment set in case of smoothness of compact operators, whenever it is non-empty. On the other hand, in view of the discussions in the previous chapters, in case the norm attainment set is empty or the concerned operator is not compact, it is not difficult to expect that the exact norm attainment set needs to be replaced by norming sequences.

6.2 Smoothness in $\mathbb{L}(\mathbb{H})$

As early as 1973, Holub [55] characterized the smoothness of compact operators on a real Hilbert space by showing that a compact operator $T \in S_{\mathbb{K}(\mathbb{H})}$ is smooth if and only if whenever $\|Tx\| = \|Ty\| = 1$ for some $x, y \in S_{\mathbb{H}}$ then $x = \pm y$. Abatzoglou [1] then used the notion of differentiability of norm to characterize smoothness of a bounded linear operator T on a real Hilbert space. Let us recall that the norm is said to be Gateaux differentiable at an element $x \neq \theta$ if for all $y \in \mathbb{X}$ the limit $\lim_{t\to 0} \frac{\|x+ty\|-\|x\|}{t}$ exists and the norm is said to be Frechet differentiable if the convergence to the limit is uniform for all $y \in \mathbb{X}$. It is well known that the norm is Gateaux differentiable at $x \in S_{\mathbb{X}}$ if and only if x is a smooth point. Abatzoglou [1, Theorem 3.1] proved that $T \in S_{\mathbb{L}(\mathbb{H})}$ is Frechet differentiable if and only if T attains norm at $\pm x$, $\|x\| = 1$ and $\sup\{\|Ty\| : x \perp_B y, \|y\| = 1\} < 1$. Later on Kittaneh and Younis [69] proved that $T \in S_{\mathbb{L}(\mathbb{H})}$ is a smooth point if and only if $\|T\|_e < 1$ and T attains norm at a unique (up to scalar multiplication) vector x (say) of $S_{\mathbb{X}}$, where $\|T\|_e$ denotes the essential spectrum of T. Recently a complete characterization of smoothness of norm one operator T was obtained for real Hilbert space in [105] using the notion of B-J orthogonality. Going in the same spirit, we here present a characterization of the same for operators on both real and complex Hilbert spaces. It should be noted that the technique of our proof is completely different from that of Abatzoglou. For our purpose, we require the following theorem from Chap. 4, which characterizes B-J orthogonality for operators on Hilbert spaces.

Theorem 6.2.1 *Let $\mathbb{H}$ be a Hilbert space and let $T \in \mathbb{L}(\mathbb{H})$. Then for any $A \in \mathbb{L}(\mathbb{H})$, $T \perp_B A \Leftrightarrow Tx \perp Ax$ for some $x \in M_T$ if and only if $M_T = S_{H_0}$, where H_0 is a finite-dimensional subspace of $\mathbb{H}$ and $\|T\|_{H_0^\perp} < \|T\|$.*

We also need the following proposition.

Proposition 6.2.2 *Suppose T is a bounded linear operator on a Hilbert space $\mathbb{H}$ that does not attain its norm, i.e., there does not exist $x \in S_{\mathbb{H}}$ such that $\|Tx\| = \|T\|$. Then there exists a sequence $\{e_n\}$ of orthonormal vectors such that $\|Te_n\| \longrightarrow \|T\|$.*

Proof We note that T^*T is a positive operator, $\|T^*T\| = \|T\|^2$ and T attains norm if and only if T^*T attains norm. Also $\|T\|^2 \in \sigma(T^*T) = \sigma_{disc}(T^*T) \cup \sigma_{ess}(T^*T)$. By Theorem VII.10 of Reed and Simon [112], $\|T\|^2$ can't belong to $\sigma_{disc}(T^*T)$. So by Theorem VII.12 (Weyl's criterion) there exists $\{e_n\}$ of orthonormal vectors such that $\|(T^*T - \|T\|^2)e_n\| \longrightarrow 0$. Now $| \langle (T^*T - \|T\|^2)e_n, e_n\rangle | \leq \|(T^*T - \|T\|^2)e_n\| \longrightarrow 0$. This implies $\langle (T^*T - \|T\|^2)e_n, e_n\rangle = \langle T^*Te_n, e_n\rangle - \langle \|T\|^2 e_n, e_n\rangle \longrightarrow 0$. Thus $\|Te_n\| \longrightarrow \|T\|$. □

By applying Theorem 6.2.1 and Proposition 6.2.2, we completely classify the smooth points of $\mathbb{L}(\mathbb{H})$.

Theorem 6.2.3 ([105, Theorem 4.6]) *Let $\mathbb{H}$ be a Hilbert space. Then $T \in \mathbb{L}(\mathbb{H})$ is a smooth point if and only if T attains norm at a unique (up to scalar multiplication) vector x (say) of $S_{\mathbb{X}}$ and* $\sup\{\|Ty\| : x \perp y, y \in S_{\mathbb{H}}\} < \|T\|$.

Proof **Necessary part**: We first show that $M_T \neq \emptyset$. If possible $M_T = \emptyset$, then by Proposition 6.2.2 there exists a sequence $\{e_n\}$ of orthonormal vectors such that $\|Te_n\| \to \|T\|$. Then there exists an orthonormal basis $\mathcal{B}$ containing $\{e_n : n = 1, 2, \ldots\}$. Define a linear operator A_1 on $\mathbb{H}$ as: $A_1 e_{2n} = Te_{2n}$ and A_1 takes every other element of $\mathcal{B}$ to 0. It is easy to verify that A_1 is bounded. Consider $A_2 = T - A_1$. Then $T \perp_B A_1, T \perp_B A_2$. Indeed, $\|Te_{2n+1}\| \to \|T\|$ and $\langle Te_{2n+1}, A_1 e_{2n+1}\rangle \to 0$ and so by Theorem 4.2.1, we get $T \perp_B A_1$. Similarly we can show that $T \perp_B A_2$. However, $T \not\perp_B (A_1 + A_2)$, which contradicts the fact that T is smooth.

We next show that T attains norm at a unique (up to scalar multiplication) vector x (say) of $S_{\mathbb{X}}$. If possible, let $\|Tx_1\| = \|Tx_2\| = \|T\|$ with $x_1, x_2 \in S_{\mathbb{H}}$, where $\{x_1, x_2\}$ is linearly independent. By Theorem 3.3.5, the norm attaining set M_T is the unit sphere of some subspace of $\mathbb{H}$. So without loss of generality, we assume that $x_1 \perp x_2$. Let $H_0 = \text{span}\{x_1, x_2\}$. Then $\mathbb{H} = H_0 \oplus H_0^{\perp}$. Define $A_1, A_2 : \mathbb{H} \longrightarrow \mathbb{H}$ as follows: $A_1(c_1x_1 + c_2x_2 + h) = c_1Tx_1$, $A_2(c_1x_1 + c_2x_2 + h) = c_2Tx_2 + Th$, where $h \in H_0^{\perp}$ and c_1, c_2 are scalars. Then, both A_1, A_2 are bounded linear operators and $T \perp_B A_1, T \perp_B A_2$. But $T = A_1 + A_2$ and so $T \not\perp_B (A_1 + A_2)$, which contradicts the fact that T is smooth.

Finally, assuming that T attains norm at a unique vector x (up to scalar multiplication) we show that $\sup\{\|Ty\| : x \perp y, y \in S_{\mathbb{H}}\} < \|T\|$. If possible, let $\sup\{\|Ty\| : x \perp y, y \in S_{\mathbb{H}}\} = \|T\|$. Let $H_0 = \text{span}\{x\}$. Then $\mathbb{H} = H_0 \oplus H_0^{\perp}$. Define $A_1, A_2 : \mathbb{H} \longrightarrow \mathbb{H}$ as follows: Let $w = u + v \in \mathbb{H}$, where $u \in H_0, v \in H_0^{\perp}$. Then $A_1 w = Tu$, $A_2 w = Tv$. It is easy to check that both A_1, A_2 are bounded linear operators and $T \perp_B A_1, T \perp_B A_2$. But $T = A_1 + A_2$ and so $T \not\perp_B (A_1 + A_2)$, which contradicts the fact that T is smooth.

Sufficient part: Let $T \perp_B A_i$ $(i = 1, 2)$. Then by Theorem 6.2.1, $Tx \perp A_1 x$ and $Tx \perp A_2 x$. As Tx is a smooth point of $\mathbb{H}$ so $Tx \perp (A_1 + A_2)x$. From this, it is elementary to observe that $T \perp_B (A_1 + A_2)$. Thus by applying Theorem 2.3.2, T is smooth. □

Remark 6.2.4 For a compact operator T on a real Hilbert space $\mathbb{H}$, if T attains norm only at $\pm x$, $\|x\| = 1$, then it easy to see that $\sup\{\|Ty\| : x \perp_B y, \|y\| = 1\} < \|T\|$ and so smoothness of T follows easily from the above theorem.

6.3 Smoothness in $\mathbb{K}(\mathbb{X}, \mathbb{Y})$

In this section, we discuss smoothness of compact operators between Banach spaces. Motivated by the result of Holub [55, Theorem 3.3], Deeb and Khalil [31] proved that a compact operator $T \in S_{\mathbb{K}(\ell_p)}$, $1 < p < \infty$, on a real ℓ_p space is smooth if and only if whenever $\|Tx\| = \|Ty\| = 1$ for some $x, y \in S_{\ell_p}$ then $x = \pm y$. Hennefield [53] obtained a sufficient condition for smoothness of compact operator assuming the space $\mathbb{X}$ to be a smooth reflexive Banach space with a Schauder basis. In 1975, Heinrich [52] further generalized the result and characterized smooth compact operators on Banach spaces. In 1993, Grząślewicz and Younis [48] provided an alternative proof of the same result, depending upon the structure of extreme linear functionals on the space of compact operators (see Theorem 4.4.1).

Having mentioned some of the relevant results in the literature, we next present the desired characterization of smooth compact operators between Banach spaces.

Theorem 6.3.1 ([48, Theorem 2]) *Let $\mathbb{X}$, $\mathbb{Y}$ be Banach spaces. Then $T \in \mathbb{K}(\mathbb{X}, \mathbb{Y})$ is a smooth operator if and only if T^* attains norm at a unique (up to scalar multiplication) vector y_0^* (say) of $S_{\mathbb{Y}^*}$ and $T^* y_0^*$ is a smooth point of $\mathbb{X}^*$.*

Proof Suppose that T is smooth. Since T^* is compact, T^* attains its norm. If possible, suppose that $y_i^* \in M_{T^*}$ for $i = 1, 2$, and $\{y_1^*, y_2^*\}$ is linearly independent. Suppose $x_i^{**} \in J(Ty_i^*)$ for $i = 1, 2$. Then $x_i^{**} \otimes y_i^*$ are two linearly independent functionals. Moreover,

$$x_i^{**} \otimes y_i^*(T) = x_i^{**}(T^* y_i^*) = \|T^* y_i^*\| = \|T^*\| = \|T\|$$

yields that $x_i^{**} \otimes y_i^* \in J(T)$, which contradicts that T is smooth. This proves that T^* attains norm at a unique (up to scalar multiplication) vector y_0^* (say). Now, suppose that $x_i^{**} \in J(T^* y_0^*)$ for $i = 1, 2$ and $x_1^{**} \neq x_2^{**}$. Then $x_i^{**} \otimes y_0^*$ are two linearly independent functionals in $J(T)$, which is again a contradiction. Therefore, $J(T^* y_0^*)$ is singleton, i.e., $T^* y_0^*$ is smooth.

Conversely, suppose that $\Phi \in Ext J(T)$. Then clearly, $\Phi \in Ext(B_{\mathbb{K}(\mathbb{X},\mathbb{Y})^*})$. Now, by Theorem 4.4.1, we get $\Phi = x^{**} \otimes y^*$, for some $x^{**} \in Ext(B_{\mathbb{X}^{**}})$, $y^* \in Ext(B_{\mathbb{Y}^*})$. Observe that

$$\|T\| = x^{**} \otimes y^*(T) = x^{**}(T^* y^*) \leq \|T^* y^*\| \leq \|T^*\| = \|T\|.$$

Therefore, $y^* \in M_{T^*}$ and $x^{**} \in J(T^* y^*)$. By hypothesis, we get $y^* = y_0^*$ and $J(T^* y_0^*) = \{x^{**}\}$, i.e., $Ext J(T) = \{x^{**} \otimes y_0^*\}$. This clearly yields that $J(T)$ is singleton, i.e., T is smooth. □

In particular, if $\mathbb{X}$ is a reflexive Banach space, then using Theorem 6.3.1, we can say that $T \in \mathbb{K}(\mathbb{X}, \mathbb{Y})$ is smooth if and only if T attains norm at a unique (up to scalar multiplication) vector x (say) of $S_{\mathbb{X}}$ and Tx is a smooth point of $\mathbb{Y}$. However, we present here an alternative proof of this fact using B-J orthogonality techniques. The proof is a slight refinement of [105, Theorems 4.1 and 4.2].

Theorem 6.3.2 *Let $\mathbb{X}$ be a reflexive Banach space and let $\mathbb{Y}$ be any Banach space. Let $T \in S_{\mathbb{K}(\mathbb{X},\mathbb{Y})}$. Then T is smooth if and only if T attains norm at a unique (up to scalar multiplication) vector x (say) of $S_{\mathbb{X}}$ and Tx is a smooth point of $\mathbb{Y}$.*

Proof **Necessary part**: Observe that T is a compact operator on a reflexive Banach space and so $M_T \neq \emptyset$. We show that if $\|Tx_1\| = \|Tx_2\| = \|T\|$ for $x_1, x_2 \in S_{\mathbb{X}}$, then x_1 is a scalar multiple of x_2. If possible, suppose x_1 is not a scalar multiple of x_2. Now, there exists a subspace H_1 of codimension 1 such that $x_1 \perp_B H_1$. There exists a scalar a with $| a |\leq 1$ such that $ax_1 + x_2 \in H_1$. Let H_2 be a subspace of H_1 with codimension 1 in H_1 such that $(ax_1 + x_2) \perp_B H_2$. Let $z \in S_{\mathbb{X}}$. Then z can be written uniquely as $z = \alpha x_1 + h_1$, for some scalar α and $h_1 \in H_1$. Again h_1 can be written uniquely as $h_1 = \beta(ax_1 + x_2) + h_2$, for some scalar β and $h_2 \in H_2$. Thus $z = (\alpha + a\beta)x_1 + \beta x_2 + h_2$. Define operators $A_1, A_2 : \mathbb{X} \longrightarrow \mathbb{Y}$ as follows:

$$A_1(z) = (\alpha + a\beta)Tx_1, \; A_2(z) = \beta Tx_2 + Th_2.$$

Clearly, both A_1, A_2 are compact linear operators. Then $T\perp_B A_1, T\perp_B A_2$ but $T = A_1 + A_2$ which shows that $T \not\perp_B (A_1 + A_2)$, which contradicts the fact that T is smooth. Hence T attains norm at a unique (up to scalar multiplication) vector $x \in S_{\mathbb{X}}$.

We next show that Tx is a smooth point in $\mathbb{Y}$. If possible let Tx be not smooth. Then following Theorem 2.3.2, there exist $y, z \in \mathbb{Y}$ such that $Tx\perp_B y, Tx\perp_B z$ but $Tx \not\perp_B (y + z)$. There exists a hyperplane H_x such that $x\perp_B H_x$. Define two operators $A_1, A_2 : \mathbb{X} \longrightarrow \mathbb{Y}$ as follows:

$$A_1(ax + h) = ay, \; A_2(ax + h) = az,$$

where a is a scalar and $h \in H_x$. Then it is easy to verify that both A_1, A_2 are compact linear operators and $T\perp_B A_1, T\perp_B A_2$. But $T \not\perp_B A_1 + A_2$. For otherwise, since $M_T = \{\alpha x : |\alpha| = 1\}$, we have by Theorem 4.3.10 for real scalars and by Theorem 4.7.6 for complex scalars, $Tx\perp_B(y + z)$, which is not possible. This contradiction shows that Tx is a smooth point.

Sufficient part: Suppose $M_T = \{\alpha x : |\alpha| = 1\}$ and Tx is a smooth point of $\mathbb{Y}$. Let $T \perp_B A_1$ and $T \perp_B A_2$. Then by Theorems 4.3.10 and 4.7.6, $Tx \perp_B A_1 x$ and $Tx \perp_B A_2 x$. Now, Tx is smooth point and so $Tx \perp_B (A_1 + A_2)x$. So for all λ, we get $\|T + \lambda(A_1 + A_2)\| \geq \|Tx + \lambda(A_1 + A_2)x\| \geq \|Tx\| = \|T\|$. This shows that $T \perp_B (A_1 + A_2)$ and so T is smooth. □

Remark 6.3.3 (i) We would like to remark that Theorem 6.3.2 generalizes and improves on both the results of Holub [55] and Hennefield [53].
(ii) Another alternative proof of Theorem 6.3.2 follows easily using Lemma 4.4.3, which states that

$$Ext\ J(T) = \{y^* \otimes x \in \mathbb{K}(\mathbb{X}, \mathbb{Y})^* : x \in M_T \cap Ext(B_{\mathbb{X}}), y^* \in Ext\ J(Tx)\},$$

where $y^* \otimes x : \mathbb{K}(\mathbb{X}, \mathbb{Y}) \to \mathcal{K}$ is defined by $y^* \otimes x(S) = y^*(Sx)$ for every $S \in \mathbb{K}(\mathbb{X}, \mathbb{Y})$.

We next proceed towards another direction of the study of smoothness in the space of compact operators. Without further mentioning, we will assume that a Banach space $\mathbb{X}$ is canonically embedded in $\mathbb{X}^{**}$. Consider the subspace $M = \{S^* \in \mathbb{K}(\mathbb{Y}^*, \mathbb{X}^*) : S \in \mathbb{K}(\mathbb{X}, \mathbb{Y})\}$ of $\mathbb{K}(\mathbb{Y}^*, \mathbb{X}^*)$. Observe that the mapping $\psi : \mathbb{K}(\mathbb{X}, \mathbb{Y}) \to M$ defined as $\psi(S) = S^*$ for all $S \in \mathbb{K}(\mathbb{X}, \mathbb{Y})$ is an isometric isomorphism. Therefore, for any $T \in \mathbb{K}(\mathbb{X}, \mathbb{Y})$, if T^* is smooth in $\mathbb{K}(\mathbb{Y}^*, \mathbb{X}^*)$, then since T^* is also smooth in M, we get that T is smooth in $\mathbb{K}(\mathbb{X}, \mathbb{Y})$. On the other hand, smoothness of T in $\mathbb{K}(\mathbb{X}, \mathbb{Y})$ does not always ensure smoothness of T^* in $\mathbb{K}(\mathbb{Y}^*, \mathbb{X}^*)$. Suppose x_0^* be a smooth point of $\mathbb{X}^*$. Let there exists a smooth point y_0 of $\mathbb{Y}$ such that y_0 is not smooth in $\mathbb{Y}^{**}$. Consider the rank one operator $T \in \mathbb{K}(\mathbb{X}, \mathbb{Y})$ defined as $T(x) = x_0^*(x)y_0$ for all $x \in \mathbb{X}$. Observe that, $T^*(y^*) = y^*(y_0)x_0^*$ for all $y^* \in \mathbb{Y}^*$. Now, $M_{T^*} = \{\alpha y_0^* : |\alpha| = 1\}$, where $J(y_0) = \{y_0^*\} \subset S_{\mathbb{Y}^*}$. Therefore, from Theorem 6.3.1, it follows that T is a smooth operator in $\mathbb{K}(\mathbb{X}, \mathbb{Y})$. Furthermore, $T^{**}(x^{**}) = x^{**}(x_0^*)y_0$ for all $x^{**} \in \mathbb{X}^{**}$. Now, $M_{T^{**}} = \{\alpha x_0^{**} : |\alpha| = 1\}$, where x_0^{**} is the unique norming linear functional of x_0^*. However, since y_0 is not smooth in $\mathbb{Y}^{**}$, again from Theorem 6.3.1, it follows that T^* is not a smooth operator in $\mathbb{K}(\mathbb{Y}^*, \mathbb{X}^*)$. Thus, although smoothness of T^* in $\mathbb{K}(\mathbb{Y}^*, \mathbb{X}^*)$ ensures the smoothness of T in $\mathbb{K}(\mathbb{X}, \mathbb{Y})$, the converse is not always true. This fact motivated Rao [109] to explore when the smoothness of T in $\mathbb{K}(\mathbb{X}, \mathbb{Y})$ ensures the smoothness of T^* in $\mathbb{K}(\mathbb{Y}^*, \mathbb{X}^*)$. In the following theorem it is proved that if $\mathbb{Y}$ is a Hahn-Banach smooth space then smoothness of T ensures smoothness of T^*. Recall that a Banach space $\mathbb{Y}$ is said to be Hahn–Banach smooth, if each functional in $\mathbb{Y}^*$ has a unique norm preserving extension in $\mathbb{Y}^{***}$.

Theorem 6.3.4 ([109, Theorem 1]) *Let $\mathbb{Y}$ be a Hahn–Banach smooth space. Then for any smooth operator $T \in \mathbb{K}(\mathbb{X}, \mathbb{Y})$, T^* is a smooth operator in $\mathbb{K}(\mathbb{Y}^*, \mathbb{X}^*)$.*

Proof Let T be a smooth operator in $\mathbb{K}(\mathbb{X}, \mathbb{Y})$ with $\|T\| = 1$. Then from Theorem 6.3.1, it follows that T^* attains norm at a unique vector (up to scalar multiplication) say y_0^*. Clearly, $y_0^* \in Ext(B_{\mathbb{Y}^*})$. Since $\mathbb{Y}$ is Hahn–Banach smooth, $y_0^* \in Ext(B_{\mathbb{Y}^{***}})$. Moreover, $T^*y_0^*$ is smooth. Let $J(T^*y_0^*) = \{x_0^{**}\}$. Since $J(T^*y_0^*)$ is an extremal subset of $B_{\mathbb{X}^{**}}$, $x_0^{**} \in Ext(B_{\mathbb{X}^{**}})$. Observe that $x_0^{**} \otimes y_0^*(T) = x_0^{**}(T^*y_0^*) = \|T^*y_0^*\| = \|T^*\| = \|T\| = 1$, i.e., $x_0^{**} \otimes y_0^* \in J(T)$. Since T is smooth, $J(T)$ is singleton, i.e., $J(T) = \{x_0^{**} \otimes y_0^*\}$. Observe that $y_0^* \otimes x_0^{**}(T^*) = 1$, i.e., $y_0^* \otimes x_0^{**} \in J(T^*)$. We show that $J(T^*) = \{y_0^* \otimes x_0^{**}\}$. Suppose that $\phi \in Ext J(T^*)$. Now, $J(T^*)$ is an extremal subset of $B_{\mathbb{K}(\mathbb{Y}^*,\mathbb{X}^*)^*}$. Therefore, from Theorem 4.4.1, it follows that $\phi = \tau \otimes x^{**}$, where $\tau \in Ext(B_{\mathbb{Y}^{***}})$ and $x^{**} \in Ext(B_{\mathbb{X}^{**}})$. Suppose that $\tau|_{\mathbb{Y}} = y^*$. Observe that since T is compact operator, $T^{**}(\mathbb{X}^{**}) \subset \mathbb{Y}$. Therefore, $\tau \otimes x^{**}(T^*) = 1$ implies that $x^{**} \otimes y^*(T) = 1$. Thus, $\|x^{**} \otimes y^*\| = 1 = \|y^*\|$, i.e., $x^{**} \otimes y^* \in J(T)$. This gives that $x^{**} \otimes y^* = x_0^{**} \otimes y_0^*$. We claim that y^*, y_0^* are linearly dependent. If y^*, y_0^* are linearly independent, then there exists $w \in \mathbb{Y}$ such that $y^*(w) = 1$ and $y_0^*(w) = 0$. Now, fix $x^* \in \mathbb{X}^*$ such that $x^{**}(x^*) = 1$. Consider the rank one operator $S \in \mathbb{K}(\mathbb{X}, \mathbb{Y})$ defined by $S(x) = x^*(x)w$ for all $x \in \mathbb{X}$. Then $x^{**} \otimes y^*(S) = 1$, whereas $x_0^{**} \otimes y_0^*(S) = 0$. This contradiction shows that $y^* = \alpha y_0^*$ for some $|\alpha| = 1$. Thus, τ is the unique extension of αy_0^* to $\mathbb{Y}^{***}$. Similarly, it can be shown that $x^{**} = \beta x_0^{**}$ for some $|\beta| = 1$. Now, from $x^{**} \otimes y^*(T) = 1 = x_0^{**} \otimes y_0^*(T)$, it follows that $\alpha\beta = 1$. Hence, $\phi = \tau \otimes x^{**} = \alpha y_0^* \otimes \beta x_0^{**} = y_0^* \otimes x_0^{**}$. Thus, $J(T^*) = \{y_0^* \otimes x_0^{**}\}$, i.e., T^* is smooth. □

6.4 Smoothness in $\mathbb{L}(\mathbb{X}, \mathbb{Y})$

The investigation of smoothness of bounded linear operators between some particular Banach spaces was initiated by many mathematicians in several papers, a few of them being [31, 43, 146]. Motivated by the characterization of smoothness for Hilbert space operators, a sufficient condition for the smoothness of operators between real Banach spaces was obtained for the first time in [105]. We here present the proof, which again depends on Theorem 4.4.8 of Chap. 4. For the convenience of the reader, we recall the result here.

Theorem 6.4.1 *Let $\mathbb{X}, \mathbb{Y}$ be real Banach spaces, let $T \in \mathbb{L}(\mathbb{X}, \mathbb{Y})$ and let $M_T = D \cup (-D)$, where D is a non-empty compact connected subset of $S_{\mathbb{X}}$. If $\sup\{\|Tx\| : x \in C\} < \|T\|$ for all closed subset C of $S_{\mathbb{X}}$ with $d(M_T, C) > 0$ then for any $A \in \mathbb{L}(\mathbb{X}, \mathbb{Y})$, $T \perp_B A$, if and only if there exists $z \in M_T$ such that $Tz \perp_B Az$.*

We now present the sufficient condition for smoothness.

Theorem 6.4.2 ([105, Theorem 4.3]) *Let $\mathbb{X}, \mathbb{Y}$ be real Banach spaces. Then $T \in \mathbb{L}(\mathbb{X}, \mathbb{Y})$ is a smooth point if T attains norm only at $\pm x_0$, Tx_0 is smooth and $\sup\{\|Tx\| : x \in C\} < \|T\|$ for all closed subsets C of $S_{\mathbb{X}}$ with $d(\pm x_0, C) > 0$.*

Proof Let $T \perp_B A_1$, $T \perp_B A_2$, where $A_1, A_2 \in \mathbb{L}(\mathbb{X}, \mathbb{Y})$. Then by Theorem 6.4.1, $Tx_0 \perp_B A_1 x_0$, $Tx_0 \perp_B A_2 x_0$. As Tx_0 is a smooth point so $Tx_0 \perp_B (A_1 + A_2)x_0$ and so $T \perp_B (A_1 + A_2)$. Thus, from Theorem 2.3.2, it follows that T is smooth. □

It is possible to obtain another sufficient condition for smoothness, by applying the following result. For a proof of the same, see Theorem 4.4.7 of Chap. 4.

Theorem 6.4.3 *Let $\mathbb{X}, \mathbb{Y}$ be real Banach spaces. Suppose $T \in \mathbb{L}(\mathbb{X}, \mathbb{Y})$ and $M_T = \{\pm x\}$ for some $x \in S_{\mathbb{X}}$. Also assume that T further satisfies the following property:*

Given any $\delta > 0$, if $\{H_\alpha : \alpha \in \Lambda\}$ is the collection of all hyperspaces such that $d(x_0, H_\alpha) > \delta$ then $\sup\{\|Tx\| : x \in (\cup_\alpha H_\alpha) \cap S_{\mathbb{X}}\} < \|T\|$. Then for any $A \in \mathbb{L}(\mathbb{X}, \mathbb{Y})$, $T \perp_B A$ if and only if $Tx \perp_B Ax$.

Theorem 6.4.4 ([126, Theorem 3.2]) *Let $\mathbb{X}, \mathbb{Y}$ be real Banach spaces and let $T \in \mathbb{L}(\mathbb{X}, \mathbb{Y})$. Suppose that the following conditions hold:*
(i) *$M_T = \{\pm x_0\}$, for some $x_0 \in S_{\mathbb{X}}$.*
(ii) *Tx_0 is a smooth point of $\mathbb{Y}$.*
(iii) *Given any $\delta > 0$, if $\{H_\alpha : \alpha \in \Lambda\}$ is the collection of hyperspaces such that $d(x_0, H_\alpha) > \delta$ then $\sup\{\|Tx\| : x \in (\cup_\alpha H_\alpha) \cap S_{\mathbb{X}}\} < \|T\|$.*
Then T is smooth.

Proof Let $T \perp_B A_i (i = 1,\ 2)$ with $A_i \in \mathbb{L}(\mathbb{X}, \mathbb{Y})$. Then by Theorem 6.4.3, $Tx_0 \perp_B A_i x_0$ $(i = 1,\ 2)$. As Tx_0 is a smooth point of $\mathbb{Y}$ so $Tx_0 \perp_B (A_1 + A_2)x_0$. Therefore, $T \perp_B (A_1 + A_2)$. So, from Theorem 2.3.2, it follows that T is smooth. □

On the other hand, we prove that the following conditions are necessary for smoothness of a bounded linear operator T, which attains its norm. Note that a part of the following theorem was proved in [126, Theorem 3.3].

Theorem 6.4.5 *Let $\mathbb{X}, \mathbb{Y}$ be Banach spaces. Let $T \in \mathbb{L}(\mathbb{X}, \mathbb{Y})$ be a smooth point such that $M_T \neq \emptyset$. Then*
(i) *$M_T = \{\alpha x_0 : |\alpha| = 1\}$, for some $x_0 \in S_{\mathbb{X}}$.*
(ii) *Tx_0 is a smooth point of $\mathbb{Y}$.*
(iii) *$\sup\{\|Tx\| : x \in H_\alpha \cap S_{\mathbb{X}}\} < \|T\|$ for all $\alpha \in \Lambda$, where $\{H_\alpha : \alpha \in \Lambda\}$ is the collection of all hyperspaces such that $d(x_0, H_\alpha) > 0$.*

Proof (i) If possible, suppose that $x_1, x_2 \in M_T$, where $\{x_1, x_2\}$ is linearly independent. Let $y_i^* \in J(Tx_i)$ for $i = 1, 2$. Then $y_i^* \otimes x_i(T) = y_i^*(Tx_i) = \|Tx_i\| = \|T\|$, i.e., $y_i^* \otimes x_i \in J(T)$. It is straightforward to check that $\{y_i^* \otimes x_i : i = 1, 2\}$ is linearly independent. This clearly contradicts that T is smooth. Therefore, $M_T = \{\alpha x_0 : |\alpha| = 1\}$, for some $x_0 \in S_{\mathbb{X}}$.
(ii) If possible, suppose that Tx_0 is not smooth. Let y_1^*, y_2^* be two distinct points in $J(Tx_0)$. Then clearly, $y_i^* \otimes x_0$, $(i = 1, 2)$ are two distinct elements in $J(T)$, which contradicts that T is smooth. Therefore, Tx_0 is a smooth point of $\mathbb{Y}$.
(iii) If possible, suppose that $\sup\{\|Tx\| : x \in H_\alpha \cap S_{\mathbb{X}}\} = \|T\|$ for some $\alpha \in \Lambda$. Then there exists a sequence $\{x_n\} \in H_\alpha \cap S_{\mathbb{X}}$ such that $\|Tx_n\| \to \|T\|$. Then each $z \in \mathbb{X}$ can be written as

$$z = ax_0 + h,$$

where a is a scalar and $h \in H_\alpha$. Suppose $d(x_0, H_\alpha) = r > 0$. So $\inf\{\|x_0 + h\| : h \in H_\alpha\} = r$. Then for any scalar $a \neq 0$, $\|x_0 + \frac{1}{a}h\| \geq r \Rightarrow \|z\| = \|ax_0 + h\| \geq r|a| \Rightarrow |a| \leq \frac{\|z\|}{r}$. Define linear operators $A_1, A_2 : \mathbb{X} \to \mathbb{Y}$ as follows:

$$A_1(z) = aTx_0, \quad A_2 z = Th.$$

It is easy to see that both A_1, A_2 are bounded. Now, for each scalar λ, $\|T + \lambda A_1\| \geq \|(T + \lambda A_1)x_n\| = \|Tx_n\| \to \|T\|$. Thus $T \perp_B A_1$. Similarly, for each scalar λ, $\|T + \lambda A_2\| \geq \|(T + \lambda A_2)x_0\| = \|Tx_0 + \lambda A_2 x_0\| = \|Tx_0\| = \|T\|$. Hence $T \perp_B A_2$. But $T = A_1 + A_2$, which shows that $T \not\perp_B (A_1 + A_2)$. This shows that T is not right additive and so from Theorem 2.3.2, it follows that T is not smooth. □

It is apparent that for real Banach spaces, the necessary and sufficient conditions for smoothness, as presented in Theorems 6.4.4 and 6.4.5, respectively, are quite close. A complete characterization of smoothness in this direction still remains elusive. However, using the notion of norming sequence and s.i.p., we have the following characterization of the smoothness of a bounded linear operator between Banach spaces.

Theorem 6.4.6 ([128, Theorem 2.1]) *Let $\mathbb{X}, \mathbb{Y}$ be Banach spaces and let $T \in S_{\mathbb{L}(\mathbb{X},\mathbb{Y})}$. Then the following two conditions are equivalent:*
(i) *T is smooth.*
(ii) *For any $A \in \mathbb{L}(\mathbb{X}, \mathbb{Y})$, $T \perp_B A \Leftrightarrow$ for any norming sequence $\{x_n\}$ for T, any subsequential limit of $\{[Ax_n, Tx_n]\}$ is 0, for any s.i.p. $[\ ,\ ]$ on $\mathbb{Y}$.*

Proof $(i) \Rightarrow (ii)$:
Let $A \in \mathbb{L}(\mathbb{X}, \mathbb{Y})$ be such that for any norming sequence $\{x_n\}$ for T, any subsequential limit of $\{[Ax_n, Tx_n]\}$ be 0, for any s.i.p. $[\ ,\ ]$ on $\mathbb{Y}$. Let $[Ax_{n_k}, Tx_{n_k}] \to 0$ for some subsequence $\{x_{n_k}\}$ of $\{x_n\}$. Then, for any scalar λ,

$$\begin{aligned}\|T+\lambda A\|\|T\| &\geq \|(T+\lambda A)x_{n_k}\|\|Tx_{n_k}\|\\ &\geq |[(T+\lambda A)x_{n_k}, Tx_{n_k}]|\\ &= |[Tx_{n_k}, Tx_{n_k}]+\lambda[Ax_{n_k}, Tx_{n_k}]|\\ &\rightarrow \|T\|^2.\end{aligned}$$

Thus, we have, $\|T+\lambda A\| \geq \|T\|$ for all scalar λ. So, $T\perp_B A$.
Now, let $T\perp_B A$. Suppose $\{x_n\}$ is a norming sequence for T and $[\ ,\ ]$ is any s.i.p. on $\mathbb{Y}$. Since $\{[Ax_n, Tx_n]\}$ is a bounded sequence of scalars, it has a convergent subsequence. We claim that every subsequential limit of $\{[Ax_n, Tx_n]\}$ is 0. On the contrary, suppose that there exists a subsequence $\{x_{n_k}\}$ such that $[Ax_{n_k}, Tx_{n_k}] \rightarrow r \neq 0$. Let us consider the operator $B = T - \frac{\|T\|^2}{r}A$. Clearly, $B \in \mathbb{L}(\mathbb{X}, \mathbb{Y})$. Now,

$$\begin{aligned}[][Bx_{n_k}, Tx_{n_k}] &= [(T-\frac{\|T\|^2}{r}A)x_{n_k}, Tx_{n_k}]\\ &= \|Tx_{n_k}\|^2 - \frac{\|T\|^2}{r}[Ax_{n_k}, Tx_{n_k}]\\ &\rightarrow \|T\|^2 - \frac{\|T\|^2}{r}r = 0.\end{aligned}$$

Thus, we have $[Bx_{n_k}, Tx_{n_k}] \rightarrow 0$. Now, proceeding as before, we can show that $T\perp_B B$. Again, as $T\perp_B A$, $T\perp_B \frac{\|T\|^2}{r}A$. Therefore, following Theorem 2.3.2, we get $T\perp_B B + \frac{\|T\|^2}{r}A$, i.e., $T \perp_B T$, which is a contradiction. Thus, every subsequential limit of $\{[Ax_n, Tx_n]\}$ is 0, in any s.i.p. $[\ ,\ ]$ on $\mathbb{Y}$.
$(ii) \Rightarrow (i)$:
Following Theorem 2.3.2, we know that T is smooth if and only if B-J orthogonality is right additive at T. Let $T\perp_B A$ and $T\perp_B B$ for some $A, B \in \mathbb{L}(\mathbb{X}, \mathbb{Y})$. We want to show that $T\perp_B(A+B)$. Let $\{x_n\}$ be any norming sequence for T and let $[\ ,\ ]$ be any s.i.p on $\mathbb{Y}$. Then $\{[Ax_n, Tx_n]\}$ has a convergent subsequence, say, $\{[Ax_{n_i}, Tx_{n_i}]\}$. So, by the hypothesis $[Ax_{n_i}, Tx_{n_i}] \rightarrow 0$. Also, $\{[Bx_{n_i}, Tx_{n_i}]\}$ has a convergent subsequence, say, $[Bx_{n_{i_k}}, Tx_{n_{i_k}}] \rightarrow 0$. Therefore, $[(A+B)x_{n_{i_k}}, Tx_{n_{i_k}}] = [Ax_{n_{i_k}}, Tx_{n_{i_k}}] + [Bx_{n_{i_k}}, Tx_{n_{i_k}}] \rightarrow 0$. Observe that, for any scalar λ,

$$\begin{aligned}\|T+\lambda(A+B)\|\|T\| &\geq \|(T+\lambda(A+B))x_{n_{i_k}}\|\|Tx_{n_{i_k}}\|\\ &\geq |[(T+\lambda(A+B))x_{n_{i_k}}, Tx_{n_{i_k}}]|\\ &= |[Tx_{n_{i_k}}, Tx_{n_{i_k}}]+\lambda[(A+B)x_{n_{i_k}}, Tx_{n_{i_k}}]|\\ &\rightarrow \|T\|^2.\end{aligned}$$

Therefore, $T\perp_B(A+B)$. Hence, T is smooth. This completes the proof. □

In general there is no guarantee that a bounded linear operator between Banach spaces will attain its norm. However, if $M_T \neq \emptyset$, then a nice characterization of smoothness of $T \in S_{\mathbb{L}(\mathbb{X},\mathbb{Y})}$ was obtained in [128, Theorem 3.3] for real Banach spaces. However, the proof can be suitably modified so that the result also holds for complex Banach spaces.

Theorem 6.4.7 ([128, Theorem 3.3]) *Let* $\mathbb{X}$, $\mathbb{Y}$ *be Banach spaces and let* $T \in S_{\mathbb{L}(\mathbb{X},\mathbb{Y})}$ *be such that* $M_T \neq \phi$. *Then* T *is smooth if and only if the following conditions hold:*
(i) $M_T = \{\alpha x_0 : |\alpha| = 1\}$, *for some* $x_0 \in S_{\mathbb{X}}$.
(ii) Tx_0 *is smooth in* $\mathbb{Y}$.
(iii) *For any* $A \in \mathbb{L}(\mathbb{X}, \mathbb{Y})$, $T \perp_B A \Leftrightarrow Tx_0 \perp_B Ax_0$.

Proof We first prove the necessary part of the theorem. Let $T \in \mathbb{L}(\mathbb{X}, \mathbb{Y})$ be smooth. Then (i) and (ii) follows from Theorem 6.4.5. However, we present here an alternative proof of (ii) using s.i.p.
(ii) Suppose on the contrary, Tx_0 is not a smooth point of $\mathbb{Y}$. Then there exist $f, g \in \mathbb{Y}^*$ such that $f \neq g$, $\|f\| = \|g\| = \|Tx_0\|$ and $f(Tx_0) = g(Tx_0) = \|Tx_0\|^2$. Now, for each $v(\neq \frac{Tx_0}{\|Tx_0\|}) \in S_{\mathbb{Y}}$, choose exactly one $f_v \in S_{\mathbb{Y}^*}$ such that $f_v(v) = \|v\| = 1$ and for $\frac{Tx_0}{\|Tx_0\|}$, choose $f_{\frac{Tx_0}{\|Tx_0\|}} = \frac{1}{\|Tx_0\|} f$. For a scalar λ, choose $f_{\lambda v} = \overline{\lambda} f_v$ for all $v \in S_{\mathbb{Y}}$. Clearly, $f_{Tx_0} = f$. Then the mapping $[\ ,\ ] : \mathbb{Y} \times \mathbb{Y} \longrightarrow \mathcal{K}$ defined by $[u, v] = f_v(u)$ for all $u, v \in \mathbb{Y}$, is a s.i.p. on $\mathbb{Y}$. Moreover, $f(y) = [y, Tx_0]$ for all $y \in \mathbb{Y}$. Clearly $\ker(f) \neq \ker(g)$. So, there exists $y \in \ker(g) \setminus \ker(f)$. Hence, $[y, Tx_0] = f(y) \neq 0$. Then for each scalar λ,

$$\|Tx_0\| \|Tx_0 + \lambda y\| = \|g\| \|Tx_0 + \lambda y\| \geq |g(Tx_0 + \lambda y)| = \|Tx_0\|^2.$$

This implies that $Tx_0 \perp_B y$. Let H be a hyperspace such that $x_0 \perp_B H$. Let $z \in \mathbb{X}$. Then z can be uniquely written as $z = \alpha x_0 + h$, where $h \in H$ and α is a scalar. Define $A : \mathbb{X} \longrightarrow \mathbb{Y}$ by $A(\alpha x_0 + h) = \alpha y$. Then it is easy to verify that $A \in \mathbb{L}(\mathbb{X}, \mathbb{Y})$. Now, for any scalar λ, we have

$$\|T + \lambda A\| \geq \|Tx_0 + \lambda A x_0\| = \|Tx_0 + \lambda y\| \geq \|Tx_0\| = \|T\|,$$

since $Tx_0 \perp_B y$ and $x_0 \in M_T$. Thus, $T \perp_B A$. Therefore, from Theorem 6.4.6, it follows that $[Ax_0, Tx_0] = 0$, since $x_0 \in M_T$. This gives that $[y, Tx_0] = f(y) = 0$, a contradiction. Hence, Tx_0 is smooth.
(iii) One part of the result, namely, $Tx_0 \perp_B Ax_0 \Rightarrow T \perp_B A$ is obvious. We prove the other part. Let $T \perp_B A$. If possible, suppose that $Tx_0 \not\perp_B Ax_0$. Let f be the unique norming linear functional of Tx_0. If $f(Ax_0) = 0$, then for any scalar λ, $\|Tx_0 + \lambda Ax_0\| \geq |f(Tx_0 + \lambda Ax_0)| = |f(Tx_0)| = \|Tx_0\|$. Hence, $Tx_0 \perp_B Ax_0$, and we are done. Therefore, assume $f(Ax_0) \neq 0$. Let $f(Ax_0) = r(\neq 0)$. Now, consider the linear operator $B = T - \frac{\|T\|}{r} A$. Clearly, $B \in \mathbb{L}(\mathbb{X}, \mathbb{Y})$. Then

$$\begin{aligned} f(Bx_0) &= f(Tx_0 - \frac{\|T\|}{r}Ax_0) \\ &= f(Tx_0) - \frac{\|T\|}{r}f(Ax_0) \\ &= \|Tx_0\| - \frac{\|T\|}{r}r \\ &= \|T\| - \|T\| = 0. \end{aligned}$$

Therefore, $Tx_0 \perp_B Bx_0$, by Theorem 2.1.15. Observe that for each scalar λ, we have

$$\|T + \lambda B\| \geq \|Tx_0 + \lambda Bx_0\| \geq \|Tx_0\| = \|T\|.$$

Hence, $T \perp_B B$. Now, T is smooth, $T \perp_B B$ and $T \perp_B \frac{\|T\|}{r} A$ implies that $T \perp_B (B + \frac{\|T\|}{r} A) = T$, which is a contradiction. Thus, $T \perp_B A \Rightarrow Tx_0 \perp_B Ax_0$. Therefore, we get $T \perp_B A \Leftrightarrow Tx_0 \perp_B Ax_0$.

Next, we prove the sufficient part. Let $A, B \in \mathbb{L}(\mathbb{X}, \mathbb{Y})$ be such that $T \perp_B A$ and $T \perp_B B$. Then by the hypothesis we have $Tx_0 \perp_B Ax_0$ and $Tx_0 \perp_B Bx_0$. Since Tx_0 is smooth, $Tx_0 \perp_B (Ax_0 + Bx_0)$. Thus, we have $T \perp_B (A + B)$, as $x_0 \in M_T$. Therefore, by applying Theorem 2.3.2, we get T is smooth. This completes the proof. □

Next, we provide an easy sufficient condition for smoothness of $T \in \mathbb{L}(\mathbb{X}, \mathbb{Y})$, which works for both real and complex Banach spaces. The real version of the following theorem is proved in [128, Theorem 3.4].

Theorem 6.4.8 *Let $\mathbb{X}$, $\mathbb{Y}$ be Banach spaces. Let $T \in \mathbb{L}(\mathbb{X}, \mathbb{Y})$ be such that the following conditions hold:*
(i) $M_T = \{\alpha x_0 : |\alpha| = 1\}$ *for some* $x_0 \in S_{\mathbb{X}}$.
(ii) Tx_0 *is a smooth point in* $\mathbb{Y}$.
(iii) *For every norming sequence* $\{x_n\}$ *for* T, $\{x_n\}$ *has a convergent subsequence converging to* ax_0, *where* $|a| = 1$.
Then T is smooth.

Proof Let $A \in \mathbb{L}(\mathbb{X}, \mathbb{Y})$ be such that $T \perp_B A$. Then from Theorem 4.5.1, it follows that either (a) or (b) holds:
(a) There exists a norming sequence $\{x_n\}$ for T such that $\|Ax_n\| \to 0$ as $n \to \infty$.
(b) For each $\alpha \in U$, there exist two norming sequences $\{x_n = x_n(\alpha)\}$, $\{y_n = y_n(\alpha)\}$ for T and two sequences of positive real numbers $\{\epsilon_n = \epsilon_n(\alpha)\}$, $\{\delta_n = \delta_n(\alpha)\}$ such that $\epsilon_n \to 0$, $\delta_n \to 0$ as $n \to \infty$, $Ax_n \in (Tx_n)_\alpha^{+\epsilon_n}$ and $Ay_n \in (Ty_n)_\alpha^{-\delta_n}$ for all $n \in \mathbb{N}$.
Suppose that (a) holds. Then $Ax_n \to 0$. From hypothesis, it follows that $\{x_n\}$ has a convergent subsequence converging to ax_0, where $|a| = 1$. Without loss of generality, assume that $x_n \to ax_0$. Therefore, $Ax_n \to aAx_0$. So, $Ax_0 = 0$. Therefore, $Tx_0 \perp_B Ax_0$.
Now, suppose that (b) holds. Choose a scalar λ arbitrarily. Then there exists $\alpha \in U$ and a real t such that $\lambda = t\alpha$. By hypothesis, $\{x_n = x_n(\alpha)\}$ and $\{y_n = y_n(\alpha)\}$ have convergent subsequences converging to ax_0 and bx_0, respectively, where

$|a| = |b| = 1$. Without loss of generality, assume that $x_n \to ax_0$ and $y_n \to bx_0$. If $t \geq 0$, then from $Ax_n \in (Tx_n)_{\alpha}^{+\epsilon_n}$, it follows that

$$\begin{aligned}
&\|Tx_n + t\alpha Ax_n\| \geq \sqrt{1-\epsilon_n^2}\|Tx_n\| \\
\Rightarrow &\lim \|Tx_n + t\alpha Ax_n\| \geq \lim \sqrt{1-\epsilon_n^2} \lim \|Tx_n\| \\
\Rightarrow &\|aTx_0 + t\alpha a Ax_0\| \geq \|aTx_0\| \\
\Rightarrow &\|Tx_0 + t\alpha Ax_0\| \geq \|Tx_0\| \\
\Rightarrow &\|Tx_0 + \lambda Ax_0\| \geq \|Tx_0\|.
\end{aligned}$$

Similarly, if $t \leq 0$, then from $Ay_n \in (Ty_n)^{-\delta_n}$ it follows that $\|Tx_0 + \lambda Ax_0\| = \|Tx_0 + t\alpha Ax_0\| \geq \|Tx_0\|$. Since λ was chosen arbitrarily, we can conclude that $Tx_0 \perp_B Ax_0$. Thus, for any $A \in \mathbb{L}(\mathbb{X}, \mathbb{Y})$, $T \perp_B A \Leftrightarrow Tx_0 \perp_B Ax_0$. Therefore, from Theorem 6.4.7, it follows that T is smooth. □

Before proving our next result that characterizes operator smoothness in a special setting, we recall that a linear projection P on $\mathbb{X}$ is called an $L-$projection if $\|x\| = \|Px\| + \|x - Px\|$ for all $x \in \mathbb{X}$. A closed subspace $J \subseteq \mathbb{X}$ is called an $L-$summand (or $L-$ideal) if it is the range of an $L-$projection. Moreover, note that if P is a contractive projection, then for $x \in range(P)$ and $y \in \ker(P)$, $x \perp_B y$. Now, if for $T \in \mathbb{L}(\mathbb{X}, \mathbb{Y})$, $M_T = \{\alpha x_0 : |\alpha| = 1\}$ and span$\{x_0\}$ is an $L-$ideal of $\mathbb{X}$, then the smoothness of T can be characterized in the following way.

Theorem 6.4.9 ([87, Theorem 4.4]) *Let $\mathbb{X}, \mathbb{Y}$ be Banach spaces. Let $T \in \mathbb{L}(\mathbb{X}, \mathbb{Y})$ be such that $M_T = \{\alpha x_0 : |\alpha| = 1\}$, where $span\{x_0\}$ is an $L-$ideal of $\mathbb{X}$. Then T is smooth if and only if the following conditions hold:*
(i) *Tx_0 is smooth in $\mathbb{Y}$.*
(ii) *$\|T\|_{ker(P)} < \|T\|$, where P is the $L-$projection with $range(P) = span\{x_0\}$.*

Proof The proof follows easily from Corollary 5.5.14, Theorems 6.4.5, 6.4.7 and the fact that $x_0 \perp_B \ker(P)$. □

Deeb and Khalil [31] proved that for real ℓ_p spaces, $T \in S_{\mathbb{L}(\ell_p)}$ is smooth if and only if T attains norm only at $\pm x \in S_{\ell_p}$ and $d(T, \mathbb{K}(\ell_p)) < 1$. In [48], Grząślewicz and Younis characterized smooth points of $\mathbb{L}(\ell^p, E)$, where $1 < p < \infty$, E is a Banach space and $\mathbb{K}(\ell^p, E)$ is an $M-$ideal in $\mathbb{L}(\ell^p, E)$. In fact, they proved the following theorem.

Theorem 6.4.10 ([48, Theorem 1]) *Suppose $1 < p < \infty$ and E is a Banach space such that $\mathbb{K}(\ell^p, E)$ is an $M-$ideal in $\mathbb{L}(\ell^p, E)$. If $T \in \mathbb{L}(\ell^p, E)$ with $\|T\| = 1$ then T is smooth if and only if the following conditions hold:*
(i) *T attains its norm at a unique (up to a constant multiple) unit vector $x \in \ell^p$.*
(ii) *Tx is a smooth point in E.*
(iii) *$dist(T, \mathbb{K}(\ell^p, E)) < 1$.*

In the following theorem, we generalize the necessary part of Theorem 6.4.10, the proof of which follows easily from Theorem 6.4.7.

Theorem 6.4.11 ([87, Theorem 4.5]) *Let $\mathbb{X}$, $\mathbb{Y}$ be Banach spaces. Let $T \in \mathbb{L}(\mathbb{X}, \mathbb{Y})$ be such that $M_T \neq \emptyset$. Suppose that T is smooth. Then the following conditions hold:*
(i) $M_T = \{\alpha x_0 : |\alpha| = 1\}$ *for some* $x_0 \in S_{\mathbb{X}}$.
(ii) Tx_0 *is smooth point in* $\mathbb{Y}$.
(iii) $dist(T, \mathbb{K}(\mathbb{X}, \mathbb{Y})) < \|T\|$.

Grząślewicz and Younis in [48, Lemma 1] proved that the converse of Theorem 6.4.11 is also true if we additionally assume that $\mathbb{X}$ is a reflexive Banach space and $\mathbb{K}(\mathbb{X}, \mathbb{Y})$ is an $M-$ideal in $\mathbb{L}(\mathbb{X}, \mathbb{Y})$. Using B-J orthogonality of operators, we give an alternative proof of [48, Lemma 1]. We would like to mention that the proof of the following theorem is in [87, Theorem 4.6] for real scalars.

Theorem 6.4.12 *Let $\mathbb{X}$ be a reflexive Banach space and let $\mathbb{Y}$ be a Banach space. Let $\mathbb{K}(\mathbb{X}, \mathbb{Y})$ be an $M-$ideal in $\mathbb{L}(\mathbb{X}, \mathbb{Y})$. Let $T \in \mathbb{L}(\mathbb{X}, \mathbb{Y})$ be such that the following conditions hold:*
(i) $M_T = \{\alpha x_0 : |\alpha| = 1\}$ *for some* $x_0 \in S_{\mathbb{X}}$.
(ii) Tx_0 *is smooth point in* $\mathbb{Y}$.
(iii) $dist(T, \mathbb{K}(\mathbb{X}, \mathbb{Y})) < \|T\|$.
Then T is smooth.

Proof Without loss of generality, assume that $\|T\| = 1$. Let $A, B \in \mathbb{L}(\mathbb{X}, \mathbb{Y})$ be such that $T \perp_B A$ and $T \perp_B B$. Then by Theorem 4.7.6, $Tx_0 \perp_B Ax_0$ and $Tx_0 \perp_B Bx_0$. Since Tx_0 is smooth, by Remark 2.3.4, $Tx_0 \perp_B (Ax_0 + Bx_0)$. Thus, $T \perp_B (A + B)$, since $x_0 \in M_T$. Hence, T is smooth. □

Combining Theorems 6.4.11 and 6.4.12, we obtain the following characterization of the smooth points of $\mathbb{L}(\mathbb{X}, \mathbb{Y})$, where $\mathbb{X}$ is a reflexive Banach space, $\mathbb{Y}$ is a Banach space, and $\mathbb{K}(\mathbb{X}, \mathbb{Y})$ is an $M-$ideal in $\mathbb{L}(\mathbb{X}, \mathbb{Y})$.

Theorem 6.4.13 ([87, Theorem 4.7]) *Let $\mathbb{X}$ be a reflexive Banach space and let $\mathbb{Y}$ be a Banach space. Suppose $\mathbb{K}(\mathbb{X}, \mathbb{Y})$ is an $M-$ideal in $\mathbb{L}(\mathbb{X}, \mathbb{Y})$. Let $T \in \mathbb{L}(\mathbb{X}, \mathbb{Y})$ be such that $M_T \neq \emptyset$. Then T is smooth if and only if the following conditions hold:*
(i) $M_T = \{\alpha x_0 : |\alpha| = 1\}$ *for some* $x_0 \in S_{\mathbb{X}}$.
(ii) Tx_0 *is smooth point in* $\mathbb{Y}$.
(iii) $dist(T, \mathbb{K}(\mathbb{X}, \mathbb{Y})) < \|T\|$.

Remark 6.4.14 We would like to remark that if $\mathbb{X}^*$ and $\mathbb{Y}$ are separable then $\mathbb{K}(\mathbb{X}, \mathbb{Y})$ is also separable. A well-known theorem of Mazur asserts that in a separable Banach space smooth points are dense (see [54, p. 171]). On the other hand, Martin [91] established the existence of compact operators that cannot be approximated by norm attaining operators. From these facts it is easy to see that there is a compact operator which is smooth but does not attain its norm. Thus, for smoothness of $T \in \mathbb{L}(\mathbb{X}, \mathbb{Y})$, it is not necessary that $M_T \neq \emptyset$. However, assuming that $M_T \neq \emptyset$, we mention the following open question.

Question. Let $\mathbb{X}$, $\mathbb{Y}$ be Banach spaces and let $T \in \mathbb{L}(\mathbb{X}, \mathbb{Y})$ be such that
(i) $M_T = \{\alpha x_0 : |\alpha| = 1\}$ for some $x_0 \in S_{\mathbb{X}}$.
(ii) Tx_0 is smooth point in $\mathbb{Y}$.
(iii) $dist(T, \mathbb{K}(\mathbb{X}, \mathbb{Y})) < \|T\|$.
Is T necessarily smooth?

6.5 $k-$Smoothness in $\mathbb{L}(\mathbb{H})$

The notion of $k-$smoothness of an element of a Banach space is a relatively new concept but it is geometrically fascinating. In [65] Khalil and Saleh have generalized the notion of smoothness and introduced the notion of $k-$smoothness or multi-smoothness. Following [65], we say that an element $x \in S_{\mathbb{X}}$ is $k-$smooth or the order of smoothness of x is k, if $J(x)$ contains exactly k linearly independent vectors, i.e., if $k = \dim(span\ J(x))$, where $J(x)$ is the set of all supporting linear functionals of x, i.e., $J(x) = \{x^* \in S_{\mathbb{X}^*} : x^*(x) = 1\}$. Of course, smooth points are same as $1-$smooth points. Note that $J(x)$ is a weak*-compact convex subset of $S_{\mathbb{X}^*}$. Following this definition, an operator $T \in \mathbb{L}(\mathbb{X}, \mathbb{Y})$ is said to be $k-$smooth if $k = \dim(span\ J(T))$, i.e., if there exist exactly k linearly independent functionals in $S_{\mathbb{L}(\mathbb{X},\mathbb{Y})^*}$ supporting the operator T. The following figure gives a nice illustration of $1-$smooth, $2-$smooth, and $3-$smooth elements in $\ell_\infty(\mathbb{R}^3)$.

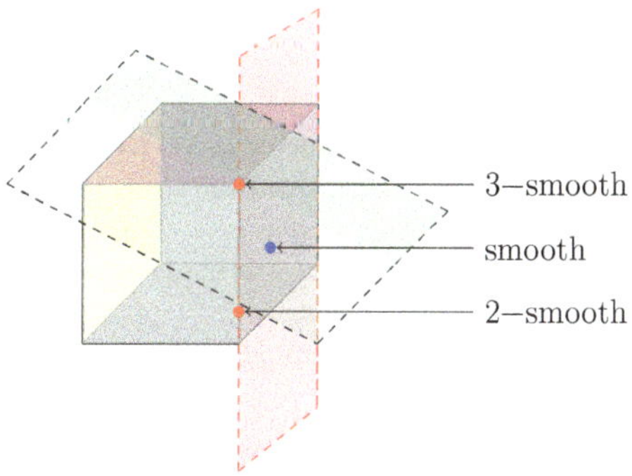

One can visualize concretely that in $\ell_2(\mathbb{R}^3)$, every point of the unit sphere is $1-$smooth.

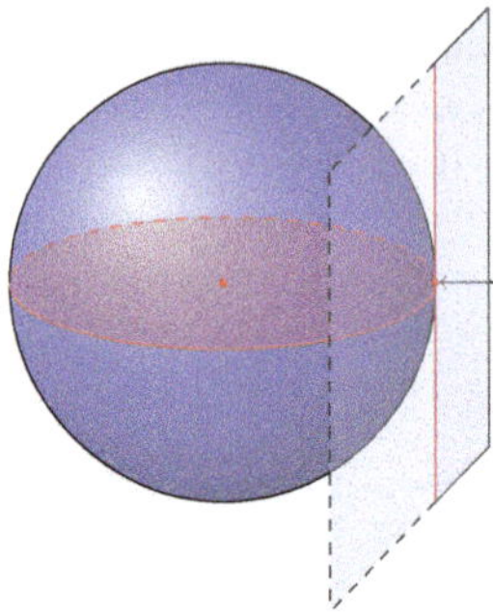

The $k-$smoothness of an element has a nice connection with geometric concepts like extreme points and exposed points. If we look at $\ell_\infty(\mathbb{R}^3)$, then it is easy to see that $(1, 1, 1)$ is an extreme point as well as multi-smooth point of order 3. This is nothing unusual. Khalil and Saleh [65] proved that if $x \in S_{\mathbb{X}}$ is an $n-$smooth point in a Banach space of dimension n, then the point x is an extreme point of the unit ball $B_{\mathbb{X}}$. Wójcik [149] went a step further to prove that if $x \in S_{\mathbb{X}}$ is an $n-$smooth point in a Banach space of dimension n, then the point x is an exposed point of the unit ball $B_{\mathbb{X}}$.

Theorem 6.5.1 ([149, Theorem 4.2]) *Let $\mathbb{X}$ be a Banach space such that* $\dim(\mathbb{X}) = n$. *Suppose that $x \in S_{\mathbb{X}}$ is $n-$smooth. Then x is an exposed point of $B_{\mathbb{X}}$.*

Proof Suppose that x is $n-$smooth and x_i^*, $(1 \leq i \leq n)$ are n linearly independent functionals in $J(x)$. Let $x^* = \sum_{i=1}^{n} \frac{1}{n} x_i^* \in J(x)$. Now, consider the hyperplane $M = \{w \in \mathbb{X} : x^*(w) = 1\}$. We claim that $M \cap S_{\mathbb{X}} = \{x\}$. Clearly, $x \in M \cap S_{\mathbb{X}}$. Suppose $y \in M \cap S_{\mathbb{X}}$. Then $x^*(y) = 1$, i.e., $\sum_{i=1}^{n} \frac{1}{n} x_i^*(y) = 1$. Moreover, $|x_i^*(y)| \leq 1$, i.e., $x_i^*(y) \in B_{\mathbb{K}}$ for all $1 \leq i \leq n$. Observe that $1 \in Ext(B_{\mathbb{K}})$. Therefore, for all $1 \leq i \leq n$, $x_i^*(y) = 1$, i.e., $x_i^*(x) = x_i^*(y)$. Now, from the linear independence of $\{x_i^* : 1 \leq i \leq n\}$ and $\dim(\mathbb{X}^*) = n$, we conclude that $x = y$. This proves that $M \cap S_{\mathbb{X}} = \{x\}$, i.e., x is an exposed point of $B_{\mathbb{X}}$. □

In general, the converse of the above result is not true, e.g., if we consider the space $\ell_2(\mathbb{R}^3)$ then $(1, 0, 0)$ is an extreme point as well as an exposed point of the unit ball but it is not $3-$smooth. However, if we consider a polyhedral Banach space then the following result holds. Recall that a finite-dimensional Banach space $\mathbb{X}$ is said to be polyhedral if $B_{\mathbb{X}}$ contains only finitely many extreme points. In particular, a two-dimensional polyhedral Banach space is said to be a polygonal Banach space.

Theorem 6.5.2 ([85, Theorem 3.5]) *Let $\mathbb{X}$ be an $n-$dimensional polyhedral Banach space. If $x \in S_{\mathbb{X}}$ is an exposed point of $B_{\mathbb{X}}$, then x is $n-$smooth.*

Proof Let $x \in S_{\mathbb{X}}$ be an exposed point of $\mathbb{X}$ and x be $k-$smooth. If possible, assume that $k < n$. Let $\{x_1^*, x_2^*, \ldots, x_k^*\}$ be a linearly independent subset of $Ext\ J(x)$. Observe that $\dim(\ker x_1^* \cap \ker x_2^* \cap \ldots \cap \ker x_k^*) = n - k > 0$. Choose $z \in \cap_{i=1}^{k} \ker x_i^*$. Let $Y = span\{x, z\}$. Then Y is a polygonal Banach space. If possible, let x be $2-$smooth in Y. Then there are linearly independent vectors $y_1^*, y_2^* \in S_{Y^*}$

such that $y_1^*(x) = y_2^*(x) = 1$. Suppose z_1^*, z_2^* are two norm preserving extensions of y_1^* and y_2^*, respectively. Then $z_1^*, z_2^* \in J(x)$. Therefore, $z_1^*, z_2^* \in span\ J(x) = span\ Ext\ J(x)$. Since $x_i^*(z) = 0$ for all $1 \leq i \leq k$, we have $z_1^*(z) = z_2^*(z) = 0$. Thus, $y_1^*(z) = y_2^*(z) = 0$, which contradicts that y_1^*, y_2^* are linearly independent. This implies that x is smooth point in Y. Hence, there exist $x_1, x_2 \in S_Y \subseteq S_{\mathbb{X}}$ such that $x = \frac{1}{2}x_1 + \frac{1}{2}x_2$. Thus, $x \notin Ext(B_{\mathbb{X}})$, i.e., x is not an exposed point of $B_{\mathbb{X}}$. This contradicts the hypothesis of the theorem. Therefore, $k = n$, completing the proof of the theorem. □

Combining this result with Theorem 6.5.1 we get the following characterization of an exposed point.

Theorem 6.5.3 *Let $\mathbb{X}$ be an n−dimensional polyhedral Banach space. Then $x \in S_{\mathbb{X}}$ is an exposed point of the unit ball $B_{\mathbb{X}}$ if and only if x is n−smooth.*

On the other hand, Lin and Rao [79] proved that if a unit vector $x \in \mathbb{X}$ is k−smooth, then there are k linearly independent extreme points in $J(x)$. In fact, they proved the following proposition.

Proposition 6.5.4 ([79, Proposition 2.1]) *A vector $x \in S_{\mathbb{X}}$ is k−smooth if and only if $span\ J(x)$ is k−dimensional. Also the basis vectors can be taken to be the extreme points of $J(x)$.*

Lin and Rao in [79] also studied when a k−smooth point in $\mathbb{X}$ remains to be k−smooth in $\mathbb{X}^{**}$, considering the canonical embedding of $\mathbb{X}$ in $\mathbb{X}^{**}$. Recall that a Banach space $\mathbb{X}$ is said to be Hahn–Banach smooth if each $x^* \in \mathbb{X}^*$ has a unique norm preserving extension in $\mathbb{X}^{***}$. Now, it is easy to check that in a Hahn–Banach smooth Banach space $\mathbb{X}$, a vector $x \in S_{\mathbb{X}}$ is k−smooth in $\mathbb{X}$ if and only if x is k−smooth in $\mathbb{X}^{**}$.

Next, we move to the space of operators. We begin with the study of k−smoothness of operators between Hilbert spaces. The study of multi-smooth points in the space of compact operators was initiated by Khalil and Saleh [65], later on Lin and Rao [79] and Wójcik [149] continued the investigation. It is well known that every element of unit sphere $S_{\mathbb{H}}$ is 1−smooth but the same is not true if we consider the elements of $S_{\mathbb{L}(\mathbb{H})}$. We first prove the following theorem which provides a sufficient condition for k−smoothness of an operator between arbitrary Hilbert spaces.

Theorem 6.5.5 ([83, Theorem 2.1]) *Suppose $\mathbb{H}_1, \mathbb{H}_2$ are Hilbert spaces. Let $T \in S_{\mathbb{L}(\mathbb{H}_1,\mathbb{H}_2)}$ be such that $M_T = S_{H_0}$, where H_0 is a finite-dimensional subspace of $\mathbb{H}_1$ with $\dim(H_0) = n$ and $\|T\|_{H_0^\perp} < 1$. Then T is n^2−smooth, if $\mathbb{H}_1, \mathbb{H}_2$ are considered to be complex and it is $\binom{n+1}{2}$−smooth, if the spaces are considered to be real.*

Proof First we show that $dist(T, \mathbb{K}(\mathbb{H}_1, \mathbb{H}_2)) < 1$. If possible, let this be not true. Then for all $S \in \mathbb{K}(\mathbb{H}_1, \mathbb{H}_2)$, $dist(T, \text{span}\{S\}) \geq dist(T, \mathbb{K}(\mathbb{H}_1, \mathbb{H}_2)) \geq 1$, i.e., for all scalars λ, $\|T - \lambda S\| \geq dist(T, \text{span}\{S\}) \geq 1$. Therefore, $T \perp_B S$. Define an operator $S : \mathbb{H} \to \mathbb{H}$ by $Sx = Tx$ whenever $x \in H_0$ and $Sx = 0$, whenever $x \in H_0^\perp$. Since H_0 is finite dimensional, $S \in \mathbb{K}(\mathbb{H}_1, \mathbb{H}_2)$. Thus, $T \perp_B S$. Using

Theorem 4.2.4, we get $x \in M_T = S_{H_0}$ such that $Tx \perp Sx$, i.e., $Tx \perp Tx$, a contradiction. This proves our claim. Observe that $\mathbb{K}(\mathbb{H}_1, \mathbb{H}_2)$ is an $M-$ideal in $\mathbb{L}(\mathbb{H}_1, \mathbb{H}_2)$ and $dist(T, \mathbb{K}(\mathbb{H}_1, \mathbb{H}_2)) < 1$. Therefore, from Lemma 4.4.3, it follows that $ExtJ(T) = \{y^* \otimes x : x \in M_T, y^* \in ExtJ(Tx)\}$, where $y^* \otimes x(S) = y^*(Sx) = \langle Sx, Tx\rangle$ for each $S \in \mathbb{L}(\mathbb{H}_1, \mathbb{H}_2)$. So we can write $ExtJ(T) = \{x \otimes Tx : x \in M_T\}$, where $x \otimes Tx(S) = \langle Sx, Tx\rangle$ for each $S \in \mathbb{L}(\mathbb{H}_1, \mathbb{H}_2)$.

Now, let $\{e_1, e_2, \ldots, e_n\}$ be an orthonormal basis of H_0. Suppose that $\mathbb{H}_1, \mathbb{H}_2$ are complex Hilbert spaces. Then

$$\begin{aligned}
&\dim(span\ ExtJ(T))\\
&= \dim(span\ \{x \otimes Tx : x \in M_T\})\\
&= \dim\left(span\left\{\sum_{i,j=1}^{n} a_i\overline{a_j}e_i \otimes Te_j : \sum_{i=1}^{n}|a_i|^2 = 1\right\}\right)\\
&= \dim(span\ \{e_i \otimes Te_j : 1 \le i, j \le n\})\\
&= n^2.
\end{aligned}$$

Next, suppose that the Hilbert spaces are real. Then

$$\begin{aligned}
&\dim(span\ ExtJ(T))\\
&= \dim\left(span\left\{\sum_{i,j=1}^{n} a_i a_j e_i \otimes Te_j : \sum_{i=1}^{n}|a_i|^2 = 1\right\}\right)\\
&= \binom{n+1}{2}.
\end{aligned}$$

Therefore, if $\mathbb{H}_1, \mathbb{H}_2$ are complex Hilbert spaces, then T is n^2-smooth, and if the spaces are real Hilbert spaces, then T is $\binom{n+1}{2}-$smooth. □

Note that for a compact operator T, $dist(T, \mathbb{K}(\mathbb{H}_1, \mathbb{H}_2)) = 0 < 1$. Therefore, we have the following corollary providing a sufficient condition for the $k-$smoothness of a compact operator between Hilbert spaces.

Corollary 6.5.6 ([149, Theorem 2.4]) *Suppose $\mathbb{H}_1, \mathbb{H}_2$ are Hilbert spaces. Let $T \in S_{\mathbb{K}(\mathbb{H}_1, \mathbb{H}_2)}$ be such that $M_T = S_{H_0}$, where H_0 is a finite-dimensional subspace of $\mathbb{H}_1$ with* $\dim(H_0) = n$. *Then T is n^2-smooth, if $\mathbb{H}_1, \mathbb{H}_2$ are considered to be complex and it is $\binom{n+1}{2}-$smooth, if the spaces are considered to be real.*

We now show that the conditions mentioned in Theorem 6.5.5 are necessary for $k-$smoothness. To do so, we require the following theorem on B-J orthogonality for operators on a complex Hilbert space. We note that in case of real Hilbert spaces, an analogous result can be obtained from Theorem 4.4.5.

Theorem 6.5.7 ([83, Theorem 2.3]) *Suppose $\mathbb{H}_1, \mathbb{H}_2$ are complex Hilbert spaces. Let $T \in S_{\mathbb{L}(\mathbb{H}_1, \mathbb{H}_2)}$ be such that $dist(T, \mathbb{K}(\mathbb{H}_1, \mathbb{H}_2)) < 1$. Then for $A \in \mathbb{L}(\mathbb{H}_1, \mathbb{H}_2)$, $T \perp_B A$ if and only if there exists $x \in M_T$ such that $Tx \perp Ax$.*

Proof Note that, if there exists $x \in M_T$ such that $Tx \perp Ax$, then clearly $T \perp_B A$. On the other hand, let $T \perp_B A$. Then from Theorem 4.7.1, it follows that there exist $\lambda_i \geq 0$, $f_i \in Ext\ J(T)$ for $1 \leq i \leq 3$ such that $\lambda_1 + \lambda_2 + \lambda_3 = 1$ and $(\lambda_1 f_1 + \lambda_2 f_2 + \lambda_3 f_3)(A) = 0$. Using (4.4.3), we get $x_i \in M_T$, $y_i^* \in Ext\ J(Tx_i)$ for $1 \leq i \leq 3$ such that $f_i = y_i^* \otimes x_i$. Since Tx_i is smooth, it is easy to check that $y_i^* \otimes x_i(A) = \langle Ax_i, Tx_i \rangle$. Thus,

$$\begin{aligned} &\sum_{i=1}^{3} \lambda_i f_i(A) = 0 \\ \Rightarrow &\sum_{i=1}^{3} \lambda_i y_i^* \otimes x_i(A) = 0 \\ \Rightarrow &\sum_{i=1}^{3} \lambda_i \langle Ax_i, Tx_i \rangle = 0. \end{aligned} \tag{6.5.1}$$

Consider the set $W = \{\langle Ax, Tx \rangle : x \in S_{H_0}\} = \{\langle T^*Ax, x \rangle : x \in S_{H_0}\}$. Since H_0 is a subspace of $\mathbb{H}_1$, W is a convex set. Using (6.5.1), we can say that 0 is in the convex hull of W, i.e., $0 \in W$. Therefore, there is $x \in M_T$ such that $\langle Ax, Tx \rangle = 0$ and so $Tx \perp Ax$. This completes the proof of the theorem. □

To prove our desired result we use a theorem proved by Lin and Rao concerning multi-smooth points of quotient space. Recall that, for $1 < p < \infty$, a Banach space $\mathbb{X}$ is said to be an (M_p) space [50, Sect. VI.5] if $\mathbb{K}(\mathbb{X} \oplus_p \mathbb{X})$ is an $M-$ideal in $\mathbb{L}(\mathbb{X} \oplus_p \mathbb{X})$.

Theorem 6.5.8 ([79, Theorem 3.8]) *Let $\mathbb{X}, \mathbb{Y}$ be separable (M_p), (M_q) spaces, respectively. Then $\mathbb{L}(\mathbb{X}, \mathbb{Y})/\mathbb{K}(\mathbb{X}, \mathbb{Y})$ has no smooth point of finite order.*

Observe that for a Hilbert space $\mathbb{H}$, $\mathbb{H} \oplus_2 \mathbb{H} \cong \mathbb{H}$ and $\mathbb{K}(\mathbb{H})$ is an $M-$ideal in $\mathbb{L}(\mathbb{H})$. Therefore, for a separable Hilbert space $\mathbb{H}$, from the last theorem we get that $\mathbb{L}(\mathbb{H})/\mathbb{K}(\mathbb{H})$ has no smooth point of finite order. Note that, we denote the annihilator of a subspace M of a Banach space $\mathbb{X}$ by $M^\dagger$, i.e., $M^\dagger = \{x^* \in \mathbb{X}^* : x^*(m) = 0 \,\forall\, m \in M\}$. In the next theorem, we identify $(\mathbb{X}/M)^*$ with $M^\dagger$, considering the isometry $\psi : M^\dagger \to (\mathbb{X}/M)^*$ defined by $\psi(x^*) = \tilde{x}^*$, where $\tilde{x}^*(x + M) = x^*(m)$ for all $x^* \in M^\dagger$ and $x \in \mathbb{X}$. Now, we are ready to prove our desired theorem.

Theorem 6.5.9 ([83, Theorem 2.4]) *Let $\mathbb{H}$ be a separable complex Hilbert space and let $T \in S_{\mathbb{L}(\mathbb{H})}$. Then T is n^2-smooth if and only if $M_T = S_{H_0}$, where H_0 is a finite-dimensional subspace of $\mathbb{H}$ with $\dim(H_0) = n$ and $\|T\|_{H_0^\perp} < 1$.*

In case the Hilbert space is real, the result holds good if we replace n^2-smoothness by $\binom{n+1}{2}-$smoothness.

Proof First we assume that $\mathbb{H}$ is a complex Hilbert space. The sufficient part follows from Theorem 6.5.5. We only prove the necessary part. Let T is n^2-smooth. Since $\mathbb{H}$ is separable, using Theorem 6.5.8, we can say that $\mathbb{L}(\mathbb{H})/\mathbb{K}(\mathbb{H})$ has no point whose order of smoothness is finite. Now, if possible, assume that $dist(T, \mathbb{K}(\mathbb{H})) = 1$, i.e., $\|T + \mathbb{K}(\mathbb{H})\| = 1$. Note that

$$\begin{aligned} J(T + \mathbb{K}(\mathbb{H})) &= \{\phi \in (\mathbb{L}(\mathbb{H})/\mathbb{K}(\mathbb{H}))^* : \|\phi\| = 1, \phi(T + \mathbb{K}(\mathbb{H})) = \|T + \mathbb{K}(\mathbb{H})\|\} \\ &= \{\phi \in \mathbb{K}(\mathbb{H})^\dagger : \|\phi\| = 1, \phi(T) = 1\} \\ &= \mathbb{K}(\mathbb{H})^\dagger \cap J(T). \end{aligned}$$

Using the above equality and the fact that T is n^2-smooth, we have that $T + \mathbb{K}(\mathbb{H})$ is a multi-smooth point of finite order. This contradiction proves that $dist(T, \mathbb{K}(\mathbb{H})) < 1$. Now, using Lemma 4.4.3, we get $M_T \neq \emptyset$. From Theorem 3.3.5, it follows that $M_T = S_{H_0}$ for some subspace H_0 of $\mathbb{H}$. Now, suppose $A \in \mathbb{L}(\mathbb{H})$ is such that $T \perp_B A$. Then applying Theorem 6.5.7, we get $x \in M_T$ such that $Tx \perp Ax$. Thus, by Theorem 4.2.4, H_0 is a finite-dimensional subspace of $\mathbb{H}$ and $\|T\|_{H_0^\perp} < 1$. Suppose that $\dim(H_0) = k$. Then from Theorem 6.5.5, we have T is k^2-smooth. Therefore, $k^2 = n^2$, i.e., $k = n$. Hence, $\dim(H_0) = n$.
Proceeding similarly and using Theorem 4.4.5, we can prove the theorem for real Hilbert spaces. □

Although a complete characterization of $k-$smoothness of operators between Hilbert spaces has been obtained here, the problem is far more complicated for bounded linear operators between Banach spaces. In the next section we deal with this, by pointing out several attributes of Banach space operators, in light of the concept of $k-$smoothness.

6.6 $k-$Smoothness in $\mathbb{L}(\mathbb{X}, \mathbb{Y})$

We begin this section with the following lemma.

Lemma 6.6.1 ([83, Lemma 3.1]) *Suppose $\mathbb{X}, \mathbb{Y}$ are Banach spaces. If $\{x_1, x_2, \ldots, x_m\}$ and $\{y_1^*, y_2^*, \ldots, y_n^*\}$ are linearly independent subsets of $\mathbb{X}$ and $\mathbb{Y}^*$, respectively, then $\{y_i^* \otimes x_j : 1 \leq i \leq n,\ 1 \leq j \leq m\}$ is a linearly independent subset of $\mathbb{L}(\mathbb{X}, \mathbb{Y})^*$.*

Proof Consider scalars c_{ij} such that

$$\sum_{1 \leq i \leq n, 1 \leq j \leq m} c_{ij} y_i^* \otimes x_j = 0. \tag{6.6.1}$$

Choose $y \in \mathbb{Y}, \phi \in \mathbb{X}^*$. Define $S \in \mathbb{L}(\mathbb{X}, \mathbb{Y})$ by $Sx = \phi(x)y$ for all $x \in \mathbb{X}$. Now, from (6.6.1), we get

$$\sum_{1\leq i\leq n,1\leq j\leq m} c_{ij}y_i^* \otimes x_j(S) = 0$$
$$\Rightarrow \sum_{1\leq i\leq n,1\leq j\leq m} c_{ij}y_i^* S(x_j) = 0$$
$$\Rightarrow \sum_{1\leq i\leq n,1\leq j\leq m} c_{ij}\phi(x_j)y_i^*(y) = 0$$
$$\Rightarrow \phi\Bigg(\sum_{1\leq i\leq n,1\leq j\leq m} c_{ij}x_j y_i^*(y)\Bigg) = 0$$
$$\Rightarrow \sum_{1\leq i\leq n,1\leq j\leq m} c_{ij}x_j y_i^*(y) = 0, \text{ (since } \phi \in \mathbb{X}^* \text{ is arbitrary)}$$
$$\Rightarrow \sum_{1\leq i\leq n} c_{ij}y_i^*(y) = 0 \text{ for all } 1 \leq j \leq m$$
$$\Rightarrow \sum_{1\leq i\leq n} c_{ij}y_i^* = 0 \text{ (since } y \in \mathbb{Y} \text{ is arbitrary)}$$
$$\Rightarrow c_{ij} = 0 \text{ for all } 1 \leq j \leq m, 1 \leq i \leq n.$$

This proves that $\{y_i^* \otimes x_j : 1 \leq i \leq n, 1 \leq j \leq m\}$ is a linearly independent subset of $\mathbb{L}(\mathbb{X}, \mathbb{Y})^*$. □

Now, we provide a sufficient condition for $k-$smoothness of operators between infinite-dimensional Banach spaces. The following theorem was proved in [83, Theorem 3.2] for real scalars.

Theorem 6.6.2 *Suppose $\mathbb{X}$ is a reflexive Banach space and $\mathbb{Y}$ is an arbitrary Banach space. Let $\mathbb{K}(\mathbb{X}, \mathbb{Y})$ be an $M-$ideal in $\mathbb{L}(\mathbb{X}, \mathbb{Y})$. Suppose that $T \in S_{\mathbb{L}(\mathbb{X},\mathbb{Y})}$ is such that $dist(T, \mathbb{K}(\mathbb{X}, \mathbb{Y})) < 1$ and $M_T \cap Ext(B_{\mathbb{X}}) = \{\alpha x_i : |\alpha| = 1, 1 \leq i \leq r\}$, where $\{x_1, x_2, \ldots, x_r\}$ is linearly independent in $\mathbb{X}$. Let Tx_i be m_i-smooth for each $1 \leq i \leq r$. Then T is $k-$smooth, where $m_1 + m_2 + \ldots + m_r = k$.*

Proof Since Tx_i is m_i-smooth, $m_i = \dim(span\ Ext\ J(Tx_i))$, for each $1 \leq i \leq r$. Suppose $\{y_{ij}^* \in Ext\ J(Tx_i) : 1 \leq j \leq m_i\}$ is a basis of $span\ Ext\ J(Tx_i)$ for each $1 \leq i \leq r$. Using similar arguments as in Lemma 6.6.1, we can show that $\{y_{ij}^* \otimes x_i : 1 \leq i \leq r, 1 \leq j \leq m_i\}$ is linearly independent. Now, by Lemma 4.4.3, we get that

$$\begin{aligned} &span\ Ext\ J(T)\\ &= span\ \{y_{ij}^* \otimes x_i : y_{ij}^* \in Ext\ J(Tx_i), 1 \leq i \leq r\}\\ &= span\ \{y_{ij}^* \otimes x_i : 1 \leq i \leq r, 1 \leq j \leq m_i\}. \end{aligned}$$

Thus, $\dim(span\ Ext\ J(T)) = m_1 + m_2 + \ldots + m_r$. Consequently, T is $k-$smooth, where $k = m_1 + m_2 + \ldots + m_r$. This completes the proof. □

Noting that $\mathbb{K}(\ell_p)$ is an $M-$ideal in $\mathbb{L}(\ell_p)$, where $1 < p < \infty$, the following corollary follows easily from Theorem 6.6.2.

Corollary 6.6.3 ([83, Corollary 3.3]) *Let $T \in S_{\mathbb{L}(\ell_p)}$, where $1 < p < \infty$. Suppose that $dist(T, \mathbb{K}(\ell_p)) < 1$ and $M_T = \{\alpha x_i : |\alpha| = 1, 1 \leq i \leq k\}$, where $\{x_1, x_2, \ldots, x_k\}$ is linearly independent in ℓ_p. Then T is $k-$smooth.*

Proof Observe that, ℓ_p is strictly convex and smooth space for $1 < p < \infty$. Therefore, $x_i \in Ext(B_{\ell_p})$ and Tx_i is smooth for each $1 \leq i \leq k$. Thus, from Theorem 6.6.2, it follows that T is $k-$smooth. □

Example 6.6.4 The above result can be used to determine the order of smoothness of operator T on infinite-dimensional $\ell_p(1 < p(\neq 2) < \infty)$ spaces. As for example consider the operator $T \in \mathbb{L}(\ell_4, \ell_4)$ defined by

$$T(a_1, a_2, a_3, a_4, \ldots) = 2^{-\frac{3}{4}}(a_1 + a_2, a_1 - a_2, 0, 0, \ldots).$$

Then it is easy to see that $M_T \cap Ext(B_{\ell_4}) = \Big\{\alpha\Big(\frac{1}{2^{\frac{1}{4}}}, \frac{1}{2^{\frac{1}{4}}}, 0, 0, 0, \ldots\Big), \alpha\Big(-\frac{1}{2^{\frac{1}{4}}}, \frac{1}{2^{\frac{1}{4}}}, 0, 0, \ldots\Big) : |\alpha| = 1\Big\}$. Since the space ℓ_4 and the operator T satisfy all the conditions of Corollary 6.6.3, we can conclude that T is $2-$smooth.

The following example shows that the converse of Corollary 6.6.3 is not true. In other words, there exists $k-$smooth operator $T \in \mathbb{L}(\ell_p)$ such that M_T is not of the form $\{\alpha x_i : |\alpha| = 1, 1 \leq i \leq k\}$, where $\{x_1, x_2, \ldots, x_k\}$ is linearly independent in ℓ_p.

Example 6.6.5 Choose the operator $T \in \mathbb{L}(\ell_p)$, $(1 < p < \infty)$ defined by $T(\sum_i a_i e_i) = a_1 e_1 + a_2 e_2$, where $\{e_i : i \in \mathbb{N}\}$ is the canonical basis of ℓ_p. Then $M_T = span\{e_1, e_2\}$. Consider $x = ae_1 + be_2$, where $a, b \neq 0$. Then $x \in M_T$ and hence by Lemma 4.4.3, we get $e_1^* \otimes e_1, e_2^* \otimes e_2, x^* \otimes x \in Ext\ J(T)$. Now, it is easy to check that $\{e_1^* \otimes e_1, e_2^* \otimes e_2, x^* \otimes x\}$ is linearly independent. Therefore, T is $k-$smooth, where $k \geq 3$. But clearly M_T does not contain 3 linearly independent vectors.

Remark 6.6.6 In [65, Theorem 2.2], Khalil and Saleh claimed that if $1 < p < \infty$ and $T \in S_{\mathbb{K}(\ell_p)}$, then T is a multi-smooth point of order k if and only if T attains its norm at exactly k linearly independent vectors. Unfortunately, this claim is not correct. Lin and Rao in [79, Remark 2.23] and Wójcik in [149, Example 2.5] pointed out the mistake in the proof of [65, Theorem 2.2]. Example 6.6.5 also shows that [65, Theorem 2.2] is not correct.

Observe that if $\mathbb{X}$ is a reflexive Banach space, then $\mathbb{L}(\mathbb{X}, \mathbb{Y}) = \mathbb{K}(\mathbb{X}, \mathbb{Y})$ and $\mathbb{K}(\mathbb{X}, \mathbb{Y})$ is trivially an $M-$ideal in $\mathbb{K}(\mathbb{X}, \mathbb{Y})$. Therefore, as an immediate consequence of Theorem 6.6.2, we obtain the following corollary. The following corollary is proved in [85, Theorem 2.2] for real scalars.

Corollary 6.6.7 *Suppose $\mathbb{X}$ is a finite-dimensional Banach space and $\mathbb{Y}$ is an arbitrary Banach space. Let $T \in S_{\mathbb{L}(\mathbb{X},\mathbb{Y})}$ be such that $M_T \cap Ext(B_{\mathbb{X}}) = \{\alpha x_i : |\alpha| = 1, 1 \leq i \leq r\}$, where $\{x_1, x_2, \ldots, x_r\}$ is linearly independent in $\mathbb{X}$. Then T is $k-$smooth if and only if Tx_i is m_i-smooth for each $1 \leq i \leq r$ and $m_1 + m_2 + \ldots + m_r = k$.*

Now, we can completely characterize the order of smoothness of a linear operator from ℓ_1^n ($n \in \mathbb{N}$) to an arbitrary Banach space.

Corollary 6.6.8 ([85, Corollary 2.3]) *Let $\mathbb{Y}$ be a Banach space and let $T \in \mathbb{L}(\ell_1^n, \mathbb{Y})$, $\|T\| = 1$. Then T is $k-$smooth if and only if $M_T \cap Ext(B_{\ell_1^n}) = \{\alpha x_i : |\alpha| = 1, 1 \leq i \leq r\}$ for some $1 \leq r \leq n$, Tx_i is m_i-smooth for each $1 \leq i \leq r$ and $m_1 + m_2 + \ldots + m_r = k$.*

Proof Note that $B_{\ell_1^n}$ contains only finitely many extreme points, and therefore if $M_T \cap Ext(B_{\ell_1^n}) = \{\alpha x_i : |\alpha| = 1, 1 \leq i \leq r\}$ for some $1 \leq r \leq n$, then $\{x_1, x_2, \ldots, x_r\}$ is always linearly independent set in ℓ_1^n. The rest of the proof follows as a consequence Corollary 6.6.7. □

We next prove the following proposition which gives a nice relation between the order of smoothness of an operator and its adjoint.

Proposition 6.6.9 ([83, Proposition 3.8]) *Let $\mathbb{X}, \mathbb{Y}$ be finite-dimensional Banach spaces and let $T \in S_{\mathbb{L}(\mathbb{X},\mathbb{Y})}$. Then T is $k-$smooth if and only if T^* is $k-$smooth.*

Proof We first prove that $Ext\ J(T) = Ext\ J(T^*)$. Let $y^* \otimes x \in Ext\ J(T)$. Then $x \in M_T \cap Ext(B_{\mathbb{X}})$ and $y^* \in Ext\ J(Tx)$. Now,

$$y^* \otimes x(T) = 1 \Rightarrow y^*(Tx) = 1 \Rightarrow x(T^*y^*) = 1 \Rightarrow \|T^*y^*\| = 1.$$

Thus, $y^* \in M_{T^*}$ and $x \in J(T^*y^*)$. Moreover, $x \in Ext(B_{\mathbb{X}})$ and $Ext\ J(Tx) \subseteq Ext(B_{\mathbb{Y}^*})$. Therefore, $x \in Ext\ J(T^*y^*)$ and $y^* \in M_{T^*} \cap Ext(B_{\mathbb{Y}^*})$. Thus, $y^* \otimes x \in Ext\ J(T^*)$. This proves that $Ext\ J(T) \subseteq Ext\ J(T^*)$. Now, replacing T by T^*, we have $Ext\ J(T^*) \subseteq Ext\ J(T)$. Thus, $Ext J(T) = Ext\ J(T^*)$, i.e., $\dim(span\ Ext\ J(T)) = \dim(span\ Ext\ J(T^*))$. Consequently, T is $k-$smooth if and only if T^* is $k-$smooth. □

Remark 6.6.10 The result holds for infinite-dimensional Banach spaces if we assume that $\mathbb{X}$ is a reflexive Banach space, $\mathbb{K}(\mathbb{X}, \mathbb{Y})$ is an $M-$ideal in $\mathbb{L}(\mathbb{X}, \mathbb{Y})$ and $T \in S_{\mathbb{L}(\mathbb{X},\mathbb{Y})}$ be such that $\text{dist}(T, \mathbb{K}(\mathbb{X}, \mathbb{Y})) < 1$.

As an immediate consequence of Proposition 6.6.9 we get the following corollary.

Corollary 6.6.11 ([83, Corollary 3.9]) *Let $\mathbb{X}$ be a finite-dimensional Banach space. Let $T \in S_{\mathbb{L}(\mathbb{X},\ell_\infty^n)}$. Then T is $k-$smooth if and only if $M_{T^*} \cap Ext(B_{\ell_1^n}) = \{\alpha x_i : |\alpha| = 1, 1 \leq i \leq r\}$ for some $1 \leq r \leq n$, T^*e_i is m_i-smooth for each $1 \leq i \leq r$ and $m_1 + m_2 + \ldots + m_r = k$.*

Proof From Proposition 6.6.9, it follows that T is $k-$smooth if and only if T^* is $k-$smooth. Now, $T^* \in S_{\mathbb{L}(\ell_1^n,\mathbb{X}^*)}$. Moreover, using Corollary 6.6.8, we have that T^* is $k-$smooth if and only if $M_{T^*} \cap Ext(B_{\ell_1^n}) = \{\alpha x_i : |\alpha| = 1, 1 \leq i \leq r\}$ for some $1 \leq r \leq n$, T^*e_i is m_i-smooth for each $1 \leq i \leq r$ and $m_1 + m_2 + \ldots + m_r = k$. This completes the proof of the corollary. □

Next, we provide another sufficient condition for $k-$smoothness of operators.

Theorem 6.6.12 ([83, Theorem 3.6]) *Let $\mathbb{X}$ be a reflexive Banach space and let $\mathbb{Y}$ be a finite-dimensional Banach space with* $\dim(\mathbb{Y}) = m$. *Suppose $T \in S_{\mathbb{L}(\mathbb{X},\mathbb{Y})}$ is such that $\{x_1, x_2, \ldots, x_r\} \subseteq M_T \cap Ext(B_{\mathbb{X}}) \subseteq span\{x_1, x_2, \ldots, x_r\}$, where $\{x_1, x_2, \ldots, x_r\}$ is linearly independent. Let Tx_i be $m-$smooth for $i = 1, 2, \ldots, r$. Then T is $mr-$smooth.*

Proof For each $1 \leq i \leq r,\ Tx_i$ is $m-$smooth. Consider a linearly independent subset $\{y_{ij}^* : 1 \leq j \leq m\}$ of $Ext\ J(Tx_i)$ for each $1 \leq i \leq r$. We first prove that $\dim(span\ J(T)) \leq mr$. Since $\mathbb{Y}$ is finite dimensional, we have $\mathbb{L}(\mathbb{X}, \mathbb{Y}) = \mathbb{K}(\mathbb{X}, \mathbb{Y})$. Hence, T is compact operator and it trivially follows that $\mathbb{K}(\mathbb{X}, \mathbb{Y})$ an $M-$ideal in $\mathbb{L}(\mathbb{X}, \mathbb{Y})$. Thus, by Lemma 4.4.3, we get

$$Ext\ J(T) = \{y^* \otimes x \in \mathbb{K}(\mathbb{X}, \mathbb{Y})^* : x \in M_T \cap Ext(B_{\mathbb{X}}),\ y^* \in Ext\ J(Tx)\}.$$

Let $x \in M_T \cap Ext(B_{\mathbb{X}})$. Then there are scalars $c_i, i = 1, 2, ..., r$ such that $x = c_1x_1 + c_2x_2 + ... + c_rx_r$. Now, assume that $y^* \in Ext\ J(Tx)$. Since $\{y_{1j}^* : 1 \leq j \leq m\}$ is linearly independent, and hence forms a basis of $\mathbb{Y}^*$, we get scalars d_j $(1 \leq j \leq m)$ such that $y^* = \sum_{1\leq j\leq m} d_j y_{1j}^*$. Thus,

$$\begin{aligned} y^* \otimes x &= y^* \otimes (c_1x_1 + c_2x_2 + \cdots + c_rx_r) \\ &= \Big(\sum_{1\leq j\leq m} d_j y_{1j}^*\Big) \otimes \Big(\sum_{1\leq i\leq r} c_i x_i\Big) \\ &= \sum_{1\leq i\leq r, 1\leq j\leq m} c_i d_j y_{1j}^* \otimes x_i \\ &\in span\{y_{1j}^* \otimes x_i : 1 \leq i \leq r, 1 \leq j \leq m\}. \end{aligned}$$

Since $x \in M_T \cap Ext(B_{\mathbb{X}})$ and $y^* \in Ext\ J(Tx)$ are arbitrary, we have $Ext\ J(T) \subseteq span\{y_{1j}^* \otimes x_i : 1 \leq i \leq r, 1 \leq j \leq m\}$. Now,

$$\begin{aligned} &\dim(span\ Ext\ J(T)) \\ &\leq \dim(span\ \{y_{1j}^* \otimes x_i : 1 \leq i \leq r, 1 \leq j \leq m\}) \\ &= mr, \text{ (by Lemma 6.6.1)}. \end{aligned}$$

Thus, T is $k-$smooth, where $k \leq mr$. Note that, $y_{ij}^* \otimes x_i \in Ext\ J(T)$ for all $1 \leq i \leq r, 1 \leq j \leq m$. Now using similar arguments as in Lemma 6.6.1, we can show that $\{y_{ij}^* \otimes x_i : 1 \leq i \leq r, 1 \leq j \leq m\}$ is linearly independent subset of $Ext\ J(T)$. Thus, $\dim(span\ Ext\ J(T)) \geq mr$, i.e., $k \geq mr$. This proves that $k = mr$ and T is $mr-$smooth, completing the proof. □

The following example illustrates the advantage of Theorem 6.6.12, for which $k-$smoothness of the operator cannot be obtained using Corollary 6.6.7 but can be obtained using above theorem.

Example 6.6.13 Let $\mathbb{X} = \ell_\infty(\mathbb{R}^4)$ and let $\mathbb{Y}$ be a two-dimensional real Banach space such that $B_{\mathbb{Y}}$ is the convex hull of $\{\pm(2, 1), \pm(2, -1)\}$. Define $T \in \mathbb{L}(\mathbb{X}, \mathbb{Y})$ by $T(x, y, z, w) = (y + w, x)$. Then

$$M_T \cap Ext(B_{\mathbb{X}}) = \{\pm(1, 1, 1, 1), \pm(1, 1, -1, 1), \pm(-1, 1, 1, 1), \pm(1, -1, 1, -1)\}.$$

Clearly, $\{(1, 1, 1, 1), (1, 1, -1, 1), (-1, 1, 1, 1), (1, -1, 1, -1)\}$ is linearly dependent. Therefore, we cannot find the order of smoothness of T using Corollary 6.6.7. Note that, $M_T \cap Ext(B_{\mathbb{X}}) \subseteq span\,\{\pm(1, 1, -1, 1), \pm(-1, 1, 1, 1), \pm(1, -1, 1, -1)\}$, where $\{(1, 1, -1, 1), (-1, 1, 1, 1), (1, -1, 1, -1)\}$ is linearly independent. Moreover, $T(1, 1, -1, 1)$, $T(-1, 1, 1, 1)$, $T(1, -1, 1, -1)$ are $2-$smooth. Therefore, using Theorem 6.6.12, we can conclude that T is $6-$smooth.

6.7 $k-$Smoothness of Operators Between Special Banach Spaces

In this section, we determine the order of smoothness of an operator $T \in \mathbb{L}(\mathbb{X}, \mathbb{Y})$, when both $\mathbb{X}, \mathbb{Y}$ are two-dimensional real Banach spaces. Note that, from Corollary 6.6.7 it follows that when $|M_T \cap Ext(B_{\mathbb{X}})| = 2$, i.e., $M_T \cap Ext(B_{\mathbb{X}}) = \{x_1, -x_1\}$, then T is smooth if Tx_1 is smooth and T is $2-$smooth if Tx_1 is non-smooth. Moreover, if $|M_T \cap Ext(B_{\mathbb{X}})| = 4$, i.e., $M_T \cap Ext(B_{\mathbb{X}}) = \{x_1, x_2, -x_1, x_2\}$, then once again, using Corollary 6.6.7, we can conclude that T is $2-$smooth when both Tx_1, Tx_2 are smooth, T is $3-$smooth when only one of Tx_1, Tx_2 is smooth and T is $4-$smooth when both Tx_1, Tx_2 are non-smooth. The situation is little bit complicated when $|M_T \cap Ext(B_{\mathbb{X}})| > 4$. In this case, we need further investigation. We separately consider the cases: $|M_T \cap Ext(B_{\mathbb{X}})| = 6$ and $|M_T \cap Ext(B_{\mathbb{X}})| \geq 8$. Let us first consider the case $|M_T \cap Ext(B_{\mathbb{X}})| = 6$.

Theorem 6.7.1 ([85, Theorem 3.1]) *Suppose $\mathbb{X}, \mathbb{Y}$ are two-dimensional real Banach spaces. Let $T \in \mathbb{L}(\mathbb{X}, \mathbb{Y})$, $\|T\| = 1$ and $M_T \cap Ext(B_{\mathbb{X}}) = \{\pm x_1, \pm x_2, \pm x_3\}$. Then the following are true.*
(i) *If Tx_i is smooth for each $1 \leq i \leq 3$, then T is $3-$smooth.*
(ii) *If Tx_1 is not smooth and either Tx_2, Tx_3 interior point of same line segment of unit sphere or $Tx_2, -Tx_3$ are interior point of same line segment of unit sphere, then T is $3-$smooth.*
(iii) *If Tx_1 is not smooth, Tx_2, Tx_3 are not interior point of the same line segment of unit sphere and $Tx_2, -Tx_3$ are not interior point of the same line segment of unit sphere, then T is $4-$smooth.*

Proof Since $\dim(\mathbb{X}) = \dim(\mathbb{Y}) = 2$, T must be $k-$smooth for some $1 \leq k \leq 4$.
(i) Assume that Tx_i is smooth for each $1 \leq i \leq 3$. Then $J(Tx_i)$ is singleton for each $1 \leq i \leq 3$. Our first claim is that Tx_1, Tx_2, Tx_3 cannot have same supporting linear functional. Suppose on the contrary that $J(Tx_i) = \{y^*\}$ for all

$i = 1, 2, 3$. Then $y^*(Tx_1) = y^*(Tx_2) = y^*(Tx_3) = 1$. Hence, for all $t \in [0, 1]$, we have $y^*(tTx_1 + (1-t)Tx_2) = 1 \Rightarrow \|tTx_1 + (1-t)Tx_2\| = 1$, since $\|y^*\| = 1$. Thus, $\|T(tx_1 + (1-t)x_2)\| = 1$ and $\|T\| = 1$ together imply that $\|tx_1 + (1-t)x_2\| = 1$ for all $t \in [0, 1]$. This shows that x_1, x_2 are on same line segment of unit sphere. Similarly, we can show that x_1, x_3 and x_2, x_3 are on same line segment of unit sphere. This contradicts the fact that x_1, x_2, x_3 are distinct points of $Ext(B_{\mathbb{X}})$. Therefore, without loss of generality, we assume that for all $i = 1, 2, 3$, $J(Tx_i) = \{y_i^*\}$ and $y_1^* \neq \pm y_2^*$. Since $\mathbb{X}$ is two-dimensional and x_1, x_2, x_3 are distinct points of $Ext(B_{\mathbb{X}})$, we get $x_3 = \gamma x_1 + \delta x_2$ for some $\gamma(\neq 0), \delta(\neq 0) \in \mathbb{R}$. Clearly, $\{y_1^*, y_2^*\}$ is linearly independent in $\mathbb{Y}^*$, since $y_1^* \neq \pm y_2^*$. Therefore, there are scalars $\alpha, \beta \in \mathbb{R}$, such that $y_3^* = \alpha y_1^* + \beta y_2^*$. Since T is $k-$smooth,

$$
\begin{aligned}
k &= \dim(span\ J(T)) \\
&= \dim(span\ Ext\ J(T)) \\
&= \dim(span\ \{y_i^* \otimes x_i : 1 \leq i \leq 3\}).
\end{aligned}
$$

To conclude (i), we only have to show that $\{y_i^* \otimes x_i : 1 \leq i \leq 3\}$ is linearly independent. Let

$$
\begin{aligned}
& a_1 y_1^* \otimes x_1 + a_2 y_2^* \otimes x_2 + a_3 y_3^* \otimes x_3 = 0, \text{ where } a_1, a_2, a_3 \in \mathbb{R}, \\
\Rightarrow\ & a_1 y_1^* \otimes x_1 + a_2 y_2^* \otimes x_2 + a_3(\alpha y_1^* + \beta y_2^*) \otimes (\gamma x_1 + \delta x_2) = 0 \\
\Rightarrow\ & (a_1 + a_3\alpha\gamma) y_1^* \otimes x_1 + (a_2 + a_3\beta\delta) y_2^* \otimes x_2 + a_3\alpha\delta y_1^* \otimes x_2 + a_3\beta\gamma y_2^* \otimes x_1 = 0.
\end{aligned}
$$

Now, applying Lemma 6.6.1, we get $a_1 + a_3\alpha\gamma = 0,\ a_2 + a_3\beta\delta = 0,\ a_3\alpha\delta = 0$ and $a_3\beta\gamma = 0$. Solving these 4 equations, we have $a_1 = a_2 = a_3 = 0$. Therefore, $\{y_i^* \otimes x_i : 1 \leq i \leq 3\}$ is linearly independent. This proves (i).

(ii) Assume that Tx_1 is not smooth. Without loss of generality, we may assume that Tx_2, Tx_3 are interior points of the same line segment of the unit sphere. Then Tx_2, Tx_3 have same unique supporting linear functional say, z^*, i.e., $J(Tx_2) = J(Tx_3) = \{z^*\}$. Since $\mathbb{Y}$ is two-dimensional and Tx_1 is not smooth, it is easy to check that $Ext\ J(Tx_1) = \{y_1^*, y_2^*\}$ for some linearly independent set $\{y_1^*, y_2^*\}$ of $\mathbb{Y}^*$. Clearly, there are scalars $a, b \in \mathbb{R} \setminus \{0\}$ such that $x_3 = ax_1 + bx_2$ and there are scalars $\alpha, \beta \in \mathbb{R}$ such that $z^* = \alpha y_1^* + \beta y_2^*$. Therefore, $z^* \otimes x_3 = (\alpha y_1^* + \beta y_2^*) \otimes (ax_1 + bx_2) = a\alpha y_1^* \otimes x_1 + a\beta y_2^* \otimes x_1 + bz^* \otimes x_2 \in span\{y_1^* \otimes x_1, y_2^* \otimes x_1, z^* \otimes x_2\}$. Thus,

$$
\begin{aligned}
k &= \dim(span\ Ext\ J(T)) \\
&= \dim(span\ \{y_1^* \otimes x_1, y_2^* \otimes x_1, z^* \otimes x_2, z^* \otimes x_3\}) \\
&= \dim(span\ \{y_1^* \otimes x_1, y_2^* \otimes x_1, z^* \otimes x_2\}).
\end{aligned}
$$

Now, we show that $\{y_1^* \otimes x_1, y_2^* \otimes x_1, z^* \otimes x_2\}$ is linearly independent. Let $a_i \in \mathbb{R}$ $(i = 1, 2, 3)$ be such that $a_1 y_1^* \otimes x_1 + a_2 y_2^* \otimes x_1 + a_3 z^* \otimes x_2 = 0$. Then

$$a_1 y_1^* S(x_1) + a_2 y_2^* S(x_1) + a_3 z^* S(x_2) = 0 \text{ for all } S \in \mathbb{L}(\mathbb{X}, \mathbb{Y}). \qquad (6.7.1)$$

Define $S_1, S_2 \in \mathbb{L}(\mathbb{X}, \mathbb{Y})$ as follows:

$$\begin{aligned} S_1 x_1 &= 0 & \qquad S_2 x_1 &= u_2 \\ S_1 x_2 &= u_1 & \qquad S_2 x_2 &= 0, \end{aligned}$$

where $u_1 \notin \ker(z^*)$ and $u_2 \in \ker(y_1^*) \setminus \ker(y_2^*)$. Now, putting S_1, S_2 in (6.7.1), we get $a_2 = a_3 = 0$. Thus, $a_1 y_1^* \otimes x_1 = 0$. Note that $a_1 = 0$, since $x_1 \neq 0$ and $y_1^* \neq 0$. Therefore, $\{y_1^* \otimes x_1, y_2^* \otimes x_1, z^* \otimes x_2\}$ is a linearly independent subset of $\mathbb{L}(\mathbb{X}, \mathbb{Y})^*$, proving that $k = 3$. Thus, T is $3-$smooth.

(iii) Assume that Tx_1 is not smooth, Tx_2, Tx_3 are not interior points of the same line segment of the unit sphere and $Tx_2, -Tx_3$ are not interior points of the same line segment of the unit sphere. Then there is a linearly independent subset $\{y_{11}^*, y_{12}^*\}$ of $\mathbb{Y}^*$ such that $Ext\ J(Tx_1) = \{y_{11}^*, y_{12}^*\}$. Moreover, there exist $y_2^* \in Ext\ J(Tx_2)$ and $y_3^* \in Ext\ J(Tx_3)$ such that $y_2^* \neq \pm y_3^*$. Now,

$$\begin{aligned} 4 \geq k &= \dim(span\ Ext\ J(T)) \\ &\geq \dim(span\ \{y_{11}^* \otimes x_1, y_{12}^* \otimes x_1, y_2^* \otimes x_2, y_3^* \otimes x_3\}). \end{aligned}$$

As in the previous case, choosing S suitably from $\mathbb{L}(\mathbb{X}, \mathbb{Y})$, we can easily show that $\{y_{11}^* \otimes x_1, y_{12}^* \otimes x_1, y_2^* \otimes x_2, y_3^* \otimes x_3\}$ is a linearly independent subset of $\mathbb{L}(\mathbb{X}, \mathbb{Y})^*$. Therefore $k = 4$, implying that T is $4-$smooth. This completes the proof. □

In the above theorem, in addition to the assumption $|M_T \cap Ext(B_{\mathbb{X}})| = 6$, if we assume the strict convexity of either $\mathbb{X}$ or $\mathbb{Y}$, then the order of smoothness of T can be determined from the following corollary.

Corollary 6.7.2 ([85, Corollary 3.2]) *Suppose $\mathbb{X}, \mathbb{Y}$ are two-dimensional real Banach spaces and either $\mathbb{X}$ or $\mathbb{Y}$ is strictly convex. Let $T \in \mathbb{L}(\mathbb{X}, \mathbb{Y})$ be such that $M_T \cap Ext(B_{\mathbb{X}}) = \{\pm x_1, \pm x_2, \pm x_3\}$. Then T is $3-$smooth if and only if Tx_i is smooth for all $i = 1, 2, 3$, otherwise T is $4-$smooth.*

Proof First assume that $\mathbb{X}$ is strictly convex. It is enough to show that case (ii) of Theorem 6.7.1 does not hold. Suppose on the contrary that Tx_2, Tx_3 are interior points of the same line segment. Then Tx_2, Tx_3 have same supporting linear functional. Then there exists $y^* \in S_{\mathbb{Y}^*}$ such that $y^*(Tx_2) = y^*(Tx_3) = 1$. Hence, for all $t \in [0, 1]$, we get $y^*((1-t)Tx_2 + tTx_3) = 1 \Rightarrow \|(1-t)x_2 + tx_3\| = 1$, contradicting the strict convexity of $\mathbb{X}$. Therefore, case (ii) of Theorem 6.7.1 does not hold. Now, the proof immediately follows from Theorem 6.7.1.
Note that, case (ii) of Theorem 6.7.1 does not arise, if $\mathbb{Y}$ is strictly convex. Therefore, the conclusion follows trivially from Theorem 6.7.1. □

To determine the order of smoothness of an operator T between two-dimensional real Banach spaces $\mathbb{X}$ and $\mathbb{Y}$, the only remaining case to consider is $|M_T \cap Ext(B_{\mathbb{X}})| \geq 8$. We take care of this case in the next theorem.

Theorem 6.7.3 ([85, Theorem 3.3]) *Suppose* $\mathbb{X}$, $\mathbb{Y}$ *are two-dimensional real Banach spaces. Let* $T \in \mathbb{L}(\mathbb{X}, \mathbb{Y})$ *be such that* $|M_T \cap Ext(B_{\mathbb{X}})| \geq 8$. *Then the following are true.*
(i) *If* Tx *is not smooth for some* $x \in M_T \cap Ext(B_{\mathbb{X}})$, *then* T *is* 4*–smooth.*
(ii) *Suppose* Tx *is smooth for each* $x \in M_T \cap Ext(B_{\mathbb{X}})$. *If there exist* $x_i \in M_T \cap Ext(B_{\mathbb{X}})$, $y_i^* \in J(Tx_i)$ *for* $i = 1, 2, 3, 4$ *such that* $x_2 = ax_1 + bx_3$, $x_4 = cx_1 + dx_3$ *and* $y_2^* = \alpha_1 y_1^* + \alpha_2 y_3^*$, $y_4^* = \beta_1 y_1^* + \beta_2 y_3^*$ *with* $\beta_1\alpha_2 ad - \beta_2\alpha_1 bc \neq 0$, *then* T *is* 4*–smooth. Otherwise* T *is* 3*–smooth.*

Proof Since $\dim(\mathbb{X}) = \dim(\mathbb{Y}) = 2$, T must be k–smooth for some $1 \leq k \leq 4$. Since $|M_T \cap Ext(B_{\mathbb{X}})| \geq 8$, we may assume that $\{\pm x_1, \pm x_2, \pm x_3, \pm x_4\} \subseteq M_T \cap Ext(B_{\mathbb{X}})$.
(i) Suppose that Tx_1 is not smooth. Without loss of generality, we may assume that $x_1 = \frac{(1-s)x_2 - sx_4}{\|(1-s)x_2 - sx_4\|}$ and $x_3 = \frac{(1-t)x_2 + tx_4}{\|(1-t)x_2 + tx_4\|}$ for some $s, t \in (0, 1)$. Let $\{y_{11}^*, y_{12}^*\}$ be a linearly independent subset of $Ext\ J(Tx_1)$. Suppose $y_2^* \in Ext\ J(Tx_2)$, $y_4^* \in Ext\ J(Tx_4)$. Note that $y_2^* \neq \pm y_4^*$. For otherwise, if $y_2^* = y_4^*$, then proceeding similarly as in Theorem 6.7.1 (i), we can show that $\|(1-t)x_2 + tx_4\| = 1$ for all $t \in [0, 1]$. This contradicts that $x_3 \in Ext(B_{\mathbb{X}})$. Thus, $y_2^* \neq y_4^*$. Similarly, we can show that $y_2^* \neq -y_4^*$. Thus, y_2^* and y_4^* are linearly independent. Since T is k–smooth, we have

$$\begin{aligned} 4 \geq k &= \dim(span\ Ext\ J(T)) \\ &\geq \dim(span\ \{y_{11}^* \otimes x_1, y_{12}^* \otimes x_1, y_2^* \otimes x_2, y_4^* \otimes x_4\}). \end{aligned}$$

Now we show that $\{y_{11}^* \otimes x_1, y_{12}^* \otimes x_1, y_2^* \otimes x_2, y_4^* \otimes x_4\}$ is linearly independent. Let there are scalars $a, b, c, d \in \mathbb{R}$ such that $ay_{11}^* \otimes x_1 + by_{12}^* \otimes x_1 + cy_2^* \otimes x_2 + dy_4^* \otimes x_4 = 0$. Then

$$ay_{11}^* S(x_1) + by_{12}^* S(x_1) + cy_2^* S(x_2) + dy_4^* S(x_4) = 0 \ \forall\ S \in \mathbb{L}(\mathbb{X}, \mathbb{Y}). \quad (6.7.2)$$

Choose $u_1 \in \ker(y_2^*) \setminus \ker(y_4^*)$ and $u_2 \in \ker(y_4^*) \setminus \ker(y_2^*)$, $u_3 \in \ker(y_{11}^*) \setminus \ker(y_{12}^*)$, $u_4 \in \ker(y_{12}^*) \setminus \ker(y_{11}^*)$. Now, for $1 \leq i \leq 4$, define $S_i \in \mathbb{L}(\mathbb{X}, \mathbb{Y})$ as follows:

$$\begin{array}{llll} S_1x_1 = 0 & S_2x_1 = 0 & S_3x_1 = u_3 & S_4x_1 = u_4 \\ S_1x_2 = u_1 & S_2x_2 = u_2 & S_3x_2 = 0 & S_4x_2 = 0. \end{array}$$

Putting S_1, S_2, S_3, S_4 in (6.7.2), we get $a = b = c = d = 0$. Therefore, $\{y_{11}^* \otimes x_1, y_{12}^* \otimes x_1, y_2^* \otimes x_2, y_4^* \otimes x_4\}$ is linearly independent, proving that $k = 4$. Hence, T is 4–smooth.

(ii) Suppose Tx is smooth for each $x \in M_T \cap Ext(B_{\mathbb{X}})$ and $\beta_1\alpha_2 ad - \beta_2\alpha_1 bc \neq 0$. Clearly $4 \geq k = \dim(span\ Ext\ J(T)) \geq \dim(span\ \{y_i^* \otimes x_i : 1 \leq i \leq 4\})$. We claim that $\{y_i^* \otimes x_i : 1 \leq i \leq 4\}$ is linearly independent.
Let $a_1 y_1^* \otimes x_1 + a_2 y_2^* \otimes x_2 + a_3 y_3^* \otimes x_3 + a_4 y_4^* \otimes x_4 = 0$, where $a_i \in \mathbb{R}$, $1 \leq i \leq 4$. Then

$a_1 y_1^* \otimes x_1 + a_2(\alpha_1 y_1^* + \alpha_2 y_3^*) \otimes (ax_1 + bx_3) + a_3 y_3^* \otimes x_3 + a_4(\beta_1 y_1^* + \beta_2 y_3^*) \otimes (cx_1 + dx_3) = 0.$
$\Rightarrow (a_1 + a_2\alpha_1 a + a_4\beta_1 c) y_1^* \otimes x_1 + (a_2\alpha_1 b + a_4\beta_1 d) y_1^* \otimes x_3 + (a_2\alpha_2 a + a_4\beta_2 c) y_3^* \otimes x_1 + (a_3 + a_2\alpha_2 b + a_4\beta_2 d) y_3^* \otimes x_3 = 0.$

Now, from Lemma 6.6.1, it follows that $\{y_1^* \otimes x_1, y_1^* \otimes x_3, y_3^* \otimes x_1, y_3^* \otimes x_3\}$ is linearly independent. Therefore, we get $a_1 + a_2\alpha_1 a + a_4\beta_1 c = 0, a_2\alpha_1 b + a_4\beta_1 d = 0, a_2\alpha_2 a + a_4\beta_2 c = 0$ and $a_3 + a_2\alpha_2 b + a_4\beta_2 d = 0$. Using $\beta_1\alpha_2 ad - \beta_2\alpha_1 bc \neq 0$ and solving these equations, we have $a_1, a_2, a_3, a_4 = 0$. Therefore, $\dim(span\ \{y_i^* \otimes x_i : 1 \leq i \leq 4\}) = 4 \Rightarrow k = 4$. Thus, T is $4-$smooth.

Now, assume that for each $\{\pm x_i : 1 \leq i \leq 4\} \subseteq M_T \cap Ext(B_{\mathbb{X}})$ and $y_i^* \in J(Tx_i)$ for $i = 1, 2, 3, 4,\ x_2 = ax_1 + bx_3, x_4 = cx_1 + dx_3$ and $y_2^* = \alpha_1 y_1^* + \alpha_2 y_3^*, y_4^* = \beta_1 y_1^* + \beta_2 y_3^* \Rightarrow \beta_1\alpha_2 ad - \beta_2\alpha_1 bc = 0$. Then the set $\{y_i^* \otimes x_i : 1 \leq i \leq 4\}$ is linearly dependent. Hence, $k < 4$. Proceeding similarly as in Theorem 6.7.1 (i) we can show that $\{y_i^* \otimes x_i : 1 \leq i \leq 3\}$ is linearly independent. Therefore, $k = 3$ and so T is $3-$smooth, completing the proof. □

Note that if $\mathbb{X}$ is a two-dimensional real Banach space such that $S_{\mathbb{X}}$ is a polygon with $|Ext(B_{\mathbb{X}})| > 6$, then the identity operator on $\mathbb{X}$ satisfies the hypothesis of Theorem 6.7.3 (i) and so it is $4-$smooth. In the following two examples we show that there exist two-dimensional real Banach spaces $\mathbb{X}, \mathbb{Y}$ and operators $T \in \mathbb{L}(\mathbb{X}, \mathbb{Y})$ such that both the cases of Theorem 6.7.3 (ii) hold.

Example 6.7.4 ([85, Example 3.4]) (i) Suppose $\mathbb{X}$ is a two-dimensional real Banach space such that $S_{\mathbb{X}}$ is a regular octagon with vertices $\pm(1, 0), \pm(\frac{1}{\sqrt{2}}, \frac{1}{\sqrt{2}}), \pm(0, 1), \pm(-\frac{1}{\sqrt{2}}, \frac{1}{\sqrt{2}})$. Define $T \in \mathbb{L}(\mathbb{X}, \mathbb{X})$ as follows:

$$T(1, 0) = \left(\frac{1}{2} + \frac{1}{2\sqrt{2}}, \frac{1}{2\sqrt{2}}\right),\ T(0, 1) = \left(-\frac{1}{2\sqrt{2}}, \frac{1}{2} + \frac{1}{2\sqrt{2}}\right).$$

Observe that

$$M_T \cap Ext(B_{\mathbb{X}}) = \left\{\pm(1, 0), \pm\left(\frac{1}{\sqrt{2}}, \frac{1}{\sqrt{2}}\right), \pm(0, 1), \pm\left(-\frac{1}{\sqrt{2}}, \frac{1}{\sqrt{2}}\right)\right\}.$$

Moreover, Tx is smooth for each $x \in M_T \cap Ext(B_{\mathbb{X}})$. In this case, it can be verified that T is $3-$smooth.

(ii) Consider two-dimensional real Banach spaces $\mathbb{X}, \mathbb{Y}$ such that $S_{\mathbb{X}}$ is a regular octagon with vertices $\pm(1, 0), \pm(\frac{1}{\sqrt{2}}, \frac{1}{\sqrt{2}}), \pm(0, 1), \pm(-\frac{1}{\sqrt{2}}, \frac{1}{\sqrt{2}})$ and $S_{\mathbb{Y}}$ is an irregular octagon with vertices $\pm(1, 0), \pm\left(\frac{17\sqrt{2}-30}{324-234\sqrt{2}}, \frac{35\sqrt{2}-56}{324-234\sqrt{2}}\right), \pm(0, 1), \pm(-\frac{1}{\sqrt{2}}, \frac{1}{\sqrt{2}})$. Define $T \in \mathbb{L}(\mathbb{X}, \mathbb{Y})$ as follows:

$$T(1, 0) = \left(\frac{5\sqrt{2}+4}{12}, \frac{2+3\sqrt{2}}{12}\right),\ T(0, 1) = \left(-\frac{\sqrt{2}}{4}, \frac{2+\sqrt{2}}{4}\right).$$

Check that $M_T \cap Ext(B_{\mathbb{X}}) = \{\pm(1,0), \pm(\frac{1}{\sqrt{2}}, \frac{1}{\sqrt{2}}), \pm(0,1), \pm(-\frac{1}{\sqrt{2}}, \frac{1}{\sqrt{2}})\}$ and Tx is smooth for each $x \in M_T \cap Ext(B_{\mathbb{X}})$. In this case, it can be verified that T is 4–smooth.

Next, we determine the order of smoothness of $T \in \mathbb{L}(\ell_\infty(\mathbb{R}^3), \mathbb{Y})$, where $\dim(\mathbb{Y}) = 2$, depending on $|M_T \cap Ext(B_{\ell_\infty(\mathbb{R}^3)})|$. Note that the order of smoothness of T can be obtained using Corollary 6.6.7, if $|M_T \cap Ext(B_{\ell_\infty(\mathbb{R}^3)})| \leq 6$. Therefore, we have to only consider the case for which $|M_T \cap Ext(B_{\ell_\infty(\mathbb{R}^3)})| = 8$, i.e., $M_T \cap Ext(B_{\ell_\infty(\mathbb{R}^3)}) = \{\pm x_1, \pm x_2, \pm x_3, \pm x_4\}$, where $x_1 = (1,1,1), x_2 = (-1,1,1), x_3 = (-1,-1,1), x_4 = (1,-1,1)$. Note that, for each $1 \leq i \leq 4$, Tx_i is either smooth or 2–smooth. For $x, y \in \mathbb{X}$, suppose that $L[x,y] = \{tx + (1-t)y : 0 \leq t \leq 1\}$ and $L(x,y) = \{tx + (1-t)y : 0 < t < 1\}$. To prove our desired result, we need the following lemma.

Lemma 6.7.5 ([86, Lemma 2.11]) *Let* $\mathbb{X} = \ell_\infty(\mathbb{R}^3)$ *and let* $\mathbb{Y}$ *be a two-dimensional Banach space. Suppose* $T \in S_{\mathbb{L}(\mathbb{X},\mathbb{Y})}$ *is such that* $Rank(T) = 2$ *and* $Ext(B_{\mathbb{X}}) \subseteq M_T$. *Then* $T(B_{\mathbb{X}})$ *is a convex set with* 4 *extreme points.*

Proof Let us consider the facets $\pm G_1$ of $B_{\mathbb{X}}$, where $G_1 = \{(1,y,z) : |y|, |z| \leq 1\}$. Note that, if we assume $x = (1,0,0)$ and $F = \{(0,y,z) : |y|, |z| \leq 1\}$, then

$$G_1 = x + F, \quad -G_1 = -x + F.$$

Clearly, $B_{\mathbb{X}}$ is the convex hull of the sets G_1 and $-G_1$ Therefore, $T(B_{\mathbb{X}})$ is the convex hull of $T(G_1) = Tx + T(F)$ and $T(-G_1) = -Tx + T(F)$. Here, F is the convex hull of $\{\pm(0,1,1), \pm(0,1,-1)\}$. Thus, $T(F)$ is the convex hull of $\{\pm T(0,1,1), \pm T(0,1,-1)\}$. Hence, $T(F)$ must be a symmetric set with at most four extreme points. If $Ext(T(F)) = \{\pm z\}$, then $Tx \neq 0$ and $Ext(T(B_{\mathbb{X}})) = \{\pm Tx \pm z\}$ and we are done.

Now, let $T(F)$ be a symmetric set with exactly four distinct extreme points $\pm T(0,1,1)$ and $\pm T(0,1,-1)$. We denote $T(0,1,1)$ and $T(0,1,-1)$ by y_1 and y_2, respectively. Clearly, $T(B_{\mathbb{X}})$ is the convex hull of $\{\pm Tx \pm y_1, \pm Tx \pm y_2\}$. Since $\dim(\mathbb{Y}) = 2$, and $\{y_1 + y_2, y_1 - y_2\}$ is linearly independent, there are $a, b \in \mathbb{R}$ such that $Tx = a(y_1 + y_2) + b(y_1 - y_2)$. We claim that either $a = 0$ or $b = 0$. Suppose on the contrary that $a \neq 0, b \neq 0$. First assume that $a > 0, b > 0$. Then

$$Tx - y_1 = \frac{2a}{2a+2b+1}(Tx + y_2) + \frac{2b}{2a+2b+1}(Tx - y_2) + \frac{1}{2a+2b+1}(-Tx - y_1).$$

Now $a > 0, b > 0$ imply that $\frac{2a}{2a+2b+1}, \frac{2b}{2a+2b+1}, \frac{1}{2a+2b+1} \in (0,1)$. Moreover, we have $\|Tx + y_2\| = \|Tx - y_2\| = \|Tx + y_1\| = 1$. Since the vectors $Tx + y_2, Tx - y_2, -Tx - y_1$ are not collinear, we get $\|Tx - y_1\| < 1$. Thus $\|T(1,-1,-1)\| < 1$, which contradicts that $Ext(B_{\mathbb{X}}) \subseteq M_T$. Similarly, considering $a < 0, b < 0$ or $a < 0, b > 0$ or $a > 0, b < 0$, we get a contradiction. This proves our claim. Now, we consider the following possible alternatives, whose discussion will end the proof. (i) $a = b = 0$. (ii) $a = 0, b \neq 0$, (iii) $a \neq 0, b = 0$.

(i) Let $a = b = 0$. Then $Tx = 0$. Thus, $T(B_{\mathbb{X}})$ is the convex hull of $\{\pm y_1, \pm y_2\}$. Thus, in this case $Ext(T(B_{\mathbb{X}})) = \{\pm y_1, \pm y_2\}$ and we are done.
(ii) Let $a = 0, b \neq 0$. Then

$$Tx - y_1 - \frac{2b}{2b+1}(Tx - y_2) + \frac{1}{2b+1}(-Tx - y_1)$$

and

$$-Tx - y_2 = \frac{1}{2b+1}(Tx - y_2) + \frac{2b}{2b+1}(-Tx - y_1).$$

Thus, if $a = 0$ and $b > 0$, then $Ext(T(B_{\mathbb{X}})) = \{\pm(Tx - y_2), \pm(Tx + y_1)\}$. Moreover, we can write

$$Tx - y_2 = \frac{2b}{2b-1}(Tx - y_1) + \frac{-1}{2b-1}(-Tx - y_2)$$

and

$$-Tx - y_1 = \frac{-1}{2b-1}(Tx - y_1) + \frac{2b}{2b-1}(-Tx - y_2).$$

Thus, if $a = 0$ and $b < 0$, then $Ext(T(B_{\mathbb{X}})) = \{\pm(Tx - y_1), \pm(Tx + y_2)\}$. Hence, we are done.
(iii) Let $a \neq 0, b = 0$. Then

$$Tx - y_1 = \frac{2a}{2a+1}(Tx + y_2) + \frac{1}{2a+1}(-Tx - y_1)$$

and

$$-Tx + y_2 = \frac{1}{2a+1}(Tx + y_2) + \frac{2a}{2a+1}(-Tx - y_1).$$

Thus, if $a > 0$ and $b = 0$, then $Ext(T(B_{\mathbb{X}})) = \{\pm(Tx + y_2), \pm(Tx + y_1)\}$. Moreover, we can write

$$Tx + y_2 = \frac{2a}{2a-1}(Tx - y_1) + \frac{-1}{2a-1}(-Tx + y_2)$$

and

$$-Tx - y_1 = \frac{-1}{2a-1}(Tx - y_1) + \frac{2a}{2a-1}(-Tx + y_2).$$

Thus, if $a < 0$ and $b = 0$, then $Ext(T(B_{\mathbb{X}})) = \{\pm(Tx - y_1), \pm(Tx - y_2)\}$. Hence, we are done. This completes the proof of the lemma.

□

We can further generalize the previous lemma for $\mathbb{X} = \ell_\infty(\mathbb{R}^n)$. We illustrate this in the next lemma and the subsequent remark.

Lemma 6.7.6 ([86, Lemma 2.12]) *Let* $\mathbb{X} = \ell_\infty(\mathbb{R}^4)$ *and let* $\mathbb{Y}$ *be a two-dimensional Banach space. Suppose* $T \in S_{\mathbb{L}(\mathbb{X},\mathbb{Y})}$ *is such that* $Rank(T) = 2$ *and* $Ext(B_{\mathbb{X}}) \subseteq M_T$. *Then* $T(B_{\mathbb{X}})$ *is a convex set with* 4 *extreme points.*

Proof Let us consider the facets $\pm G_1$ of $B_{\mathbb{X}}$, where $G_1 = \{(1, y, z, w) : |y|, |z|, |w| \leq 1\}$. For $x = (1, 0, 0, 0)$ and $F = \{(0, y, z, w) : |y|, |z|, |w| \leq 1\}$,

$$G_1 = x + F, \quad -G_1 = -x + F.$$

Note that there are $x_i \in F(1 \leq i \leq 4)$ such that $\pm x + x_i \in Ext(B_{\mathbb{X}})$ and F is the convex hull of $\{\pm x_i : 1 \leq i \leq 4\}$. Clearly, $B_{\mathbb{X}}$ is the convex hull of the sets $G_1 = x + F$ and $-G_1 = -x + F$. Therefore, $T(B_{\mathbb{X}})$ is the convex hull of $T(x + F) = Tx + T(F)$ and $T(-x + F) = -Tx + T(F)$. Here, F is a symmetric cube about the origin and $|Ext(T(F))| \leq 8$.

If possible, suppose that either $|Ext(T(F))| = 8$ or $|Ext(T(F))| = 6$ and the two remaining points (say $\pm y_4$, where $y_j = Tx_j$, for $j = 1, 2, 3, 4$) belong to the boundary of $T(F)$. Suppose $\mathbb{Y}_0$ is the space $\mathbb{Y}$ endowed with a norm whose unit ball is $T(F)$. Then the operator $T_0 : \ell_\infty^3 \to \mathbb{Y}_0$ given by $T_0(t_1, t_2, t_3) = T(0, t_1, t_2, t_3)$ satisfies the hypothesis of Lemma 6.7.5. However, $T_0(B_{\ell_\infty^3})$ is not a convex set with four extreme points. Since this is not possible, we must have $|Ext(T(F))| \leq 6$ and if the equality holds, the remaining two points (say, as before, $\pm y_4$) necessarily belong to the interior of $T(F)$. However, the latter implies that $Tx + y_4$ belongs to the interior of $B_{\mathbb{Y}}$, which is a contradiction to the hypothesis of the lemma. Therefore, $|Ext(T(F))| \leq 4$. Now, we can argue similarly as in Lemma 6.7.5 and show that $|Ext(T(B_{\mathbb{X}})| = 4$. This completes the proof of the lemma. □

Remark 6.7.7 ([86, Remark 2.13]) Observe that following the same line of arguments we can show that Lemma 6.7.6 holds for $\mathbb{X} = \ell_\infty(\mathbb{R}^n)$, i.e., if $T \in S_{\mathbb{L}(\mathbb{X},\mathbb{Y})}$ with $Rank(T) = 2$ and $Ext(B_{\mathbb{X}}) \subseteq M_T$, then $T(B_{\mathbb{X}})$ is a convex set with 4 extreme points.

Now, we are in a position to determine the order of smoothness of $T \in \mathbb{L}(\ell_\infty(\mathbb{R}^3), \mathbb{Y})$ when $|M_T \cap Ext(B_{\ell_\infty(\mathbb{R}^3)})| = 8$.

Theorem 6.7.8 ([83, Theorem 3.10]) *Let* $\mathbb{X} = \ell_\infty(\mathbb{R}^3)$ *and let* $\mathbb{Y}$ *be a two-dimensional real Banach space. Suppose* $T \in S_{\mathbb{L}(\mathbb{X},\mathbb{Y})}$ *is such that* $M_T \cap Ext(B_{\mathbb{X}}) = \{\pm x_1, \pm x_2, \pm x_3, \pm x_4\}$. *Let* $S_1 = \{x_i : 1 \leq i \leq 4,\ Tx_i \text{ is smooth}\}$. *Then the following hold:*
(I) *Let* $|S_1| = 4$.
(a) *If either* $Rank(T) = 1$ *or for each* $i, j(1 \leq i \neq j \leq 4)$ *either* Tx_i, Tx_j *or* $Tx_i, -Tx_j$ *belong to the same straight line contained in* $S_{\mathbb{Y}}$, *then* T *is* 3*–smooth.*
(b) *Otherwise,* T *is* 4*–smooth.*
(II) *If* $|S_1| = 3$, *then* T *is* 4*–smooth.*
(III) *If* $|S_1| = 2$, *then* T *is* 5*–smooth.*
(IV) *If* $|S_1| < 2$, *then* $S_1 = \emptyset$ *and* T *is* 6*–smooth.*

Proof Since $\dim(\mathbb{X}) = 3$ and $\dim(\mathbb{Y}) = 2$, T must be $k-$smooth for $1 \leq k \leq 6$.

(I) Suppose $|S_1| = 4$, i.e., Tx_i is smooth for $1 \leq i \leq 4$.

(a) If the given condition is satisfied, then clearly there exists $y^* \in S_{\mathbb{Y}^*}$ such that for all $i(1 \leq i \leq 4)$, $J(Tx_i) = \{y^*\}$ or $\{-y^*\}$. Now,

$$\begin{aligned} &\dim(span\ J(T)) \\ &= \dim(span\ Ext\ J(T)) \\ &= \dim(span\ \{y^* \otimes x_1, y^* \otimes x_2, y^* \otimes x_3, y^* \otimes x_4\}) \\ &= \dim(span\ \{y^* \otimes x_1, y^* \otimes x_2, y^* \otimes x_3\}) \\ &= 3, \text{ (by Lemma 6.6.1)}. \end{aligned}$$

Therefore, T is $3-$smooth.

(b) Suppose the condition (a) is not satisfied. Thus, there exist $1 \leq i, j \leq 4$ such that $Tx_i \neq \pm Tx_j$ and neither Tx_i, Tx_j nor $Tx_i, -Tx_j$ belong to the same straight line contained in $S_{\mathbb{Y}}$. Without loss of generality, we may assume that $i = 1$, $j = 2$. Let $J(Tx_i) = \{y_i^*\}$, $(1 \leq i \leq 4)$. Then it is easy to check that $\{y_1^*, y_2^*\}$ is linearly independent. Let $y_3^* = ay_1^* + by_2^*$ and $y_4^* = cy_1^* + dy_2^*$, where $a, b, c, d \in \mathbb{R}$. Since $\|y_3^*\| = 1$, a and b cannot be zero simultaneously. Similarly, c and d cannot be zero simultaneously. Now, suppose that T is $k-$smooth. Then

$$\begin{aligned} k &= \dim(span\ J(T)) \\ &= \dim(span\ Ext\ J(T)) \\ &= \dim(span\ \{y_1^* \otimes x_1, y_2^* \otimes x_2, y_3^* \otimes x_3, y_4^* \otimes x_4\}). \end{aligned}$$

We show that $\{y_1^* \otimes x_1, y_2^* \otimes x_2, y_3^* \otimes x_3, y_4^* \otimes x_4\}$ is linearly independent. Let $c_i (1 \leq i \leq 4) \in \mathbb{R}$ be such that

$$c_1 y_1^* \otimes x_1 + c_2 y_2^* \otimes x_2 + c_3 y_3^* \otimes x_3 + c_4 y_4^* \otimes x_4 = 0.$$

Then

$c_1 y_1^* \otimes x_1 + c_2 y_2^* \otimes x_2 + c_3(ay_1^* + by_2^*) \otimes x_3 + c_4(cy_1^* + dy_2^*) \otimes (x_1 - x_2 + x_3) = 0.$

$\Rightarrow (c_1 + c_4 c) y_1^* \otimes x_1 + (c_2 - c_4 d) y_2^* \otimes x_2 + (c_3 a + c_4 c) y_1^* \otimes x_3 + (c_3 b + c_4 d) y_2^* \otimes x_3 - c_4 c y_1^* \otimes x_2 + c_4 d y_2^* \otimes x_1 = 0.$

Since $\{x_1, x_2, x_3\}$ is linearly independent subset of $\mathbb{X}$ and $\{y_1^*, y_2^*\}$ is linearly independent subset of $\mathbb{Y}^*$, from Lemma 6.6.1 it follows that $\{y_1^* \otimes x_1, y_2^* \otimes x_1, y_1^* \otimes x_2, y_2^* \otimes x_2, y_1^* \otimes x_3, y_2^* \otimes x_3\}$ is linearly independent subset of $\mathbb{L}(\mathbb{X}, \mathbb{Y})^*$. Therefore,

$$c_1 + c_4 c = 0,\ c_2 - c_4 d = 0,\ c_3 a + c_4 c = 0,\ c_3 b + c_4 d = 0,\ c_4 c = 0,\ c_4 d = 0.$$

Now, solving these equations, we get $c_i = 0$ for all $1 \leq i \leq 4$. Hence, $\{y_1^* \otimes x_1, y_2^* \otimes x_2, y_3^* \otimes x_3, y_4^* \otimes x_4\}$ is linearly independent. Thus, $k = 4$, i.e., T is $4-$smooth.

(II) Without loss of generality, we may assume that $S_1 = \{x_2, x_3, x_4\}$, i.e., Tx_2, Tx_3, Tx_4 are smooth points of $S_{\mathbb{Y}}$ and Tx_1 is $2-$smooth point of $S_{\mathbb{Y}}$. Clearly, $Rank(T) = 2$. Now, from Lemma 6.7.5 it follows that $T(B_{\mathbb{X}})$ is a polygon with 4 extreme points. Since Tx_1 is a $2-$smooth point of $S_{\mathbb{Y}}$, from Theorem 6.5.1, we get that $\pm Tx_1$ are exposed points of $B_{\mathbb{X}}$, i.e., $\pm Tx_1 \in Ext(B_{\mathbb{Y}})$. Since $\|T\| = 1$, $T(B_{\mathbb{X}}) \subseteq B_{\mathbb{Y}}$, i.e., $Ext(B_{\mathbb{Y}}) \cap T(B_{\mathbb{X}}) \subseteq Ext(T(B_{\mathbb{X}}))$. Therefore, $\pm Tx_1 \in Ext(T(B_{\mathbb{X}}))$. Suppose that $\pm Tx_2$ are the other two extreme points of the polygon $T(B_{\mathbb{X}})$. Then $L[Tx_2, -Tx_1]$ is an edge of the polygon $T(B_{\mathbb{X}})$. Now, $L[x_2, -x_1] \cap L[x_3, -x_4] = \{(-1, 0, 0)\}$. Therefore, $T(-1, 0, 0) \in L[Tx_2, -Tx_1] \cap L[Tx_3, -Tx_4]$. This implies that $Tx_3, -Tx_4 \in L[Tx_2, -Tx_1]$. Since Tx_3, Tx_4 are smooth points and Tx_1 is $2-$smooth point, $Tx_3 \neq -Tx_1$ and $-Tx_4 \neq -Tx_1$. Now, from $Rank(T) = 2$, we get either $Tx_3 \neq Tx_2$ or $-Tx_4 \neq Tx_2$. Without loss of generality, we assume that $Tx_3 \neq Tx_2$. Then $Tx_3 \in L(Tx_2, -Tx_1)$. Since $\|Tx_3\| = 1$, $L[Tx_2, -Tx_1] \subseteq S_{\mathbb{Y}}$. Now, let $J(Tx_3) = \{y^*\}$. Then it is easy to check that $J(Tx_2) = \{y^*\}$ and $J(Tx_4) = \{-y^*\}$. Since Tx_1 is $2-$smooth, clearly $Ext\ J(Tx_1) = \{y_1^*, y_2^*\}$, where $\{y_1^*, y_2^*\}$ is a linearly independent subset of $\mathbb{Y}^*$. Let $y^* = ay_1^* + by_2^*$. Suppose T is $k-$smooth. Then

$$\begin{aligned}
k &= \dim(span\ J(T)) \\
&= \dim(span\ Ext\ J(T)) \\
&= \dim(span\ \{y_1^* \otimes x_1, y_2^* \otimes x_1, y^* \otimes x_2, y^* \otimes x_3, y^* \otimes x_4\}) \\
&= \dim(span\ \{y_1^* \otimes x_1, y_2^* \otimes x_1, y^* \otimes x_2, y^* \otimes x_3\}).
\end{aligned}$$

Now, using Lemma 6.6.1 we can prove that $\{y_1^* \otimes x_1, y_2^* \otimes x_1, y^* \otimes x_2, y^* \otimes x_3\}$ is a linearly independent subset of $\mathbb{L}(\mathbb{X}, \mathbb{Y})^*$. Hence, $k = 4$, i.e., T is $4-$smooth.

(III) Without loss of generality, we assume that $S_1 = \{x_3, x_4\}$. Then $\pm Tx_3, \pm Tx_4$ are smooth points of $S_{\mathbb{Y}}$ and $\pm Tx_1, \pm Tx_2$ are $2-$smooth points of $S_{\mathbb{Y}}$. Clearly, $Rank(T) = 2$. It follows from Theorem 6.5.1 that $\pm Tx_1, \pm Tx_2$ are exposed points of $B_{\mathbb{X}}$, i.e., $\pm Tx_1, \pm Tx_2 \in Ext(B_{\mathbb{Y}})$. Since $\|T\| = 1$, $T(B_{\mathbb{X}}) \subseteq B_{\mathbb{Y}}$. Thus $Ext(B_{\mathbb{Y}}) \cap T(B_{\mathbb{X}}) \subseteq Ext(T(B_{\mathbb{X}}))$ and so $\pm Tx_1, \pm Tx_2 \in Ext(T(B_{\mathbb{X}}))$. If possible, let $Tx_1 = -Tx_2$. Then from $x_4 = x_1 - x_2 + x_3$, we get $Tx_1 = \frac{Tx_4 - Tx_3}{2}$. Since $Tx_1 \in Ext(B_{\mathbb{Y}})$, we must have $Tx_1 = Tx_4 = -Tx_3$. Thus, $Rank(T) = 1$, a contradiction. Hence $Tx_1 \neq -Tx_2$. First consider that $Tx_1 = Tx_2$. Then from $x_4 = x_1 - x_2 + x_3$, we get $Tx_3 = Tx_4$. Let $J(Tx_3) = J(Tx_4) = \{y^*\}$ and $Ext\ J(Tx_1) = Ext\ J(Tx_2) = \{y_1^*, y_2^*\}$. Then it is easy to observe that T is $5-$smooth. Now, assume that $Tx_1 \neq Tx_2$. Then $\pm Tx_1, \pm Tx_2$ are 4 distinct extreme points of the polygon $T(B_{\mathbb{X}})$ and $L[Tx_2, -Tx_1]$ is an edge of $T(B_{\mathbb{X}})$. Now, as in **(II)**, we can show that $Tx_3, -Tx_4 \in L[Tx_2, -Tx_1]$. Since $Tx_3, -Tx_4$ are smooth points, $Tx_3, -Tx_4 \in L(Tx_2, -Tx_1)$. Using $\|Tx_3\| = 1$, we can prove that $L[Tx_2, -Tx_1] \subseteq S_{\mathbb{Y}}$. Let $J(Tx_4) = \{y^*\}$. Then $J(Tx_3) = \{-y^*\}$. Let $Ext\ J(Tx_1) = \{y_1^*, y_2^*\}$ and $Ext\ J(Tx_2) = \{y_3^*, y_4^*\}$. Clearly, $\{y_1^*, y_2^*\}$ and $\{y_3^*, y_4^*\}$ are linearly independent. Now, suppose T is $k-$smooth. Then

$$
\begin{aligned}
k &= \dim(span\ J(T)) \\
&= \dim(span\ Ext\ J(T)) \\
&= \dim(span\ \{y_1^* \otimes x_1, y_2^* \otimes x_1, y_3^* \otimes x_2, y_4^* \otimes x_2, y^* \otimes x_3, y^* \otimes x_4\}) \\
&= \dim(span\ \{y_1^* \otimes x_1, y_2^* \otimes x_1, y_3^* \otimes x_2, y_4^* \otimes x_2, y^* \otimes x_3\}) \\
&= \dim(span\ \{y_1^* \otimes x_1, y_2^* \otimes x_1, y_1^* \otimes x_2, y_2^* \otimes x_2, y^* \otimes x_3\}) \\
&= 5, \text{ by simple calculation.}
\end{aligned}
$$

Hence, T is $5-$smooth.

(IV) Since $|S_1| < 2$, at least 3 points of $Tx_i,\ 1 \leq i \leq 4$ are $2-$smooth. Without loss of generality, assume that Tx_1, Tx_2, Tx_3 are $2-$smooth. If $Rank(T) = 1$, then it is easy to observe that Tx_4 is $2-$smooth. Let $Rank(T) = 2$. Then it follows from Lemma 6.7.5 that $T(B_{\mathbb{X}})$ is a polygon with 4 extreme points. First let $Ext\ (T(B_{\mathbb{X}})) = \{\pm Tx_1, \pm Tx_2\}$. Then proceeding similarly as in **(III)** we can show that $Tx_3, -Tx_4 \in L[Tx_2, -Tx_1]$. Since Tx_3 is $2-$smooth, we must have either $Tx_3 = Tx_2$ or $Tx_3 = -Tx_1$. If $Tx_3 = Tx_2$, then using $x_4 = x_1 - x_2 + x_3$, we get $Tx_4 = Tx_1$. If $Tx_3 = -Tx_1$, then similarly, we get $Tx_4 = -Tx_2$. In each case, Tx_4 is $2-$smooth. Similarly, considering other cases, we can show that if $|S_1| < 2$, then Tx_i are $2-$smooth for all $1 \leq i \leq 4$, i.e., $S_1 = \emptyset$. It now follows from Theorem 6.6.12 that T is $6-$smooth. This completes the proof of the theorem. □

Then next corollary is an immediate consequence of Theorem 6.7.8.

Corollary 6.7.9 ([83, Corollary 3.11]) *Let* $\mathbb{X} = \ell_\infty(\mathbb{R}^3)$ *and let* $\mathbb{Y}$ *be a two-dimensional real strictly convex and smooth Banach space. Suppose* $T \in S_{\mathbb{L}(\mathbb{X},\mathbb{Y})}$ *and* $M_T \cap Ext(B_{\mathbb{X}}) = \{\pm x_1, \pm x_2, \pm x_3, \pm x_4\}$. *Then the following hold:*
(i) *If* $Rank(T) = 1$ *then* T *is* $3-$*smooth.*
(ii) *If* $Rank(T) = 2$, *then* T *is* $4-$*smooth.*

Proof The proof follows from case **(I)** of Theorem 6.7.8 by observing that $S_{\mathbb{Y}}$ does not contain any non-trivial straight line segment, since $\mathbb{Y}$ is strictly convex and Tx_i is smooth for all $1 \leq i \leq 4$, since $\mathbb{Y}$ is smooth. □

The results in this section depend heavily on the geometries of the underlying real Banach spaces. It is not known whether the results also hold for complex Banach spaces. Therefore, it might be fruitful and interesting to continue the study in the setting of complex Banach spaces. We end this section with the following open question:

Question 2 *Suppose that* $\mathbb{X}$ *and* $\mathbb{Y}$ *are Banach spaces and* $T \in S_{L(\mathbb{X},\mathbb{Y})}$. *Find a tractable necessary and sufficient conditions for* T *to be a multi-smooth point of finite order. Specifically, one can consider the case* $\mathbb{X} = \ell_\infty^n$, $\mathbb{Y} = \ell_1^n$, *where* $n \geq 3$. *We note that analogous questions raised in the setting of other polyhedral Banach spaces remain unexplored. Indeed, we would like to emphasize that it should be possible to unify the strong connection between multi-smoothness of operators and* *the geometries of the corresponding underlying spaces within a single theoretical framework. It might be a good opportunity for the reader to explore this direction of research from a geometric perspective, especially in the setting of finite-dimensional polyhedral Banach spaces.*

Our goal of exploring the geometry of spaces of linear operators between Banach spaces in the light of orthogonality has multiple facets. In addition to the question on smoothness, the issue of symmetry of the B-J orthogonality is also of fundamental importance to our cause and this will be taken up in detail in the next chapter. As the reader has surely appreciated by now, the asymmetric nature of the B-J orthogonality is hugely responsible for the more complicated nature of the Banach space geometry, in contrast to its Hilbert space counterpart. Our approach to the symmetry of B-J orthogonality is essentially local and we will see that it allows us to establish the connection with geometric niceties of the underlying space(s) from the perspective of orthogonality in a more convenient manner.

Chapter 7
Symmetry of the B-J Orthogonality

7.1 Introduction

While working with the usual orthogonality relation in Hilbert spaces, little do we realize that the symmetry of the orthogonality is a precious boon! Indeed, one of the fundamental differences between the usual orthogonality in Hilbert spaces and B-J orthogonality in Banach spaces is that unlike the former one, the later one is, in general, asymmetric. In support of our claim, we may mention the result proved by James [60] and Day [29] which implies that a Banach space having dimension greater than or equal to 3 is a Hilbert space if and only if B-J orthogonality is symmetric in that Banach space. On the other hand, as we will see in this chapter, certain local symmetries of B-J orthogonality may indeed be present in a Banach space whose norm is not induced by an inner product. Motivated by such observations, the notion of symmetric points was introduced in [37, 135]. Although the concept is quite recent, it has received considerable attention, perhaps even more so because of the connections between the geometry of the space and the structure of symmetric points in it. One may see [23], where a new method is proposed to compute the isometry groups of real ℓ_p^n spaces by exploiting the said connection. The content of this chapter is based on some of the recent developments in this new yet greatly advancing area of research taking shape in the geometric theory of Banach spaces. We will see that the concept of symmetric points in Banach spaces allows us to obtain generalized versions of several results in the framework of norms, which were previously known only in the specialized setting of inner products. Moreover, the significance of the existence of such special points is apparent from the fact that symmetric points allow us to have a better understanding of the geometric structure of the concerned Banach space. Starting with the basic geometric theory of symmetric points in a global sense, we obtain a complete characterization of the same, which is also handy from a computational point of view. The duality of the two kinds of symmetries has been discussed in detail, with a special emphasis on its connection with well-known

A. Mal et al., *Birkhoff–James Orthogonality and Geometry of Operator Spaces*, Infosys Science Foundation Series, https://doi.org/10.1007/978-981-99-7111-4_7

geometric attributes of a Banach space such as strict convexity and smoothness. As mentioned in the very beginning of this monograph, we are particularly interested in considering the space of operators between Banach spaces and observing some geometric consequences arising out of such a study on it. Accordingly, the symmetric operators between Banach spaces have been studied in detail, and whenever possible, we have obtained complete characterizations of the same. It is worth mentioning that some of the problems in this area remain open, to the best of our knowledge. It is our hope that the reader will be motivated by such challenges and we opine that further progress in this area will contribute to better understanding of several fundamentally important class of operators, including isometries and extreme contractions.

We recall that an element x in a Banach space $\mathbb{X}$ is said to be left symmetric (right symmetric) if $x \perp_B y$ implies $y \perp_B x$ ($y \perp_B x$ implies $x \perp_B y$) for all $y \in \mathbb{X}$. An element $x \in \mathbb{X}$ is said to be symmetric if x is both left symmetric and right symmetric. In an inner product space, $x \perp_B y$ is equivalent to $\langle x, y \rangle = 0$ and so clearly $x \perp_B y \Leftrightarrow y \perp_B x$. Thus, in an inner product space, every element is both right symmetric and left symmetric. The same does not necessarily hold true, in general, in a Banach space. Consider the elements $x = (1, 1)$ and $y = (1, 0)$ in the Banach space $\ell_\infty^2(\mathbb{R})$. Then $x \perp_B y$ but $y \not\perp_B x$ whereas if we consider $x = (1, 0)$ and $y = (0, 1)$ then $x \perp_B y$ as well as $y \perp_B x$. This is illustrated in the following diagram:

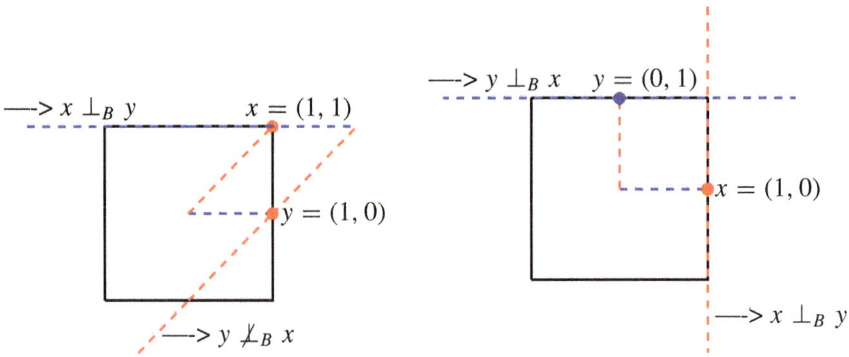

Also, if we consider the space ℓ_p^2, $(1 < p < \infty, p \neq 2)$, then using elementary calculus one can easily verify that $x = (\frac{1}{2^{1/p}}, \frac{1}{2^{1/p}})$ is a non-zero symmetric element. As mentioned before, the symmetry of B-J orthogonality characterizes Hilbert spaces among Banach space if the dimension of the space is strictly greater than 2. However, in the two-dimensional case, there are Banach spaces where B-J orthogonality is symmetric but the norm is not induced by an inner product. A two-dimensional real Banach space where B-J orthogonality is symmetric is also known as a Radon plane. An interesting topic of study in the geometric theory of Banach spaces is the characterization of Radon planes which are not Euclidean. We provide a geometric characterization of polygonal Radon planes. The asymmetric nature of B-J orthogonality makes it interesting to study left symmetric and right symmetric operators in the space of bounded linear operators between Hilbert spaces as well as Banach spaces. We characterize these distinguished operators, first in the framework of Hilbert spaces. Then we investigate them in the setting of some special Banach spaces like ℓ_1^n, ℓ_∞^n, etc. and finally we study them in the setting of general Banach spaces. We consciously follow a style that involves gradual build-up of the concepts, so as to give the reader some idea about the increasing complications in our study. In particular, this illustrates that symmetric points are important objects of study towards understanding the norm structure which can be surprisingly general. The role played by geometric concepts such as strict convexity in determining the symmetric points of operator spaces is certainly a case in point. Throughout the chapter, we assume that the scalars are real unless otherwise specified.

7.2 Characterization of Radon Plane

We recall from [14] that a centrally symmetric closed convex curve γ is said to be a Radon curve if a suitable translate of it can be used as the unit sphere of a normed plane in which the B-J orthogonality is symmetric. We would like to refer the readers to some of the excellent recent studies [14, 92, 93] for more information on various aspects of Radon curves. Let us emphasize here that the background of Minkowski geometry (i.e., the geometry of finite-dimensional real Banach spaces) is affine in nature. In particular, the image of a Radon curve under an affine transformation is again a Radon curve. Authors in [73] proved that every Radon plane is isometrically isomorphic to a special Day-James space generated by a pair of an absolute norm and its dual norm. We here focus on polygonal Radon planes. A two dimensional Banach space $\mathbb{X}$ is said to be polygonal if $B_{\mathbb{X}}$ contains only finitely many extreme points or, equivalently, if $S_{\mathbb{X}}$ is a polygon. We say that $S_{\mathbb{X}}$ is a regular polygon if all the edges of $S_{\mathbb{X}}$ have the same length with respect to the usual metric on $\mathbb{R}^2$ and all interior angles are equal in measure. In order to obtain a complete geometric characterization of Radon planes, we introduce the following definitions.

Definition 7.2.1 Let $\mathbb{X}$ be a polygonal Banach space. Let L be an edge of $S_{\mathbb{X}}$. A functional $f \in S_{\mathbb{X}^*}$ is said to be a supporting functional corresponding to L if the following two conditions are satisfied:
(1) f attains norm at some $v \in L$.
(2) $L = (v + \ker(f)) \cap S_{\mathbb{X}}$.

Definition 7.2.2 (*Translated vertex property of an edge* [90, Definition 1.1]) Let $\mathbb{X}$ be a two-dimensional polygonal Banach space. An edge L of $S_{\mathbb{X}}$ is said to satisfy the translated vertex property (TVP) if $0 \in x + L$ for some $x \in \mathbb{X}$ and the line determined by $x + L$ intersects $S_{\mathbb{X}}$ in two vertices $\pm v$. In that case, $\pm v$ are called the translated vertices of $S_{\mathbb{X}}$ corresponding to the edge L. Equivalently, an edge L satisfies TVP if and only if the kernel of the extreme supporting functional f corresponding to L contains a pair of vertices of $S_{\mathbb{X}}$. The space $\mathbb{X}$ is said to satisfy TVP if every edge of $S_{\mathbb{X}}$ satisfies TVP.

In this section, for $u, v \in \mathbb{X}$, by the symbol $\overline{uv}$, we mean the closed straight line segment joining u and v. Thus, $\overline{uv} = \{(1-t)u + tv : t \in [0, 1]\}$.

Definition 7.2.3 (*Translated edge property of two adjacent edges* [90, Definition 1.2]). Let $\mathbb{X}$ be a two-dimensional polygonal Banach space and let L, M be any two adjacent edges of $S_{\mathbb{X}}$ meeting at $x \in S_{\mathbb{X}}$. We say that the consecutive edges L and M satisfy the translated edge property (TEP) if and only if
(i) Both L and M satisfy TVP with translated vertices $\pm v$ and $\pm w$, respectively.
(ii) Either $\overline{vw}$ or $\overline{v(-w)}$ is an edge with the translated vertices $\pm x$.
The space $\mathbb{X}$ is said to satisfy TEP if every pair of adjacent edges of $S_{\mathbb{X}}$ satisfies the same.

The importance of these two geometric concepts, in our context of characterizing the Radon planes, will become evident in due course of our study. Let us now state two simple observations that relate smooth points and extreme points of the unit ball of a two-dimensional polygonal Banach space and describe the B-J orthogonality set of any unit vector. We omit the proofs; that of the first proposition is rather obvious, and the second one follows directly from Theorem 5.2.7.

Proposition 7.2.4 *Let $\mathbb{X}$ be a two-dimensional polygonal Banach space. Let $x \in S_{\mathbb{X}}$. Then x is an extreme point of $B_{\mathbb{X}}$ if and only if x is not a smooth point.*

Proposition 7.2.5 *Let $\mathbb{X}$ be a two-dimensional Banach space and $x \in S_{\mathbb{X}}$. Then the set $x^{\perp} = \{y \in \mathbb{X} : x \perp_B y\} = K \cup -K$, where K is a normal cone in $\mathbb{X}$.*

Note that a subset K of $\mathbb{X}$ is said to be a normal cone in $\mathbb{X}$ if
(i) $K + K \subset K$, (ii) $\alpha K \subset K$ for all $\alpha \geq 0$, and (iii) $K \cap (-K) = \{\theta\}$.

The following proposition is useful in studying polygonal Radon planes.

Proposition 7.2.6 ([90, Proposition 2.3]) *Let $\mathbb{X}$ be a two-dimensional polygonal Banach space such that $\mathbb{X}$ is a Radon plane. Suppose L is an edge of $S_{\mathbb{X}}$ and v is an extreme point of $S_{\mathbb{X}}$ such that $L \perp_B v$. Then $v \perp_B u$ if and only if $u \in \pm L$.*

Proof Suppose L is an edge of $S_{\mathbb{X}}$ such that $L \perp_B v$, where $v \in Ext(B_{\mathbb{X}})$. Since $\mathbb{X}$ is a Radon plane, $u \in \pm L$ gives that $v \perp_B u$.

Conversely, assume that $L \perp_B v$ and $v \perp_B u$. We want to prove that $u \in \pm L$. Since $L \perp_B v$, there exists a linear functional $f \in S_{\mathbb{X}^*}$ such that $f(x) = 1$ for all $x \in L$ and $f(v) = 0$. Again, $\mathbb{X}$ being a Radon plane, we get $u \perp_B v$. Thus, there exists a linear functional $g \in S_{\mathbb{X}^*}$ such that $g(u) = 1$ and $g(v) = 0$. Hence, $v \in \ker(g)$. We also have $v \in \ker(f)$. Since $\dim(\mathbb{X}) = 2$, we get $\ker(f) = \ker(g)$. This proves that $f = \lambda g$. As $f, g \in S_{\mathbb{X}^*}$, we get $f = \pm g$. Therefore, $f(u) = \pm 1$, and hence $u \in \pm L$. □

The next result illustrates the significance of the extreme points of the unit ball in studying Radon planes.

Theorem 7.2.7 ([90, Theorem 2.4]) *Let $\mathbb{X}$ be a two-dimensional polygonal Banach space. Then $\mathbb{X}$ is a Radon plane if and only if B-J orthogonality is left symmetric at each extreme point of $B_{\mathbb{X}}$.*

Proof Suppose $\mathbb{X}$ is a Radon plane. Hence, B-J orthogonality is symmetric at each point of $S_{\mathbb{X}}$. In particular, B-J orthogonality is left symmetric at each extreme point of $B_{\mathbb{X}}$.

Conversely, assume that B-J orthogonality is left symmetric at each extreme point of $B_{\mathbb{X}}$. Let $x, y \in S_{\mathbb{X}}$ and $x \perp_B y$. We prove that $y \perp_B x$. If $x \in Ext(B_{\mathbb{X}})$, then we are done. In a two-dimensional polygonal space, we know that each point of $S_{\mathbb{X}}$ is either an extreme point of $B_{\mathbb{X}}$ or a smooth point. Hence we consider x to be a smooth point. Note that, x can be written as $x = (1-t)v_1 + tv_2$ for some $v_1, v_2 \in Ext(B_{\mathbb{X}})$ and $t \in (0, 1)$. As $x \perp_B y$, by Theorem 2.1.14 there exists $f \in S_{\mathbb{X}^*}$ such that $f(x) = 1$ and $f(y) = 0$. Thus, $(1-t)f(v_1) + tf(v_2) = f(x) = 1$ implies that $f(v_1) = f(v_2) = 1$. This proves that $v_1 \perp_B y$ and $v_2 \perp_B y$. Again, since B-J orthogonality is left symmetric at each extreme point of $B_{\mathbb{X}}$, we get $y \perp_B v_1$ and $y \perp_B v_2$. Now, from Proposition 7.2.5, it follows that the set $x^{\perp} = \{v \in \mathbb{X} : y \perp_B v\} = K \cup (-K)$, where K is a normal cone in $\mathbb{X}$. Thus, $y \perp_B (1-t)v_1 + tv_2 = x$. Therefore, $x \perp_B y$ implies that $y \perp_B x$. Consequently, B-J orthogonality is symmetric, completing the proof. □

If $\mathbb{X}$ is a two-dimensional polygonal Banach space, then the number of vertices of $S_{\mathbb{X}}$ is either $4n$ or $4n + 2$, for some $n \in \mathbb{N}$. The major aim of the present section is to show that in the first case $\mathbb{X}$ is *never* a Radon plane, while in the second case, $\mathbb{X}$ is a Radon plane provided some additional conditions are satisfied. We begin by proving the following result that addresses our first query.

Theorem 7.2.8 ([90, Theorem 2.5]) *Let $\mathbb{X}$ be a two-dimensional Banach space such that $S_{\mathbb{X}}$ is a polygon with $4n$ vertices, where $n \in \mathbb{N}$. Then $\mathbb{X}$ is not a Radon plane.*

Proof If possible, suppose that $\mathbb{X}$ is a Radon plane. Without any loss of generality, we consider an edge L of $S_{\mathbb{X}}$. Suppose $f \in S_{\mathbb{X}^*}$ is the extreme supporting linear functional corresponding to the edge L. Then $f(L) = 1$ and $f(v) = 0$ for some $v \in S_{\mathbb{X}}$. Clearly, $L \perp_B v$. Since $\mathbb{X}$ is a Radon plane, we get $v \perp_B L$. Now, L being an edge, B-J orthogonality at v is not right unique. Thus, using Remark 2.3.4, we get v is not a smooth point of $\mathbb{X}$. Using Proposition 7.2.4, we get that $v \in Ext(B_{\mathbb{X}})$, and hence v is a vertex of $S_{\mathbb{X}}$. Now, suppose that u is the midpoint of L. Since $v \perp_B u$, there exists a supporting linear functional $g \in S_{\mathbb{X}^*}$ at v such that $g(v) = 1$ and $g(u) = 0$. The situation is illustrated in Fig. 7.1.

Note that, the plane $\mathbb{X}$ is divided into the four closed sets:

$$S_1 = \{x \in \mathbb{X} : f(x) \geq 0, g(x) \geq 0\}, S_2 = \{x \in \mathbb{X} : f(x) \leq 0, g(x) \geq 0\},$$

$$S_3 = \{x \in \mathbb{X} : f(x) \leq 0, g(x) \leq 0\}, S_4 = \{x \in \mathbb{X} : f(x) \geq 0, g(x) \leq 0\}.$$

We observe that $\ker(f)$ contains two vertices of $S_{\mathbb{X}}$, namely, $\pm v$, and $\ker(g)$ contains two smooth points of $S_{\mathbb{X}}$, namely, $\pm u$. We claim that each closed set S_i, $i = 1, 2, 3, 4$

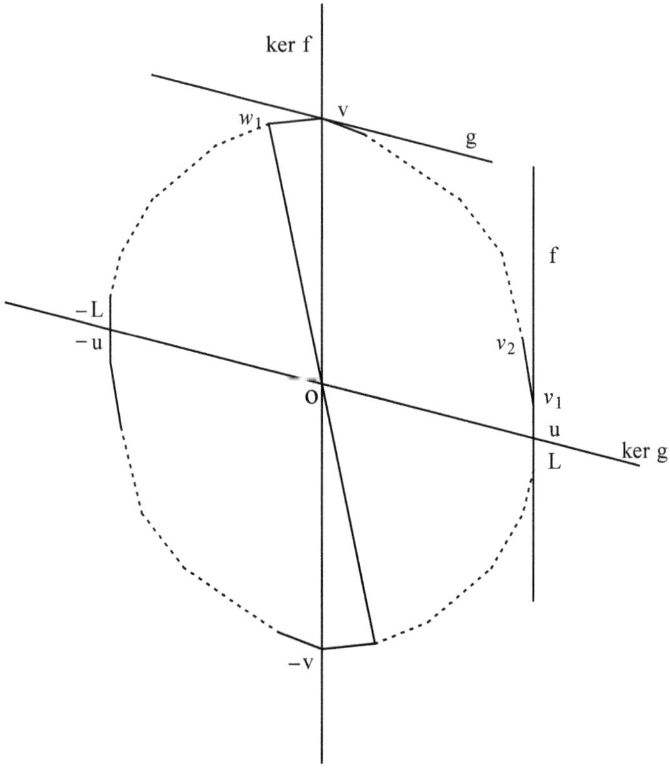

Fig. 7.1 .

(excluding $\ker(f)$), contains exactly the same number of vertices of $S_{\mathbb{X}}$. Suppose k_0 denotes the number of vertices of $S_{\mathbb{X}}$ in S_1 (excluding $\pm v$). To establish our claim, let us consider two adjacent vertices v_1 and v_2 (which may not be equal to v) of $S_{\mathbb{X}}$ in S_1 such that $v_1 \in S_{\mathbb{X}} \cap L$. Since $B_{\mathbb{X}}$ is convex, v_2 can be written as $v_2 = (1-t)v_1 + t(\alpha v)$ for some $\alpha > 0$ and $t \in (0, 1)$. This proves that $f(v_2) < f(v_1)$. Similarly, v_2 can be written as $v_2 = (1-s)v + s(\beta u)$ for some $s \in (0, 1)$ and $\beta > 0$. This proves that $g(v_2) < g(v)$. Proceeding similarly, we can prove that $g(v_1) < g(v_2)$. Now, $\overline{v_1 v_2}$ is an edge of $S_{\mathbb{X}}$ lying in S_1 and $\overline{v_1 v_2} \perp_B v_2 - v_1$. Suppose $w_1 = \frac{v_2 - v_1}{\|v_2 - v_1\|}$. Since $\mathbb{X}$ is a Radon plane, it follows that $w_1 \perp_B \overline{v_1 v_2}$. As before, the B-J orthogonality is not right unique at w_1 and hence w_1 is a vertex of $S_{\mathbb{X}}$. Since $f(w_1) < 0$ and $g(w_1) > 0$, we get that w_1 lies in S_2 (excluding $\ker(f)$ and $\ker(g)$). Continuing this process for every edge in S_1, we conclude that, corresponding to every edge of $S_{\mathbb{X}}$ lying entirely in S_1, there exists a vertex of $S_{\mathbb{X}}$ in S_2 (excluding $\ker(f)$). Thus, the number of vertices of $S_{\mathbb{X}}$ in S_2 (excluding $\ker(f)$) is greater than or equal to k_0. Similarly, corresponding to each edge of $S_{\mathbb{X}}$ lying entirely in S_2, there exists a vertex of $S_{\mathbb{X}}$ in S_1 (excluding $\ker(f)$). Thus, the number of vertices of $S_{\mathbb{X}}$ in S_1 (excluding $\ker(f)$) is greater than or equal to the number of vertices of $S_{\mathbb{X}}$ in S_2

(excluding $\ker(f)$). This shows that $S_{\mathbb{X}}$ has the same number of vertices in S_1 and S_2 (excluding $\ker(f)$). Moreover, by symmetry of $S_{\mathbb{X}}$ about the origin, the number of vertices of $S_{\mathbb{X}}$ in S_1 and S_3 (S_2 and S_4) must be equal. Thus, the number of vertices of $S_{\mathbb{X}}$ in each S_i, for $i = 1, 2, 3, 4$ (excluding $\ker(f)$), is k_0. This completes the proof of our claim. Now, we count the total number of vertices of $S_{\mathbb{X}}$. Clearly, this turns out to be $4k_0 + 2$ (k_0 in each closed set S_i, $i = 1, 2, 3, 4$ except $\ker(f)$ and 2 in $\ker(f)$). However, this contradicts the fact that $S_{\mathbb{X}}$ is a polygon with $4n$ vertices, completing the proof. □

Now we obtain a complete characterization of polygonal Radon planes in terms of the translated edge property (TEP).

Theorem 7.2.9 ([90, Theorem 2.6]) *Let $\mathbb{X}$ be a two-dimensional polygonal Banach space. Then $\mathbb{X}$ is a Radon plane if and only if $\mathbb{X}$ satisfies TEP.*

Proof First we prove the necessary part. Suppose L is an edge of $S_{\mathbb{X}}$. Let the extreme supporting linear functional corresponding to the edge L be $f \in S_{\mathbb{X}^*}$. Then $f(L) = 1$ and for some $v \in S_{\mathbb{X}}$, $f(v) = 0$. Clearly, $L \perp_B v$. Since $\mathbb{X}$ is a Radon plane, we get $v \perp_B L$. Since L is an edge, the B-J orthogonality is not right unique. Therefore, from Remark 2.3.4, it follows that v is not a smooth point of $\mathbb{X}$. Again, from Proposition 7.2.4, it follows that v is an extreme point of $B_{\mathbb{X}}$. Thus, v is a vertex of $S_{\mathbb{X}}$. This proves that each edge satisfies the TVP. Suppose L and M are any two adjacent edges of $S_{\mathbb{X}}$ with a common vertex u. Also, assume that f and g are the extreme supporting functionals corresponding to the edges L and M, respectively. Thus, $f(u) = g(u) = 1$. On the other hand, by the TVP, we have $\ker(f) \cap S_{\mathbb{X}} = \{\pm v\}$ and $\ker(g) \cap S_{\mathbb{X}} = \{\pm w\}$ for some extreme points v and w of $B_{\mathbb{X}}$. The situation is illustrated in Fig. 7.2.

We prove that $\overline{vw}$ or $\overline{v(-w)}$ is an edge of $S_{\mathbb{X}}$. Clearly $L \perp_B v$ and $M \perp_B w$. Since B-J orthogonality is symmetric, we have $v \perp_B L$, $w \perp_B M$. Now u, v, w are vertices with the property that $u \perp_B v$ and $u \perp_B w$. Suppose $y_s = (1-s)v + sw$ and $z_s = (1-s)v - sw$, where $s \in (0, 1)$. We claim that $u \perp_B y_s$ for all $s \in (0, 1)$ or $u \perp_B z_s$ for all $s \in (0, 1)$. From Proposition 7.2.5, it follows that the set $u^{\perp} = \{x \in \mathbb{X} : u \perp_B x\} = K \cup (-K)$, where K is a normal cone in $\mathbb{X}$. Thus, the claim is established. We assume that $u \perp_B y_s$ for all $s \in (0, 1)$. Then $y_s \perp_B u$ for all $s \in (0, 1)$, i.e., $vw \perp_B u$. Clearly, for all $s \in (0, 1)$, $\|y_s\| \leq 1$. If for all $s \in (0, 1)$, $\|y_s\| = 1$, then $\overline{vw}$ is an edge and we are done. If it is possible that $\overline{vw}$ is not an edge, then we can find vertices v_1, w_1, not necessarily unequal, so that $\overline{vv_1}$ and $\overline{ww_1}$ are edges, and proceeding similarly we can conclude that $\overline{vv_1} \perp_B u$, $\overline{ww_1} \perp_B u$. This is not possible, and so $\overline{vw}$ is an edge. This completes the proof of the necessary part.

Now we prove the sufficient part using Theorem 7.2.7. It is sufficient to show that $x \perp_B y$ implies $y \perp_B x$, whenever x is an extreme point of $B_{\mathbb{X}}$ and $y \in S_{\mathbb{X}}$. Suppose L and M are the two edges of $S_{\mathbb{X}}$ containing x. Assume that $f, g \in S_{\mathbb{X}^*}$ are the extreme supporting linear functionals corresponding to the edges L and M such that $\ker(f) \cap S_{\mathbb{X}} = \{\pm v\}$ and $\ker(g) \cap S_{\mathbb{X}} = \{\pm w\}$. Thus either $\overline{vw}$ or $\overline{v(-w)}$ is an edge of $S_{\mathbb{X}}$. Without any loss of generality, we assume that $\overline{vw}$ is an edge of $S_{\mathbb{X}}$. Then, using the TVP, we have $\overline{vw} \perp_B x$. We want to prove that $y \in \overline{vw}$. Following

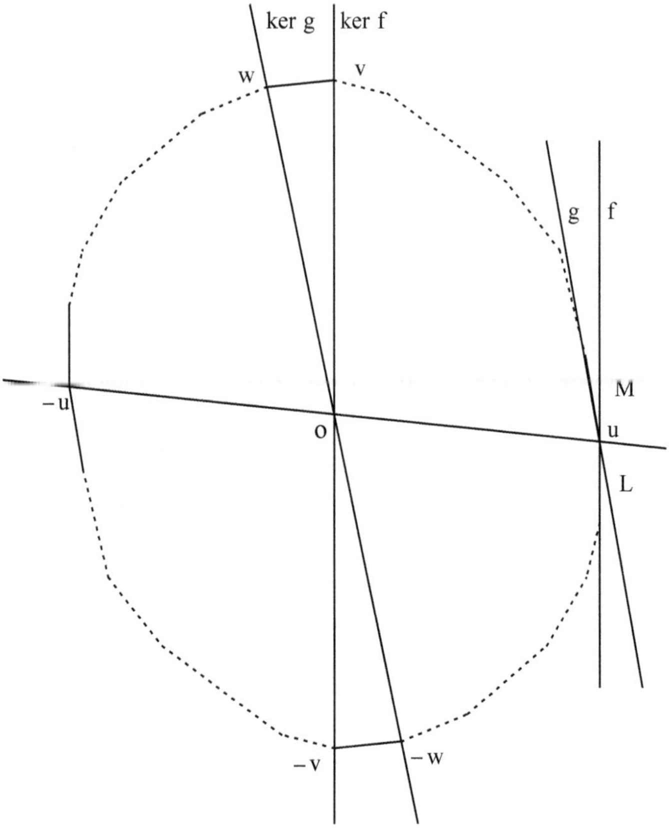

Fig. 7.2 .

the proof of [121, Lemma 2.1] we observe that every supporting linear functional at x is a convex combination of f and g. Since $x \perp_B y$, there is some $t \in [0, 1]$ such that $[(1-t)f + tg](x) = 1$ and $[(1-t)f + tg](y) = 0$. This gives that $f(y)g(y) < 0$. Thus, either $f(y) > 0, g(y) < 0$ or $f(y) < 0, g(y) > 0$. Since $\|y\| = 1$, we can conclude that either $y \in \overline{vw}$ or $-y \in \overline{vw}$. Therefore, we have $y \perp_B x$, completing the proof. □

As an application of the above theorem, it is easy to observe that regular polygons with $4n + 2$ number of vertices, where $n \in \mathbb{N}$, are the unit spheres of Radon planes.

Theorem 7.2.10 ([90, Theorem 2.7]) *Let $\mathbb{X}$ be a two-dimensional Banach space such that $S_{\mathbb{X}}$ is a regular polygon with $4n + 2$ vertices, where $n \in \mathbb{N}$. Then $\mathbb{X}$ is a Radon plane.*

Proof We prove that $\mathbb{X}$ satisfies the TEP. Suppose L and M are two adjacent edges of $S_{\mathbb{X}}$ with common vertex v_1. Then $-L$ is also an edge of $S_{\mathbb{X}}$, parallel to L. We

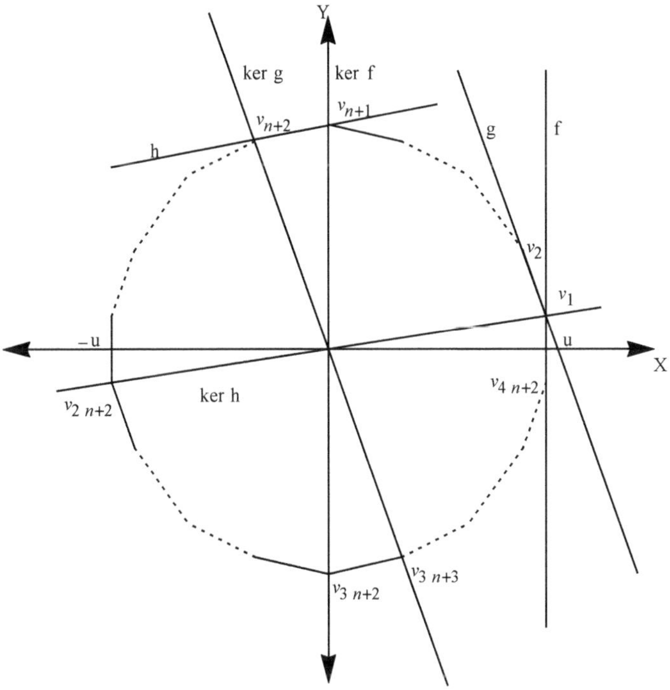

Fig. 7.3 .

choose the X-axis to be the line through the origin that bisects both L and $-L$. Note that, there is no vertex of $S_{\mathbb{X}}$ on the X-axis. Recall that for a regular polygon with $2n$ vertices, the lines of symmetries are the following straight lines:

(i) The straight line passing through the midpoints of any two parallel edges of the polygon.
(ii) The straight line passing through any two opposite vertices v_i and $-v_i$ of the polygon.

We choose the Y-axis to be the straight line through the origin, which is parallel to L and $-L$. Because of regularity of $S_{\mathbb{X}}$ it is easy to check that both the X and Y-axes are lines of symmetry. Suppose the Y-axis meets $S_{\mathbb{X}}$ at two points $\pm w = \pm(0, \beta)$, where $\beta \in \mathbb{R}$. The two axes divide the plane into four quadrants. Since $S_{\mathbb{X}}$ is symmetric about the two axes, each quadrant (excluding the two axes) contains exactly the same number of vertices. Since there is no vertex of $S_{\mathbb{X}}$ on the X-axis and the number of vertices of $S_{\mathbb{X}}$ is $4n + 2$, we conclude that $\pm w$ are two vertices of $S_{\mathbb{X}}$. Suppose the vertex v_1 is in the first quadrant. We choose the other vertices of $S_{\mathbb{X}}$ anticlockwise and denote by $v_2, v_3, \ldots, v_{4n+2}$, where $v_j = (\cos \frac{(2j-1)\pi}{4n+2}, \sin \frac{(2j-1)\pi}{4n+2})$, $j = 1, 2, \ldots, 4n + 2$. Note that the following holds (Fig. 7.3):

(1) The other vertex on L is v_{4n+2}.
(2) v_i and v_{i+1} are adjacent, $i = 1, 2, \ldots, 4n+1$.
(3) $w = v_{n+1}$ and $-w = v_{3n+2}$.
(4) The vertices of $S_{\mathbb{X}}$ on $-L$ are v_{2n+1} and v_{2n+2} in the third and fourth quadrants, respectively.
Without any loss of generality, we take the two adjacent edges $L = \overline{v_1 v_{4n+2}}$ and $M = \overline{v_1 v_2}$ of $S_{\mathbb{X}}$. Suppose $f, g \in S_{\mathbb{X}^*}$ are the extreme supporting linear functionals at v_1 corresponding to edges L and M, respectively, such that $f(v_1) = g(v_1) = 1$. It is easy to check that $\ker(f) \cap S_{\mathbb{X}} = \{v_{n+1}, v_{3n+2}\}$ and $\ker(g) \cap S_{\mathbb{X}} = \{v_{n+2}, v_{3n+3}\}$. Then $\overline{v_{n+1} v_{n+2}}$ and $\overline{v_{3n+2} v_{3n+3}}$ are the edges of $S_{\mathbb{X}}$ and $\overline{v_{n+1} v_{n+2}} = -\overline{v_{3n+2} v_{3n+3}}$. Now, suppose $h \in S_{\mathbb{X}^*}$ is the extreme supporting linear functional corresponding to edge $\overline{v_{n+1} v_{n+2}}$ such that $h(x) = 1$, where $x \in \overline{v_{n+1} v_{n+2}}$. It is easy to check that $\ker(h) \cap S_{\mathbb{X}} = \{v_1, v_{2n+2}\}$ and $v_1 = -v_{2n+2}$. Hence, the edges L and M satisfy the TEP. Therefore, from Theorem 7.2.9, it follows that $\mathbb{X}$ is a Radon plane.

□

Combining Theorems 7.2.8 and 7.2.10, we get the following complete characterization of Radon planes having regular polygons as unit spheres.

Theorem 7.2.11 ([90, Theorem 2.8]) *Let $\mathbb{X}$ be a two-dimensional Banach space such that $S_{\mathbb{X}}$ is a regular polygon. Then $\mathbb{X}$ is a Radon plane if and only if the number of vertices of $S_{\mathbb{X}}$ is $4n+2$, where $n \in \mathbb{N}$. In particular, the affine image of any regular polygon with $4n+2$ vertices is always a Radon curve.*

Remark 7.2.12 We would like to emphasize that Heil [51] first observed that if $\mathbb{X}$ is a two-dimensional polygonal Banach space such that $S_{\mathbb{X}}$ is a regular polygon, then $\mathbb{X}$ is a Radon plane if and only if the number of vertices of $S_{\mathbb{X}}$ is $4n+2$ for some natural number n. In this chapter, we have provided an alternative geometric approach to prove the same. Furthermore, our results in this direction generalize the relevant observations made by Heil in [51] by taking into consideration all possible convex polygons, and not just regular ones.

Continuing in the spirit of Remark 7.2.12, we now present a family of Radon planes whose unit spheres are polygons, but not *regular* ones. We would like to mention here that the sufficient part of the following result is a consequence of the more general fact that the property of being a Radon curve is invariant under affine transformations.

Theorem 7.2.13 ([90, Theorem 2.10]) *Let $\mathbb{X} = \mathbb{R}^2$ be endowed with a norm so that $S_{\mathbb{X}}$ is a hexagon with four vertices $\pm(1, \alpha), \pm(1, -\alpha)$, where $\alpha > 0$. Then $\mathbb{X}$ is a Radon plane if and only if the other two vertices of $S_{\mathbb{X}}$ are $\pm(0, 2\alpha)$ or $\pm(2, 0)$.*

Proof First we prove the sufficient part. From Theorem 7.2.11, we get that a regular hexagon with vertices $\pm(1, 0), \pm(\frac{1}{2}, \frac{\sqrt{3}}{2}), \pm(-\frac{1}{2}, \frac{\sqrt{3}}{2})$ is a Radon curve. Thus, an affine transformation of this Radon curve under the linear map T given by $T(1, 0) = (1, -\alpha)$, $T(\frac{1}{2}, \frac{\sqrt{3}}{2}) = (1, \alpha)$ is again a Radon curve. It is easy to observe that the

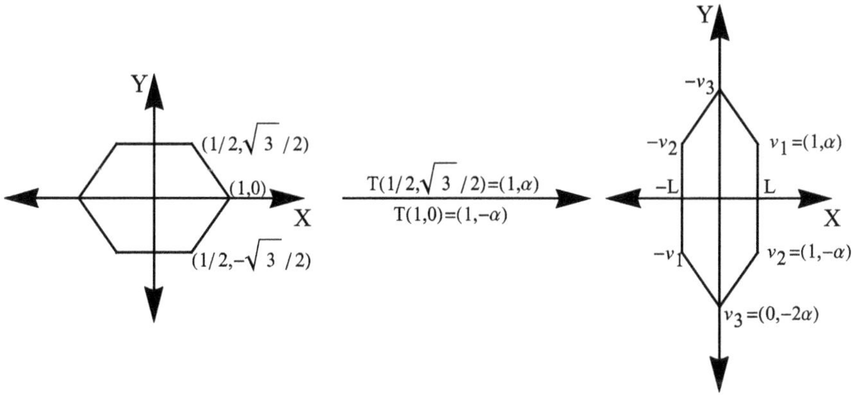

Fig. 7.4 .

vertices of the transformed Radon curve are given by $\pm(1, \alpha)$, $(1, -\alpha)$, $\pm(0, 2\alpha)$. This is illustrated in Fig. 7.4. Similarly, considering the transformation $T(-\frac{1}{2}, \frac{\sqrt{3}}{2}) = (1, \alpha)$, $T(1, 0) = (1, -\alpha)$ we can show that the hexagon with the vertices $\pm(1, \alpha)$, $\pm(1, -\alpha)$, $\pm(2, 0)$ is also a Radon curve.

Now we prove the necessary part. Suppose $\mathbb{X}$ is a two-dimensional Banach space such that $S_{\mathbb{X}}$ is a hexagon with two vertices $v_1 = (1, \alpha)$, $v_2 = (1, -\alpha)$. Then $-v_1 = (-1, -\alpha)$, $-v_2 = (-1, \alpha)$ are also vertices of $S_{\mathbb{X}}$. Assume that $\mathbb{X}$ is a Radon plane. Now, one of the following two cases must be true:

(1) The X-axis contains no vertex of $S_{\mathbb{X}}$.
(2) The X-axis contains exactly two vertices of $S_{\mathbb{X}}$.

First suppose that there are no vertices of $S_{\mathbb{X}}$ on the X-axis. Denote the edge between the vertices v_1 and v_2 by L. Then $-L$ is the edge between the vertices $-v_1$ and $-v_2$. Now, it is obvious that the X-axis meets L at the point $(1, 0)$ and $-L$ at the point $(-1, 0)$, which are smooth points of $S_{\mathbb{X}}$. The Y-axis meets $S_{\mathbb{X}}$ at two points $\pm v = \pm(0, \beta)$, where $\beta \in \mathbb{R}$. It is easy to check that for any $y \in L$, we have $y \perp_B v$. Since B-J orthogonality is symmetric in $\mathbb{X}$, we must have that $v \perp_B y$ for any $y \in L$. This proves that B-J orthogonality is not right unique at v, and so from Remark 2.3.4 it follows that v must be a non-smooth point of $\mathbb{X}$. Thus, the other two vertices of $S_{\mathbb{X}}$ are of the form $\pm(0, \beta)$.

Denote the vertex $(0, -\beta)$ by v_3. Then the vertex $(0, \beta)$ is $-v_3$. Now, by the symmetry of B-J orthogonality in $\mathbb{X}$, we must have that the edges $\overline{v_2 v_3}$ and $\overline{v_3(-v_1)}$ are parallel to $\overline{v_1(-v_1)}$ and $\overline{v_2(-v_2)}$, respectively. It is easy to check that this is true only when $\beta = 2\alpha$. The situation is illustrated in right-hand side of Fig. 7.4.

Now, suppose that the X-axis contains exactly two vertices of $S_{\mathbb{X}}$. Then the other two vertices of $S_{\mathbb{X}}$ are given by $\pm w = \pm(\eta, 0)$, where $\eta \in \mathbb{X}$. The rest of the proof of the fact that $\eta = 2$ in this case can be completed proceeding similarly as above.

This completes the proof of the necessary part and establishes the theorem completely. □

Remark 7.2.14 Using the fact that affine transformation of a Radon curve is a Radon curve, from Theorem 7.2.11 it follows that for each natural number n, there exist non-regular $4n + 2$-gons which are Radon curves.

7.3 Left Symmetric and Right Symmetric Elements in a Banach Space

The previous section dealt with the geometry of two-dimensional polygonal Banach spaces, from the viewpoint of symmetricity of B-J orthogonality. Clearly, such spaces are neither smooth nor strictly convex. Therefore, it is worth investigating the local symmetry of B-J orthogonality in connection with strict convexity and smoothness. We begin this section with an easy proposition which explores some of these interrelations.

Proposition 7.3.1 ([102, Proposition 2.1]) *Let $\mathbb{X}$ be a Banach space.*
(i) If $x \in S_{\mathbb{X}}$ is right symmetric and smooth then x is left symmetric.
(ii) If $\mathbb{X}$ is strictly convex and $x \in S_{\mathbb{X}}$ is left symmetric then x is right symmetric.
(iii) If $\mathbb{X}$ is strictly convex and $x \in S_{\mathbb{X}}$ is left symmetric then x is smooth.

Proof (i) Suppose x is right symmetric and smooth. Let $x \perp_B y$. Then using Theorem 2.1.13, we get $a \in \mathbb{R}$ such that $(ax + y) \perp_B x$. Since x is right symmetric, we have $x \perp_B (ax + y)$. On the other hand, x being smooth, from Theorem 2.3.3, it follows that there exists unique $a \in \mathbb{R}$ such that $x \perp_B (ax + y)$. Therefore, we must have $a = 0$ and so $y \perp_B x$. This proves that x is left symmetric.

(ii) Suppose $y \perp_B x$. Now, from Theorem 2.1.12, it follows that there exists $b \in \mathbb{R}$ such that $x \perp_B (bx + y)$. Since x is left symmetric, we have $(bx + y) \perp_B x$. From strict convexity of $\mathbb{X}$ and Theorem 2.2.1, it follows that there exists unique scalar b such that $(bx + y) \perp_B x$. Thus, we must have $b = 0$ and so $x \perp_B y$. Therefore, x is right symmetric.

(iii) If possible, suppose that x is not smooth, i.e., B-J orthogonality at x is not right unique. Hence, there exists $y \in \mathbb{X}$ and distinct scalars a and b such that $x \perp_B (ax + y)$ and $x \perp_B (bx + y)$. Since x is left symmetric, we have $(ax + y) \perp_B x$ and $(bx + y) \perp_B x$. Therefore, B-J orthogonality is not left unique, contradicting that $\mathbb{X}$ is strictly convex. Thus, x is smooth. □

We next provide a complete characterization of the left symmetric points of a Banach space, without any additional assumptions. We would like to mention that the next two theorems initially appeared in [130]. The Proofs of Theorems 7.3.3 and 7.3.4 presented here vary from those given in [130]. The authors would like to acknowledge Dr. Roy, the second author of [130] for supplying us with this elegant proof.

Lemma 7.3.2 *Let $\mathbb{X}$ be a Banach space and let $x, u \in \mathbb{X}$ be non-zero. Then the following hold:*

(i) $u \in x^+ \setminus x^-$ if and only if $f(u) > 0$ for all $f \in J(x)$.
(ii) $u \in x^- \setminus x^+$ if and only if $f(u) < 0$ for all $f \in J(x)$.

Proof We only prove (i) and note that the proof of (ii) can be completed similarly. Without loss of generality, assume that $\|x\| = \|u\| = 1$. We first prove the necessary part. Suppose on the contrary there exists $f \in J(x)$ such that $f(u) \leq 0$. Clearly, $f(u) \neq 0$, as $x \not\perp_B u$. Let $f(u) = -\varepsilon$ for some $\varepsilon > 0$. Since $u \notin x^-$, there exists $\lambda_0 < 0$ such that $\|x\| > \|x + \lambda_0 u\|$. On the other hand,

$$\|x\| > \|x + \lambda_0 u\| \geq f(x + \lambda_0 u) = f(x) + \lambda_0 f(u) > \|x\|,$$

which is a contradiction.

We now prove the sufficient part. Let $\lambda \geq 0$ be arbitrary. Then for any $f \in J(x)$,

$$\|x + \lambda u\| \geq f(x + \lambda u) \geq \|x\|.$$

This proves that $u \in x^+$. Moreover, $u \notin x^-$, as this would imply $u \in x^+ \cap x^-$. Then $x \perp_B u$, and by the James characterization of B-J orthogonality, there exists a member g of $J(x)$ such that $g(u) = 0$, which is clearly a contradiction. □

Using the preceding lemma, we obtain a complete characterization of left symmetric points in a Banach space.

Theorem 7.3.3 ([130, Theorem 2.1]) *Let $\mathbb{X}$ be a Banach space. Then $x \in \mathbb{X}$ is a left symmetric point in $\mathbb{X}$ if and only if given any $u \in \mathbb{X}$, the following two conditions hold true:*
(i) $u \in x^-$ implies that $x \in u^-$,
(ii) $u \in x^+$ implies that $x \in u^+$.

Proof We first prove the necessary part. Without loss of generality, let $\|x\| = \|u\| = 1$. We only prove (i) and the proof of (ii) can be completed similarly. If $x \perp_B u$, then $u \in x^+ \cap x^-$. By the left symmetry of x, $u \perp_B x$, in particular, $x \in u^+ \cap u^-$.

So assume that $x \not\perp_B u$. Suppose on the contrary that $x \notin u^-$. Then we have the following:

- $u \in x^- \setminus x^+$.
- $x \in u^+ \setminus u^-$.

We first show that x and u are linearly independent. If $x = \lambda u$ for some non-zero scalar λ, then λ must be negative. Otherwise, for any $f \in J(x)$, $f(u) > 0$, which is a contradiction to Lemma 7.3.2, since $u \in x^- \setminus x^+$. However, if $\lambda < 0$, then for any $g \in J(u)$, $g(x) < 0$, which is again a contradiction to Lemma 7.3.2, as $x \in u^+ \setminus u^-$. Therefore, x and u are linearly independent.

Fix any $f \in J(x)$. Then $f(u) < 0$. Now, consider the vector $(-f(u)x + u)$, which is non-zero and lies in the kernel of f. Thus, $x \perp_B (-f(u)x + u)$. Due to the left symmetry of x, $(-f(u)x + u) \perp_B x$. By James characterization of B-J orthogonality, there exists $h \in J(-f(u)x + u)$ such that $h(x) = 0$. Since $x \in u^+$ and $f(u) < 0$,

$$\| - f(u)x + u\| \geq \|u\| = 1.$$

On the other hand,

$$h(-f(u)x + u) = h(u) = \| - f(u)x + u\| \geq 1.$$

However, since $\|h\| = \|u\| = 1$, we must have $h(u) = 1$, and therefore $h \in J(u)$. This proves that $u \perp_B x$, which is a contradiction.

We now prove the sufficient part. Let $x \perp_B u$. In particular, $u \in x^+ \cap x^-$. It follows from the hypothesis that $x \in u^+ \cap u^-$. In other words, $u \perp_B x$ and x is left symmetric.

□

A similar treatment is indeed possible for obtaining an analogous characterization of the right symmetric points of a Banach space.

Theorem 7.3.4 ([130, Theorem 2.2]) *Let $\mathbb{X}$ be a Banach space. Then $x \in \mathbb{X}$ is a right symmetric point in $\mathbb{X}$ if and only if given any $u \in \mathbb{X}$, the following two conditions hold true:*
(i) $x \in u^-$ implies that $u \in x^-$,
(ii) $x \in u^+$ implies that $u \in x^+$.

Proof We first prove the necessary part. Without loss of generality, let $\|x\| = \|u\| = 1$. We only prove (i) and the proof of (ii) can be completed similarly. If $u \perp_B x$, then $x \in u^+ \cap u^-$. By the right symmetry of x, $x \perp_B u$, in particular, $u \in x^+ \cap x^-$.

So assume that $u \not\perp_B x$. Suppose on the contrary that $u \notin x^-$. Then we have the following:

- $u \in x^+ \setminus x^-$.
- $x \in u^- \setminus u^+$.

We first show that x and u are linearly independent. If $x = \lambda u$ for some non-zero scalar λ, then λ must be positive. Otherwise, for any $f \in J(x)$, $f(u) < 0$, which is a contradiction to Lemma 7.3.2, since $u \in x^+ \setminus x^-$. However, if $\lambda > 0$, then for any $g \in J(u)$, $g(x) > 0$, which is again a contradiction to Lemma 7.3.2, as $x \in u^- \setminus u^+$. Therefore, x and u are linearly independent.

Find a non-zero $(\alpha_0 x + \beta_0 u) \in span\{u, x\}$ such that $\|\alpha_0 x + \beta_0 u\| = 1$ and $(\alpha_0 x + \beta_0 u) \perp_B x$. The existence of such an element is always ensured by the left existence of B-J orthogonality. Note that α_0, β_0 are non-zero. Now, there exists $h \in J(\alpha_0 x + \beta_0 u)$ such that $h(x) = 0$. We claim that α_0 and β_0 can be chosen of the same sign.

Without loss of generality, let $\alpha_0 > 0$ and $\beta_0 < 0$. Since $x \in u^-$, $x \in (\beta_0 u)^+$, and $\alpha_0 x \in (\beta_0 u)^+$. Therefore,

$$1 = \|\alpha_0 x + \beta_0 u\| \geq |\beta_0|.$$

On the other hand, $x \notin u^+$. So $x \notin (\beta_0 u)^-$, and there exists $\lambda_0 < 0$ such that

$$\|\lambda_0 x + \beta_0 u\| < |\beta_0| \leq 1.$$

Since $\|\lambda x + \beta_0 u\| \to \infty$ for $\lambda \to -\infty$, by the continuity of the function $\lambda \mapsto \|\lambda x + \beta_0 u\|$, we can find $\mu_0 \leq 0$ such that $\|\mu_0 x + \beta_0 u\| = 1$.

If $\mu_0 = 0$, then $|\beta_0| = 1$. Since $\beta_0 < 0$, we must have $\beta_0 = -1$. Therefore, $h(\mu_0 x + \beta_0 u) = h(-u) = 1$ and $-h \in J(u)$. Since $x \in \ker h$, we have $u \perp_B x$, which is a contradiction to our initial assumption. Thus, $\mu_0 < 0$. Now, put $\alpha_0 = \mu_0$. Therefore, α_0 and β_0 are of the same sign, as desired.

Now, $(\alpha_0 x + \beta_0 u) \perp_B x$, with α_0 and β_0 having the same sign. By the right symmetry of x, we have $x \perp_B (\alpha_0 x + \beta_0 u)$. So there exists $f \in J(x)$ such that

$$f(\alpha_0 x + \beta_0 u) = \alpha_0 + \beta_0 f(u) = 0.$$

Thus, $f(u) < 0$, which is a contradiction to Lemma 7.3.2.

We now prove the sufficient part. Let $u \perp_B x$. Then $x \in u^+ \cap u^-$. It follows from the hypothesis that $u \in x^+ \cap x^-$. Therefore, $x \perp_B u$ and x is right symmetric. □

We observe that as an immediate consequence of Theorems 7.3.3 and 7.3.4, it is possible to completely characterize symmetric points of a Banach space. We state this useful observation in the following theorem. The proof is omitted as it is now obvious.

Theorem 7.3.5 ([130, Theorem 2.3]) *Let $\mathbb{X}$ be a Banach space. Then $x \in \mathbb{X}$ is a symmetric point in $\mathbb{X}$ if and only if given any $u \in \mathbb{X}$, the following four conditions hold true:*
(i) $u \in x^-$ implies that $x \in u^-$,
(ii) $u \in x^+$ implies that $x \in u^+$,
(iii) $x \in u^-$ implies that $u \in x^-$,
(iv) $x \in u^+$ implies that $u \in x^+$.

Next, we turn our attention towards obtaining an explicit characterization of the left symmetric points and the right symmetric points of the infinity sum of two given Banach spaces. Since B-J orthogonality is homogeneous, it is clearly sufficient to describe the left symmetric points and the right symmetric points on the unit sphere of the concerned space.

We first characterize the left symmetric points of the infinity sum of two Banach spaces.

Theorem 7.3.6 ([130, Theorem 2.4]) *Let $\mathbb{X}, \mathbb{Y}$ be Banach spaces and let $\mathbb{Z} = \mathbb{X} \bigoplus_{\infty} \mathbb{Y}$. Then $(x, y) \in S_{\mathbb{Z}}$ is a left symmetric point in $\mathbb{Z}$ if and only if either of the following is true:*
(i) $x \in S_{\mathbb{X}}$ is a left symmetric point in $\mathbb{X}$ and $y = 0$,
(ii) $x = 0$ and $y \in S_{\mathbb{Y}}$ is a left symmetric point in $\mathbb{Y}$.

Proof **Sufficient part**: First assume that (i) holds true. Suppose $(x, 0) \perp_B (u, v)$ in $\mathbb{X} \bigoplus_{\infty} \mathbb{Y}$, i.e., $\max\{\|x + \lambda u\|, \|\lambda v\|\} \geq \|x\| = 1$. Then for each λ with sufficiently small absolute value, $\|x + \lambda u\| \geq \|x\|$. Now, using the convexity of the norm function, it is easy to observe that $\|x + \lambda u\| \geq \|x\|$ for each λ, i.e., $x \perp_B u$. Since x is a left symmetric point in $\mathbb{X}$, we get $u \perp_B x$. Thus, for each $\lambda \in \mathbb{R}$, we have the following:

$$\|(u, v) + \lambda(x, 0)\| = \max\{\|u + \lambda x\|, \|v\|\} \geq \max\{\|u\|, \|v\|\} = \|(u, v)\|.$$

Therefore, $(u, v) \perp_B (x, 0)$ whenever $(x, 0) \perp_B (u, v)$. Thus we get $(x, 0)$ is a left symmetric point in $\mathbb{Z}$. If (ii) holds true, then proceeding similarly we can prove that $(0, y)$ is a left symmetric point in $\mathbb{Z}$.

Necessary part: Suppose $(x, y) \in S_{\mathbb{Z}}$ is a left symmetric point in $\mathbb{Z}$. Clearly, either $\|x\| = 1$ or $\|y\| = 1$. Without any loss of generality, we assume that $\|x\| = 1$. Observe that $\|y\| \leq 1$. We claim that $y = 0$. If possible, suppose that $y \neq 0$. Since $\max\{\|x\|, \|y + \lambda y\|\} \geq \|x\| = \|(x, y)\| = 1$, we get $(x, y) \perp_B (0, y)$. However, for each $\lambda < 0$ with sufficiently small absolute value, we have $\|(0, y) + \lambda(x, y)\| = \max\{\|\lambda x\|, \|y + \lambda y\|\} < \|y\| = \|(0, y)\|$. Thus, $(0, y) \not\perp_B (x, y)$, contradicting our assumption that (x, y) is a left symmetric point in $\mathbb{Z}$. Hence, we conclude that if $(x, y) \in S_{\mathbb{Z}}$ is a left symmetric point in $\mathbb{Z}$ with $\|x\| = 1$ then $y = 0$. Moreover, in this case, we claim that x is a left symmetric point in $\mathbb{X}$. If possible, suppose that there exists $u \in S_{\mathbb{X}}$ such that $x \perp_B u$ but $u \not\perp_B x$. It is easy to observe that either $x \notin u^+$ or $x \notin u^-$. Without any loss of generality, we assume that $x \notin u^+$. Hence, there exists $\lambda > 0$ such that $\|u + \lambda x\| < \|u\| = 1$. Now, it is easy to check that $(x, 0) \perp_B (u, 0)$ but $(u, 0) \not\perp_B (x, 0)$, contradicting our assumption that $(x, 0)$ is a left symmetric point in $\mathbb{Z}$. This contradiction completes the proof of our claim that x is a left symmetric point in $\mathbb{X}$. Therefore, all the conditions stated in (i) hold true. Similarly, assuming $\|y\| = 1$, we can prove that all the conditions stated in (ii) are satisfied. □

Next, we characterize the right symmetric points of the infinity sum of two Banach spaces. We need the following lemma to achieve our goal.

Lemma 7.3.7 ([130, Lemma 2.1]) *Suppose $\mathbb{X}, \mathbb{Y}$ are Banach spaces and $\mathbb{Z} = \mathbb{X} \bigoplus_{\infty} \mathbb{Y}$. Let $(u, v) \in S_{\mathbb{Z}}$ be such that $\|u\| = \|v\| = 1$. If $(x, y) \in \mathbb{Z}$ is such that $(u, v) \perp_B (x, y)$ then either of the following is true:*
(i) $x \in u^+$ and $y \in v^-$,
(ii) $x \in u^-$ and $y \in v^+$.

Proof From Proposition 3.2.2, it follows that either $x \in u^+$ or $x \in u^-$. Without any loss of generality, assume that $x \in u^+$. Consider the following two cases:

Case I: Suppose $x \in u^-$. By the same argument, either $y \in v^+$ or $y \in v^-$. In each of these two sub-cases, it is easy to check that the statement of the lemma holds true.

Case II: Suppose $x \notin u^-$. Then $\|u + \lambda x\| < \|u\| = 1$, whenever $\lambda < 0$ is of sufficiently small absolute value. From $(u, v) \perp_B (x, y)$, it follows that $\|(u, v) + \lambda(x, y)\| = \max\{\|u + \lambda x\|, \|v + \lambda y\|\} \geq \|(u, v)\| = 1$ for each $\lambda \in \mathbb{R}$. Thus, if $\lambda < 0$ is of sufficiently small absolute value, we must have $\|v + \lambda y\| \geq \|v\| = 1$. Now, using the convexity of the norm function, it is easy to check that $\|v + \lambda y\| \geq \|v\| = 1$ for all $\lambda < 0$. Therefore, $y \in v^-$. □

Now the promised characterization:

Theorem 7.3.8 ([130, Theorem 2.5]) *Suppose* $\mathbb{X}, \mathbb{Y}$ *are Banach spaces and* $\mathbb{Z} = \mathbb{X} \bigoplus_\infty \mathbb{Y}$. *Then* $(x, y) \in S_\mathbb{Z}$ *is a right symmetric point in* $\mathbb{Z}$ *if and only if* $\|x\| = \|y\| = 1$ *and* x, y *are right symmetric points in* $\mathbb{X}$ *and* $\mathbb{Y}$, *respectively.*

Proof **Sufficient part**: Suppose $(u, v) \in \mathbb{Z}$ is such that $(u, v) \perp_B (x, y)$. Consider the following two cases:

Case I: Let $\|u\| \neq \|v\|$. Without any loss of generality, assume that $\|u\| > \|v\|$. Since $(u, v) \perp_B (x, y)$, we get $\max\{\|u + \lambda x\|, \|v + \lambda y\|\} \geq \|(u, v)\| = \|u\|$ for all $\lambda \in \mathbb{R}$. Therefore, if $\lambda \in \mathbb{R}$ is of sufficiently small absolute value, it is easy to check that $\|u + \lambda x\| \geq \|u\|$. Now, using the convexity of the norm function, we can prove that $\|u + \lambda x\| \geq \|u\|$ for all $\lambda \in \mathbb{R}$, i.e., $u \perp_B x$. Since x is a right symmetric point in $\mathbb{X}$, we get $x \perp_B u$. Therefore, for each $\lambda \in \mathbb{R}$:

$$\|(x, y) + \lambda(u, v)\| = \max\{\|x + \lambda u\|, \|y + \lambda v\|\} \geq \|x + \lambda u\| \geq \|x\| = \|(x, y)\| = 1.$$

This proves that $(x, y) \perp_B (u, v)$.

Case II: Let $\|u\| = \|v\|$. If $\|u\| = \|v\| = 0$, then clearly $(x, y) \perp_B (0, 0) = (u, v)$. Without any loss of generality, assume that $\|u\| = \|v\| = 1$. Using Lemma 7.3.7, we get that either of the following must hold true:

(i) $x \in u^+$ and $y \in v^-$, (ii) $x \in u^-$ and $y \in v^+$. First assume that (i) holds true. Since x, y are right symmetric points in $\mathbb{X}$ and $\mathbb{Y}$, respectively, using Theorem 7.3.4, we get $u \in x^+$ and $v \in y^-$. Therefore, for each $\lambda \geq 0$, we have

$$\|(x, y) + \lambda(u, v)\| = \max\{\|x + \lambda u\|, \|y + \lambda v\|\} \geq \|x + \lambda u\| \geq \|x\| = \|(x, y)\| = 1.$$

On the other hand, for each $\lambda \leq 0$, we have

$$\|(x, y) + \lambda(u, v)\| = \max\{\|x + \lambda u\|, \|y + \lambda v\|\} \geq \|y + \lambda v\| \geq \|y\| = \|(x, y)\| = 1.$$

Thus, we get $(x, y) \perp_B (u, v)$. Similarly, if (ii) holds true, then we can deduce the same conclusion. Hence, $(x, y) \perp_B (u, v)$, whenever $(u, v) \perp_B (x, y)$, i.e., (x, y) is a right symmetric point in $\mathbb{Z}$.

Necessary part: Suppose $(x, y) \in S_{\mathbb{Z}}$ is a right symmetric point in $\mathbb{Z}$. Clearly, $\|y\| \leq 1$. We claim that $\|y\| = 1$. If possible, let $\|y\| < 1$. Then we have $\|x\| = 1$. Choose $w \in S_{\mathbb{Y}}$ such that $y \in w^+$. Now, for each $\lambda \geq 0$, we get $\|(-x, w) + \lambda(x, y)\| \geq \|w + \lambda y\| \geq \|w\| = 1 = \|(-x, w)\|$. On the other hand, for each $\lambda < 0$, we get $\|(-x, w) + \lambda(x, y)\| \geq \| - x + \lambda x\| = 1 - \lambda > \|(-x, w)\|$. Thus, $(-x, w) \perp_B (x, y)$. However, if $\lambda > 0$ is of sufficiently small absolute value, then $\|(x, y) + \lambda(-x, w)\| = \max\{\|x - \lambda x\|, \|y + \lambda w\|\} = \|(1 - \lambda)x\| = 1 - \lambda < 1 = \|(x, y)\|$. Therefore, $(x, y) \not\perp_B (-x, w)$, contradicting our assumption that (x, y) is a right symmetric point in $\mathbb{Z}$. Thus, $\|y\| = 1$. Similarly, we can prove that $\|x\| = 1$. Next, we claim that x is a right symmetric point in $\mathbb{X}$. If possible, suppose that x is not a right symmetric point in $\mathbb{X}$. Then there exists $u \in S_{\mathbb{X}}$ such that $u \perp_B x$ but $x \not\perp_B u$. Now, from Proposition 3.2.2, it follows that either $u \in x^+ \setminus x^-$ or $u \in x^- \setminus x^+$. Without any loss of generality, assume that $u \in x^+ \setminus x^-$. Choose $v \in S_{\mathbb{Y}}$ such that $v \in y^+ \setminus y^-$. Then there exists $\lambda_0 < 0$ such that $\|x + \lambda_0 u\| < \|x\| = 1$ and $\|y + \lambda_0 v\| < \|y\| = 1$. This proves that $(x, y) \not\perp_B (u, v)$. On the other hand, since $u \perp_B x$, it is easy to observe that $(u, v) \perp_B (x, y)$, which clearly contradicts our assumption that (x, y) is a right symmetric point in $\mathbb{Z}$. Hence, x must be a right symmetric point in $\mathbb{X}$. Similarly, we can prove that y is a right symmetric point in $\mathbb{Y}$. □

Our next aim is to present a relationship between left symmetry (right symmetry) of a smooth point $x \in S_{\mathbb{X}}$ and right symmetry (left symmetry) of the supporting functional of $B_{\mathbb{X}}$ at x. For $g \in S_{\mathbb{X}^*}$, we define $D_g = \{z \in S_{\mathbb{X}} : g(z) = \|g\|\}$. We will use Theorem 4.3.10 in the proof of the next result. Moreover, we require the following easy observation:

Proposition 7.3.9 ([66]) *(i) Let $\mathbb{X}$ be a reflexive and Kadets–Klee Banach space. Then for $f \in \mathbb{X}^* \setminus \{0\}$, $M_f = \pm D$, where D is a compact and connected subset of $S_{\mathbb{X}}$.*
(ii) Let $\mathbb{X}$ be a Banach space and let $f \in \mathbb{X}^*$ be such that $M_f \neq \emptyset$. Then $M_f = \pm D$, where D is a closed and connected subset of $S_{\mathbb{X}}$.

We are now ready to obtain the desired result in form of the following:

Theorem 7.3.10 ([66, Theorem 2.2]) *Suppose $\mathbb{X}$ is a reflexive Banach space and $x \in S_{\mathbb{X}}$ is a smooth point. Let f_x be the unique supporting functional of $B_{\mathbb{X}}$ at x. Then the following are true:*
(i) If x is a left symmetric point in $\mathbb{X}$, then f_x is right symmetric for the collection

$$\mathcal{A} = \{g \in S_{\mathbb{X}^*} : D_g \text{ is compact and contains only smooth points}\}.$$

(ii) If x is an exposed point of $B_{\mathbb{X}}$ and f_x is right symmetric in $\mathbb{X}^*$, then x is a left symmetric point in $\mathbb{X}$.

(iii) If x is an exposed point of $B_{\mathbb{X}}$ and x is a right symmetric point in $\mathbb{X}$, then f_x is a left symmetric point in $\mathbb{X}^*$.
(iv) If f_x is a left symmetric point in $\mathbb{X}^*$, then x is right symmetric for the collection

$$\mathcal{A} = \{y \in S_{\mathbb{X}} : y \text{ is a smooth and exposed point of } B_{\mathbb{X}}\}.$$

Proof (i) Suppose $g \in \mathcal{A}$ is such that $g \perp_B f_x$. Then from Theorem 4.3.10, we get $g(y) \perp_B f_x(y)$ for some $y \in M_g$. Thus $f_x(y) = 0$ and hence $x \perp_B y$. Now, from left symmetry of x, we get $y \perp_B x$. Since y is a smooth point, we have $J(y) = \{g\}$. This clearly proves $g(x) = 0$. Therefore, $f_x(x) \perp_B g(x)$ and so $f_x \perp_B g$.

(ii) Suppose $y \in S_{\mathbb{X}}$ is such that $x \perp_B y$. Then $f_x(y) = 0$. Let $g \in J(y)$. Then $g(y) \perp_B f_x(y)$, which implies that $g \perp_B f_x$. Since f_x is right symmetric, we have, $f_x \perp_B g$. Since x is an exposed point of $B_{\mathbb{X}}$, we get $M_{f_x} = \{\pm x\}$. Therefore, from Theorem 4.3.10, it follows that $f_x(x) \perp_B g(x)$. Hence $g(x) = 0$, and so $y \perp_B x$.

(iii) Suppose $g \in S_{\mathbb{X}^*}$ is such that $f_x \perp_B g$. Since x is an exposed point of $B_{\mathbb{X}}$, we get that $M_{f_x} = \{\pm x\}$. Now, from Theorem 4.3.10 it follows that $f_x(x) \perp_B g(x)$. Thus $g(x) = 0$, and so $y \perp_B x$ for some $y \in M_g$. Since x is right symmetric, we have $x \perp_B y$. This proves that $f_x(y) = 0$. Therefore, $g(y) \perp_B f_x(y)$, and consequently $g \perp_B f_x$.

(iv) Suppose $y \in \mathcal{A}$ is such that $y \perp_B x$. Since $y \in \mathcal{A}$, there exists $g \in S_{\mathbb{X}^*}$ such that $J(y) = \{g\}$ and $M_g = \{\pm y\}$. This implies that $g(x) = 0$ and $f_x(x) \perp_B g(x)$. Hence, $f_x \perp_B g$. Since f_x is left symmetric, we have $g \perp_B f_x$. Since $M_g = \{\pm y\}$, so from Theorem 4.3.10 it follows that $g(y) \perp_B f_x(y)$. Therefore, $f_x(y) = 0$ and so $x \perp_B y$. □

As an immediate application of the above theorem, we get the following result, which gives a complete characterization of a left symmetric (right symmetric) element $x \in S_{\mathbb{X}}$ in $\mathbb{X}$ in terms of right symmetry (left symmetry) of the supporting linear functional f_x of $B_{\mathbb{X}}$ at x, under some additional nice conditions on $\mathbb{X}$.

Corollary 7.3.11 ([66, Corollary 2.2.1]) *Let $\mathbb{X}$ be a reflexive, strictly convex and smooth Banach space. Then the following are true:*
(i) $x \in S_{\mathbb{X}}$ is a left symmetric point in $\mathbb{X}$ if and only if the supporting functional f_x of $B_{\mathbb{X}}$ at x is a right symmetric point in $\mathbb{X}^*$.
(ii) $x \in S_{\mathbb{X}}$ is a right symmetric point in $\mathbb{X}$ if and only if the supporting functional f_x of $B_{\mathbb{X}}$ at x is a left symmetric point in $\mathbb{X}^*$.

Proof (i) follows from (i) and (ii) of Theorem 7.3.10.
(ii) follows from (iii) and (iv) of Theorem 7.3.10. □

Remark 7.3.12 From Proposition 7.3.1, it follows that any non-zero left symmetric or right symmetric point in a strictly convex and smooth Banach space is a symmetric point. Therefore, if $\mathbb{X}$ is a reflexive, strictly convex, and smooth Banach space, then from Corollary 7.3.11 we can conclude that $x \in S_{\mathbb{X}}$ is a symmetric point in $\mathbb{X}$ if and only if the supporting linear functional of $B_{\mathbb{X}}$ at x is a symmetric point in $\mathbb{X}^*$.

For the rest of this section we consider Banach spaces which are not necessarily smooth. We begin with the following simple observation.

Lemma 7.3.13 ([131, Theorem 3.2]) *Suppose $\mathbb{X}$ is a reflexive Banach space. Then for any $f, g \in \mathbb{X}^* \setminus \{0\}$,*

$$f \perp_B g \Leftrightarrow M_f \cap ker(g) \neq \emptyset.$$

Proof We first assume that $f \perp_B g$. From [131, Prop. 3.1], we note that $M_f = D \cup (-D)$, where D is a connected subset of $S_{\mathbb{X}}$. Then applying Theorem 4.3.10, we have $f(x) \perp_B g(x)$, for some $x \in M_f$. This implies that $g(x) = 0$, i.e., $x \in \ker(g)$. Therefore, $M_f \cap \ker(g) \neq \emptyset$.

The sufficient part of the theorem can be proved using similar arguments as given above.

□

The following proposition shows that the sufficient part of (ii) of Corollary 7.3.11 holds even without the assumption of smoothness on $\mathbb{X}$.

Proposition 7.3.14 ([66, Proposition 2.3]) *Suppose $\mathbb{X}$ is a reflexive and strictly convex Banach space. If $x \in S_{\mathbb{X}}$ and there exists a left symmetric supporting functional of $B_{\mathbb{X}}$ at x then x is a right symmetric point in $\mathbb{X}$.*

Proof Suppose f is a left symmetric supporting linear functional of $B_{\mathbb{X}}$ at x. Consider any $y \in S_{\mathbb{X}}$ such that $y \perp_B x$. Then, there exists $g \in J(y)$ such that $g(x) = 0$. Therefore, $f(x) \perp_B g(x)$ and so $f \perp_B g$. Since f is left symmetric, we have $g \perp_B f$ and since $\mathbb{X}$ is strictly convex, we have $M_g = \{\pm y\}$. Thus, from Lemma 7.3.13, it follows that $f(y) = 0$. Therefore, $x \perp_B y$, completing the proof of the
proposition. □

The next two results are analogous to (ii) of Corollary 7.3.11 for reflexive and strictly convex Banach spaces. For a two-dimensional strictly convex space, the result holds without any additional assumption and for higher dimensions it is true under some additional assumptions. Given an element $x \subset \mathbb{X}$, let us define ${}^{\perp}x = \{y \in \mathbb{X} : y \perp_B x\}$.

Theorem 7.3.15 ([66, Theorem 2.3]) *Let $\mathbb{X}$ be a two-dimensional strictly convex Banach space. Then $x \in S_{\mathbb{X}}$ is a right symmetric point in $\mathbb{X}$ if and only if there exists a left symmetric supporting linear functional of $B_{\mathbb{X}}$ at x.*

Proof The sufficient part follows from Proposition 7.3.14. We only prove the necessary part. Suppose $x \in S_{\mathbb{X}}$ is a right symmetric point in $\mathbb{X}$. Since $\mathbb{X}$ is two-dimensional, there exists a unique linear functional $g \in S_{\mathbb{X}^*}$ such that $\ker(g) \cap S_{\mathbb{X}} = \{\pm x\}$. Since $\mathbb{X}$ is strictly convex, we have $M_g = \{\pm y\}$ for some $y \in S_{\mathbb{X}}$. Therefore, ${}^{\perp}x \cap S_{\mathbb{X}} = \{\pm y\}$. Since x is right symmetric, we have $x \perp_B y$. It is easy to observe that there exists a unique linear functional $f \in S_{\mathbb{X}^*}$ such that $\ker(f) \cap S_{\mathbb{X}} = \{\pm y\}$. Moreover, since f is the only norm one linear functional such that $\ker(f) \cap S_{\mathbb{X}} = \{\pm y\}$, either $f \in J(x)$ or $-f \in J(x)$. Without any loss of generality, assume that $f \in J(x)$. Suppose, $f \perp_B h$ for some $h \in \mathbb{X}^*$. Therefore, it follows from Lemma 7.3.13 that

$M_f \cap \ker(h) \neq \emptyset$. Once again, using strict convexity of $\mathbb{X}$, we obtain $M_f = \{\pm x\}$. Since $\mathbb{X}$ is two-dimensional, this h is unique up to scalar multiples. Thus, there exists a scalar μ such that $h = \mu g$. Hence, $f^{\perp} = \{\lambda g : \lambda \in \mathbb{R}\}$. Therefore, to complete the proof, it is enough to show that $g \perp_B f$. However, since $M_g \cap \ker(f) \neq \emptyset$, using Lemma 7.3.13 we immediately get that $g \perp_B f$. Consequently, f is a left symmetric supporting linear functional of $B_{\mathbb{X}}$ at x and this completes the proof. □

Theorem 7.3.16 ([66, Theorem 2.4]) *Suppose $\mathbb{X}$ is a reflexive and strictly convex Banach space. Let $x \in S_{\mathbb{X}}$ and let ${}^{\perp}x$ be a subspace having codimension one in $\mathbb{X}$. Then x is a right symmetric point in $\mathbb{X}$ if and only if there exists a left symmetric supporting linear functional of $B_{\mathbb{X}}$ at x.*

Proof The sufficient part follows from Proposition 7.3.14. We only prove the necessary part. Suppose $x \in S_{\mathbb{X}}$ is a right symmetric point in $\mathbb{X}$ and ${}^{\perp}x$ is a subspace of codimension one in $\mathbb{X}$. Note that, each element $y \in \mathbb{X}$ can be uniquely expressed as $y = \alpha x + h$ for some $\alpha \in \mathbb{R}$ and $h \in {}^{\perp}x$. Define a linear functional $f : \mathbb{X} \longrightarrow \mathbb{R}$ by

$$f(\alpha x + h) = \alpha.$$

It is easy to observe that $\|f\| = 1$ and $x \in M_f$. On the other hand, since $\mathbb{X}$ is strictly convex, $M_f = \{\pm x\}$. We now show that f is a left symmetric point in $\mathbb{X}^*$. Let $g \in S_{\mathbb{X}^*}$ be such that $f \perp_B g$. Then from Lemma 7.3.13 it follows that $g(x) = 0$. Suppose $M_g = \{\pm y\}$ for some $y \in S_{\mathbb{X}}$. Hence, $y \in^{\perp} x$ and so $f(y) = 0$. Now, it follows from Lemma 7.3.13 that $g \perp_B f$. Thus, f is a left symmetric supporting linear functional of $B_{\mathbb{X}}$ at x, completing the proof of the theorem. □

We now extend (i) of Theorem 7.3.10 for a strictly convex Banach space under some additional assumptions on $x^{\perp}$.

Theorem 7.3.17 ([66, Theorem 2.5]) *Suppose $\mathbb{X}$ is a reflexive and strictly convex Banach space. Let $x \in S_{\mathbb{X}}$ be a left symmetric point in $\mathbb{X}$ and let $x^{\perp} \cap S_{\mathbb{X}}$ contain only smooth points. Then there exists a right symmetric supporting functional of $B_{\mathbb{X}}$ at x.*

Proof Since x is a left symmetric point in $\mathbb{X}$, from Proposition 7.3.1 we get that x is a smooth point. Therefore, $x^{\perp}$ is a subspace of codimension one in $\mathbb{X}$. Note that, each element $y \in \mathbb{X}$ can be uniquely expressed as $y = \alpha x + h$ for some $\alpha \in \mathbb{R}$ and $h \in x^{\perp}$. Define a linear functional $f_x : \mathbb{X} \longrightarrow \mathbb{R}$ by

$$f_x(\alpha x + h) = \alpha,$$

where $\alpha \in \mathbb{R}$ and $h \in x^{\perp}$. It is easy to observe that $\|f_x\| = 1$ and $M_{f_x} = \{\pm x\}$. On the other hand, since $x \in S_{\mathbb{X}}$ is a smooth point, we have $J(x) = \{f_x\}$. Applying Lemma 7.3.13 together with the facts that $\mathbb{X}$ is strictly convex and $x^{\perp} \cap S_{\mathbb{X}}$ contain only smooth points it is easy to check that f_x is a right symmetric supporting linear functional of $B_{\mathbb{X}}$ at x. This ends the proof of the theorem. □

In the next example, we show that if $x^{\perp}$ contains non-smooth point, then Theorem 7.3.17 fails.

Example 7.3.18 ([66, Example 2.2]) Suppose $\mathbb{X} = \mathbb{R}^2$ and $\|\cdot\|$ is a norm on $\mathbb{X}$ such that the unit sphere of $\mathbb{X}$, $S_{\mathbb{X}}$ is given by $S_{\mathbb{X}} = \{(x, y) \in \mathbb{X} : x^2 + 2|y| = 1\}$. From the description of the unit sphere, it is clear that $\mathbb{X}$ is strictly convex and $\pm(1, 0)$ are the only non-smooth point of $S_{\mathbb{X}}$. The situation is illustrated in the following figure.

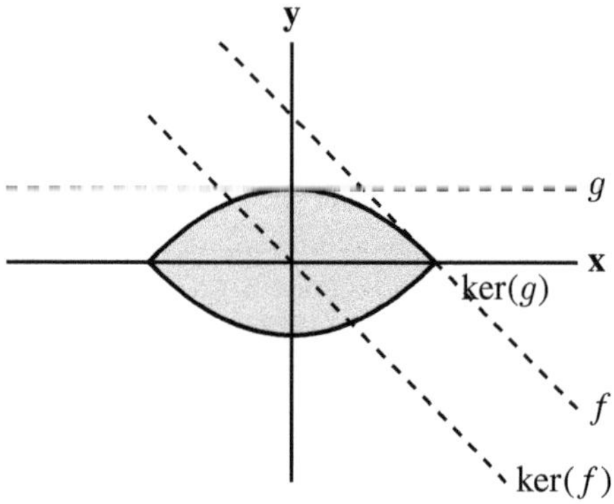

Note that $(0, \frac{1}{2}) \in S_{\mathbb{X}}$ is a smooth point and the unique norm one supporting linear functional of $B_{\mathbb{X}}$ at $(0, \frac{1}{2})$ is given by

$$g(x, y) = 2y \text{ for all } (x, y) \in \mathbb{X}.$$

Clearly, $\ker(g) = \{\lambda(1, 0) : \lambda \in \mathbb{R}\}$ and so $(0, \frac{1}{2}) \perp_B \lambda(1, 0)$ for all $\lambda \in \mathbb{R}$. Since $\mathbb{X}$ is two-dimensional and $(0, \frac{1}{2})$ is a smooth point, $(0, \frac{1}{2})^{\perp}$ is an one-dimensional subspace of $\mathbb{X}$. Therefore, $(0, \frac{1}{2})^{\perp} = \{\lambda(1, 0) : \lambda \in \mathbb{R}\}$. We now show that $(0, \frac{1}{2})$ is a left symmetric point in $\mathbb{X}$. Since B-J orthogonality is homogeneous, to prove $(0, \frac{1}{2})$ is a left symmetric point in $\mathbb{X}$, it is enough to show that $(1, 0) \perp_B (0, \frac{1}{2})$. Observe that, $h(x, y) = x$, for all $(x, y) \in \mathbb{X}$, is a supporting linear functional of $B_{\mathbb{X}}$ at $(1, 0)$. Since $\ker(h) = \{\lambda(0, 1) : \lambda \in \mathbb{R}\}$, we get $(1, 0) \perp_B (0, \frac{1}{2})$, as desired. Next, we prove that g is not right symmetric. Suppose

$$f(x, y) = x + y \text{ for all } (x, y) \in \mathbb{X}.$$

It is easy to check that $M_f = \{\pm(1, 0)\}$. Hence, $M_f \cap \ker(g) \neq \emptyset$. Thus, from Lemma 7.3.13, we get that $f \perp_B g$. However, $g \not\perp_B f$, as $M_g \cap \ker(f) = \emptyset$.

We now show that if $\mathbb{X}$ is a reflexive, strictly convex Banach space with $x \in S_{\mathbb{X}}$ (not necessarily a left symmetric point in $\mathbb{X}$) then under some additional assumptions on ${}^{\perp}x$, there exists a right symmetric supporting linear functional of $B_{\mathbb{X}}$ at x.

Theorem 7.3.19 ([66, Theorem 2.6]) *Suppose $\mathbb{X}$ is a reflexive, strictly convex Banach space and let $x \in S_{\mathbb{X}}$. Suppose H is a subspace having codimension one in $\mathbb{X}$ such that $H \subseteq {}^{\perp}x$ and $H \cap S_{\mathbb{X}}$ contains only left symmetric points. Then there exists a right symmetric supporting linear functional of $B_{\mathbb{X}}$ at x.*

Proof Define a linear functional $f : \mathbb{X} \longrightarrow \mathbb{R}$ by

$$f(\alpha x + h) = \alpha,$$

where $\alpha \in \mathbb{R}$ and $h \in H$. Since the elements of $H \cap S_{\mathbb{X}}$ are left symmetric, we get $|f(\alpha x + h)| \leq 1$ for $(\alpha x + h) \in S_{\mathbb{X}}$ and $h \neq 0$. From $f(x) = 1$, we have $\|f\| = 1$. Now, strict convexity of $\mathbb{X}$ implies that $M_f = \{\pm x\}$. Now, we show that f is a right symmetric supporting functional at x. Let $g \in S_{\mathbb{X}^*}$ be such that $g \perp_B f$. From strict convexity of $\mathbb{X}$, we get that $M_g = \{\pm y\}$ for some $y \in S_{\mathbb{X}}$. Thus from Lemma 7.3.13, it follows that $y \in \ker(f)$ and this implies $y \in H \cap S_{\mathbb{X}}$. By our assumption, we have y is a left symmetric point in $\mathbb{X}$. Therefore, from Proposition 7.3.1 we get y is a smooth point. Hence, g is the unique supporting linear functional of $B_{\mathbb{X}}$ at y. Now, $y \perp_B x$ implies that $x \in \ker(g)$. Therefore, $M_f \cap \ker(g) \neq \emptyset$. Now, using Lemma 7.3.13 we conclude that f is right symmetric. □

In the next example, we show the existence of a right symmetric supporting linear functional for a non-left symmetric point.

Example 7.3.20 ([66, Example 2.3]) Consider the same two-dimensional Banach space as in Example 7.3.18. First we show that there exists a right symmetric supporting linear functional of $B_{\mathbb{X}}$ at $(1, 0)$ and then we will show that $(1, 0)$ is not a left symmetric point in $\mathbb{X}$. Suppose

$$f(x, y) = x, \text{ for all } (x, y) \in \mathbb{X}.$$

It is easy to check that f is a supporting linear functional of $B_{\mathbb{X}}$ at $(1, 0)$ and $M_f = \{\pm(1, 0)\}$. Now, from Lemma 7.3.13, we get that for each $g \in \mathbb{X}^*$, $g \perp_B f$ if and only if $M_g \cap \ker(f) \neq \emptyset$. It is easy to observe that $\ker(f) \cap S_{\mathbb{X}} = \{\pm(0, \frac{1}{2})\}$. Since $\pm(0, \frac{1}{2}) \in S_{\mathbb{X}}$ are smooth points, the lone norm one supporting linear functional, which attains its norm at $\pm(0, \frac{1}{2})$, is given by $g(x, y) = 2y$ for all $(x, y) \in \mathbb{X}$. Therefore, ${}^{\perp}f = \{\lambda g : \lambda \in \mathbb{R}\}$. Now, to conclude f is right symmetric, it is enough to prove that $f \perp_B g$. Clearly, $f \perp_B g$, as $(0, 1) \in M_f \cap \ker(g)$. Therefore, f is right symmetric, as desired. We now claim that $(1, 0) \in S_{\mathbb{X}}$ is not a left symmetric point in $\mathbb{X}$. Suppose

$$h(x, y) = x + y, \text{ for all } (x, y) \in \mathbb{X}.$$

Note that $M_h = \{\pm(1, 0)\}$ and $\ker(h) \cap S_{\mathbb{X}} = \{\pm(1 - \sqrt{2}, -1 + \sqrt{2})\}$. Therefore, $(1, 0) \perp_B (1 - \sqrt{2}, -1 + \sqrt{2})$. On the other hand, since $(1 - \sqrt{2}, -1 + \sqrt{2}) \in S_{\mathbb{X}}$ is a smooth point, the lone norm one supporting linear functional of $B_{\mathbb{X}}$ at

$(1-\sqrt{2}, -1+\sqrt{2}) \in S_{\mathbb{X}}$ is given by $p(x, y) = \frac{(y-(\sqrt{2}-1)x)}{2-\sqrt{2}}$ for all $(x, y) \in \mathbb{X}$. However, since $p(1, 0) \neq 0$, we have $(1-\sqrt{2}, -1+\sqrt{2}) \not\perp_B (1, 0)$. We illustrate the situation with the following figure.

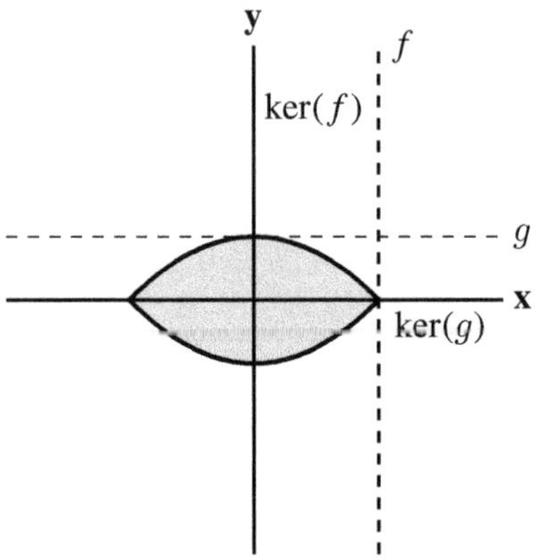

7.4 Left Symmetric and Right Symmetric Elements in $\mathbb{L}(\mathbb{H})$

Having studied some significant geometric attributes of the local symmetry of B-J orthogonality in a general Banach space, we next proceed to identify the left and right symmetric elements in the space $\mathbb{L}(\mathbb{H})$ of all bounded linear operators on a Hilbert space $\mathbb{H}$. As an elementary example of the specialty of such points, consider $T = \begin{pmatrix} 1 & 0 & 0 \\ 0 & 1/2 & 0 \\ 0 & 0 & 1/2 \end{pmatrix}$ and $A = \begin{pmatrix} 0 & 0 & 0 \\ 0 & 1 & 0 \\ 0 & 0 & 0 \end{pmatrix}$ on $(\mathbb{R}^3, \|.\|_2)$. Then $T \perp_B A$ but $A \not\perp_B T$. Similarly, we can exhibit operators T, A for which $A \perp_B T$ but $T \not\perp_B A$. In [144], Turnšek initiated the study of symmetric operators in the setting of Hilbert spaces and then continued it in [143]. Using a different approach, the authors in [39] studied left and right symmetric compact operators on Hilbert spaces. We now discuss the approach developed in [143] to characterize both left and right symmetric operators defined on Hilbert spaces, real or complex. The following definitions were introduced in [143].

Definition 7.4.1 ([143]) Let $\mathbb{X}$ be a complex Banach space. An element $x \in \mathbb{X}$ is said to be r-orthogonal to $y \in \mathbb{X}$ if $\|x + ry\| \geq \|x\|$ for all real scalars r and is written as $x \perp_r y$. In general, $x \perp_r y$ does not imply $x \perp_B y$, but other way round is always true.

Definition 7.4.2 ([143]) Let $\mathbb{X}$ be a complex Banach space. An element $x \in \mathbb{X}$ is said to be r-left symmetric if $x \perp_B y \Rightarrow y \perp_r x$. Similarly x is said to r-right symmetric if $y \perp_B x \Rightarrow x \perp_r y$. Clearly if an element is left symmetric (right symmetric) then it is r-left symmetric (r-right symmetric) but the converse is not necessarily true.

We first aim to characterize r-left symmetric operators on Hilbert spaces. For this purpose, we require the following theorem which follows from the same line of arguments as in [17, 99].

Theorem 7.4.3 ([143, Theorem 2.4]) *Let $\mathbb{H}$ be a real or complex Hilbert space and $T, A \in \mathbb{L}(\mathbb{H})$. Then $T \perp_r A$ if and only if there exists $\{x_n\} \subset S_{\mathbb{H}}$ such that $\|Tx_n\| \to \|T\|$ and $\Re\langle Tx_n, Ax_n\rangle \to 0$.*

As a consequence, the following corollary is obvious.

Corollary 7.4.4 ([143, Corollary 2.6]) *Let $\mathbb{H}$ be a real or complex Hilbert space and $T, A \in \mathbb{L}(\mathbb{H})$ and in addition T is compact. Then $T \perp_r A$ if and only if there exists $x \in M_T$ such that $\Re\langle Tx, Ax\rangle = 0$.*

In the next result we show that a non-zero positive (self-adjoint in real case) operator is never r-left symmetric.

Theorem 7.4.5 ([143, Proposition 3.2]) *Let $\mathbb{H}$ be a real or complex Hilbert space and $0 \neq T \in \mathbb{L}(\mathbb{H})$ be positive (self-adjoint in the real case). Then there exists $A \in \mathbb{L}(\mathbb{H})$ such that $T \perp_B A$ but $A \not\perp_r T$.*

Proof Assume that $\|T\| = 1$. Let $\{x_n\}$ be a sequence in $S_{\mathbb{H}}$ such that $\|Tx_n\| \to 1$. Let $H_0 = \overline{span}\ \{x_n\}$. Without loss of generality we assume that $H_0 \neq \mathbb{H}$. Consider the following two cases.

Case I: $T(H_0^{\perp}) \neq 0$. Then there exists $y \in H_0^{\perp}$ such that $\|y\| = 1$ and $Ty \neq 0$. Define $A : \mathbb{H} \to \mathbb{H}$ by $Az = \langle z, y\rangle y$ for all $z \in \mathbb{H}$. Since $Ax_n = 0, \forall n$ and $\|Tx_n\| \to \|T\|$ so it follows from Theorem 4.2.1 that $T \perp_B A$. Next, we show that $A \not\perp_r T$. If possible let $A \perp_r T$. Since A is compact so by Corollary 7.4.4 there exists $z \in M_A$ such that $\Re\langle Az, Tz\rangle = 0$. As $z \in M_A$ so $\|Az\| = |\langle z, y\rangle| = 1$. From the equality condition of Cauchy–Schwarz inequality it follows that $z = \lambda y, |\lambda| = 1$. Then

$$\Re\langle Az, Tz\rangle = \Re\langle \lambda y, \lambda Ty\rangle = \langle y, Ty\rangle = \|T^{1/2}y\|^2 \neq 0,$$

a contradiction. Thus $A \not\perp_r T$.

Case II: $T(H_0^{\perp}) = 0$. Let $x \in H_0$, $\|x\| = 1$, $Tx \neq 0$ and $y \in H_0^{\perp}$, $\|y\| = 1$. Let $S =$ span $\{x, y\}$. Define $A : \mathbb{H} \to \mathbb{H}$ by

$$Ax = y,\ Ay = x + y \text{ and } Az = 0,\ z \in S^{\perp}.$$

It is easy to see that $\langle Tx_n, Ax_n\rangle = 0$ for all n and so by Theorem 4.2.1, $T \perp_B A$. As before we can see that $\Re\langle Az, Tz\rangle \neq 0$ for each $z \in M_A$. Since A is compact it follows from Corollary 7.4.4 that $A \not\perp_r T$. This completes the proof. □

We now present a characterization of left symmetric operators. The following theorem was first proved in [143, Theorem 3.3, Corollary 3.4].

Theorem 7.4.6 *Let $\mathbb{H}$ be a real or complex Hilbert space and $T \in \mathbb{L}(\mathbb{H})$. Then T is left symmetric if and only if $T = 0$.*

Proof The sufficient part is trivial. Conversely, suppose $0 \neq T$ be left symmetric. Then T is r-left symmetric. Clearly T^* is also r-left symmetric. Let $T = U|T|$ and $T^* = V|T^*|$ be the polar decompositions of T and T^*, respectively. Note that the operators U and V can be chosen in such a way that either U or V is an isometry. Let us assume without loss of generality that U is an isometry. Now by Theorem 7.4.5 there exists an $A \in \mathbb{L}(\mathbb{H})$ such that $T \perp_B A$ but $A \not\perp_r T$. As U is an isometry, so $T \perp_B UA$ but $UA \not\perp_r T$. Hence we get an operator $S = UA$ such that $T \perp_B S$ but $S \not\perp_B T$. This contradicts the fact that T is left symmetric. So $T = 0$. □

We next characterize r-right symmetric operators. We begin with the following two lemmas.

Lemma 7.4.7 ([144, Lemma 2.1]) *Let $\mathbb{H}$ be a real or complex Hilbert space. Let $T \in \mathbb{L}(\mathbb{H})$ be positive (self-adjoint in real case). Suppose $\|T\| = 1$. If $\{x_n\}$ is a sequence of unit vectors such that $\|Tx_n\| \to 1$, then $(T - I)x_n \to 0$.*

Proof Clearly, $\|T^2x_n - x_n\|^2 = \|T^2x_n\|^2 + 1 - 2\langle T^2x_n, x_n\rangle \geq 0$ and so

$$1 \geq \|T^2x_n\|^2 \geq 2\langle T^2x_n, x_n\rangle - 1 = 2\|Tx_n\|^2 - 1.$$

Letting $n \to \infty$ we get $\|T^2x_n\| \to 1$ and so $\|T^2x_n - x_n\| \to 0$. Thus $(T + I)(T - I)x_n \to 0$ and so $(T - I)x_n \to 0$. □

Lemma 7.4.8 ([143, Lemma 4.2]) *Let $T = \begin{pmatrix} a & b \\ b & c \end{pmatrix}$, $b \in \mathbb{R}$ be positive, $T \neq \lambda I$ and let $A = \begin{pmatrix} \cos\phi & -\sin\phi \\ \sin\phi & \cos\phi \end{pmatrix}$. Then there exists $\phi \in [0, \frac{\pi}{2})$ such that $A \perp_B T$.*

Proof The matrix A is unitary and so it attains norm on each unit vector. We exhibit a unit vector $x = (x_1, x_2)^t \in \mathbb{R}^2$ such that $\langle Ax, Tx\rangle = 0$. Clearly

$$\langle Ax, Tx\rangle = (a\cos\phi + b\sin\phi)x_1^2 + (2b\cos\phi - (a - c)\sin\phi)x_1x_2 + (c\cos\phi - b\sin\phi)x_2^2.$$

Let $x_2 \neq 0$ and $t = \frac{x_1}{x_2}$. Then $\langle Ax, Tx\rangle = 0$ if and only if

$$(a\cos\phi + b\sin\phi)t^2 + (2b\cos\phi - (a - c)\sin\phi)t + (c\cos\phi - b\sin\phi) = 0.$$

This quadratic equation has real roots if the discriminant

$$(4b^2 + (a - c)^2)\sin^2\phi - 4(ac - b^2)\cos^2\phi$$

is non-negative. So it suffices to take $\phi \in [0, \frac{\pi}{2})$ such that $\tan^2\phi \geq \frac{4(ac-b^2)}{4b^2+(a-c)^2}$. □

Next, we prove the following result.

Proposition 7.4.9 ([143, Proposition 4.3]) *Let $\mathbb{H}$ be a real or complex Hilbert space. Let $T \in \mathbb{L}(\mathbb{H})$ be positive (self-adjoint in real case). If $T \neq \lambda I$ then there exists $A \in \mathbb{L}(\mathbb{H})$ such that $A \perp_B T$ but $T \not\perp_r A$.*

Proof Assume $\|T\| = 1$. Let $x \in S_{\mathbb{H}}$ be such that $Tx \neq 0$. Then $Tx = x$ and $Ty = y$ for each y with $\langle x, y\rangle = 0$ cannot hold simultaneously. So there exists a unit vector $y \in \mathbb{H}$ such that $\langle x, y\rangle = 0$ for which either $Tx \neq x$ or $Ty \neq y$. Let $\mu \in \mathbb{C}$, $|\mu| = 1$, be such that $\mu\langle Ty, x\rangle \in \mathbb{R}$. Let $z = \mu y$. Consider $\mathcal{M} = span\ \{x, z\}$ and let P be the orthogonal projection onto $\mathcal{M}$. Then $PTP - \begin{pmatrix} a & b \\ b & c \end{pmatrix}$, $b \in \mathbb{R}$, with respect to the orthonormal basis $\{x, z\}$. Clearly, PTP is positive and $PTP \neq I$ because of $Tx \neq x$ or $Tz \neq z$. Define $A : \mathbb{H} \to \mathbb{H}$ as follows:

$$Ax = x \cos\phi + z \sin\phi,\ Az = -x \sin\phi + z \cos\phi \text{ and } Av = v,\ v \in \mathcal{M}^{\perp}.$$

Then $PAP = \begin{pmatrix} \cos\phi & -\sin\phi \\ \sin\phi & \cos\phi \end{pmatrix}$, with respect to the orthonormal basis $\{x, z\}$. Using Lemma 7.4.7, there exists $\phi \in [0, \frac{\pi}{2})$ such that $PAP \perp_B PTP$ and so $A \perp_B T$. We claim that $T \not\perp_r A$. If not, then there exists $\{x_n\} \subset S_{\mathbb{H}}$ such that $\|Tx_n\| \to 1$ and $\Re\langle Tx_n, Ax_n\rangle \to 0$. By Lemma 7.4.7, we get $Tx_n - x_n \to 0$ and so $\Re\langle x_n, Ax_n\rangle \to 0$. Let $x_n = u_n + v_n$, where $u_n \in \mathcal{M}$ and $v_n \in \mathcal{M}^{\perp}$. Clearly, $\Re\langle u_n, Au_n\rangle = \|u_n\|^2 \cos\phi$ and so $\Re\langle x_n, Ax_n\rangle = (\cos\phi - 1)\|u_n\|^2 + 1$. This leads to a contradiction as $\cos\phi \neq 0$. This completes the proof. □

Now, we are in a position to characterize right symmetric operators.

Theorem 7.4.10 ([143, Corollary 4.5]) *Let $\mathbb{H}$ be a real or complex Hilbert space and $T \in \mathbb{L}(\mathbb{H})$. Then T is right symmetric if and only if T is a scalar multiple of an isometry or coisometry.*

Proof We first prove the sufficient part. Let T be an isometry and $A \perp_B T$. Then there exists $\{x_n\} \subset S_{\mathbb{H}}$ such that $\|Ax_n\| \to \|A\|$ and $\langle Ax_n, Tx_n\rangle \to 0$. Since T is an isometry the same sequence shows that $T \perp_B A$. Thus T is right symmetric. Similarly, if T^* is an isometry, i.e., T is a coisometry, then T^* is right symmetric and so is T.
Next, we prove the necessary part. Let T be right symmetric. If possible, suppose neither T nor T^* is an isometry. Let $T = U|T|$ and $T^* = V|T^*|$ be the polar decompositions of T and T^*, respectively, where $|T| \neq I$, $|T^*| \neq I$. Note that, the operators U and V can be chosen in such a way that either U or V is an isometry. Let us assume without loss of generality that U is an isometry. Now, by Proposition 7.4.9, there exists an $A \in \mathbb{L}(\mathbb{H})$ such that $A \perp_B |T|$ but $|T| \not\perp_r A$. As U is an isometry, so $UA \perp_B T$ but $T \not\perp_r UA$ and thus $T \not\perp_B UA$. Hence, we get an operator $S = UA$ such that $S \perp_B T$ but $T \not\perp_B S$. This contradicts the fact that T is right symmetric. This completes the proof. □

We end this section with the following two simple observations:

Remark 7.4.11 (i) Let $0 \neq T \in \mathbb{L}(\mathbb{H})$ be a compact operator on an infinite-dimensional real or complex Hilbert space $\mathbb{H}$. Then T is not right symmetric, as the compactness of T will force the unit ball to be compact.
(ii) Let $\mathbb{H}$ be a real or complex finite-dimensional Hilbert space and $0 \neq T \in \mathbb{L}(\mathbb{H})$. Then T is right symmetric if and only if $M_T = S_{\mathbb{H}}$.

7.5 Left Symmetric and Right Symmetric Operators on ℓ_1^n and ℓ_∞^n

In the previous section, we have obtained complete descriptions of the left symmetric and right symmetric linear operators on a finite-dimensional Hilbert space. These characterizations are not true, in general, in the framework of Banach spaces. In this section, we study the symmetric properties of B-J orthogonality, in the space of linear operators on ℓ_1^n and ℓ_∞^n. We find a necessary and sufficient condition for a linear operator T to be left symmetric and right symmetric both on ℓ_1^n and ℓ_∞^n. We write e_i for the element $(0, 0, \ldots, 0, 1, 0, \ldots, 0)$ where the i-th coordinate is 1 and all other coordinates are zero, for any $i \in \{1, 2, \ldots, n\}$. We first note that the extreme points (or their scalar multiples) of the closed unit ball of ℓ_1^n are the only right symmetric points of ℓ_1^n (see [37, Lemma 2.1]). The next theorem characterizes left symmetric operators on ℓ_1^n.

Theorem 7.5.1 ([37, Theorem 2.3]) *Let T be a linear operator on ℓ_1^n. For $n = 2$, $(0 \neq)T$ is left symmetric if and only if T attains norm at only one extreme point (up to scalar multiple), image of which is a left symmetric point of ℓ_1^n and images of other extreme points are zero. For $n \geq 3$, then T is left symmetric if and only if T is the zero operator.*

Proof **Sufficient part**: Note that, the zero operator is trivially left symmetric. Therefore, we consider that $n = 2$ and $T \neq 0$. Suppose T attains its norm at only one extreme point and images of other extreme points are zero. Without loss of generality, assume that T attains its norm only at $\pm e_1$. Let A be an operator such that $T \perp_B A$. Then from Theorem 4.3.3, it follows that $Te_1 \perp_B Ae_1$. Since Te_1 is left symmetric, we get that $Ae_1 \perp_B Te_1$. Moreover, $Ae_2 \perp_B Te_2 = 0$. Clearly, either e_1 or e_2 is in M_A. Thus, we get $A \perp_B T$. Hence, T is left symmetric.

Necessary part: We first prove that for $n \geq 2$, if $(0 \neq)T$ is left symmetric, then T attains norm at only one extreme point (up to scalar multiple), image of which is a left symmetric point of ℓ_1^n and images of other extreme points are zero. Clearly, T attains its norm at an extreme point, say at e_{i_0}.
Claim: $Te_i = 0$ for all $i \in \{1, 2, ..., n\} - \{i_0\}$.
If possible, suppose that $Te_j \neq 0$ for some $j \neq i_0$. Define a linear operator A on ℓ_1^n as follows:

$$Ae_j = Te_j, \ Ae_i = 0, \ i \neq j.$$

Observe that A attains norm only at $\pm e_j$. Moreover, $T \perp_B A$, as $Te_{i_0} \perp_B Ae_{i_0}$ and $\|Te_{i_0}\| = \|T\|$. From Theorem 4.3.3, it follows that $A \not\perp_B T$ as $Ae_j \not\perp_B Te_j$ and $M_A = \{\pm e_j\}$. Therefore, $Te_i = 0$ for all $i \neq i_0$. Next, we claim that Te_{i_0} is left symmetric. If possible, let Te_{i_0} be not left symmetric, i.e., there exists w such that $Te_{i_0} \perp_B w$ but $w \not\perp_B Te_{i_0}$. Define a linear operator A on ℓ_1^n as follows:

$$Ae_{i_0} = w, \ Ae_i = 0, \ i \neq i_0.$$

Observe that A attains norm only at $\pm e_{i_0}$. Moreover, $T \perp_B A$, as $Te_{i_0} \perp_B Ae_{i_0}$ and $\|Te_{i_0}\| = \|T\|$. But $A \not\perp_B T$ as $Ae_{i_0} \not\perp_B Te_{i_0}$. Therefore, $T \perp_B A$ but $A \not\perp_B T$, this contradiction proves our claim. This completes the proof of the necessary part for the case $n = 2$.
Now, from [23, Theorem 2.8], it follows that for $n \geq 3$, there is no non-zero left symmetric point in ℓ_1^n. Therefore, we can conclude that for $n \geq 3$, there is no non-zero left symmetric operator on ℓ_1^n. This completes the proof. □

Remark 7.5.2 Any operator T on ℓ_1^2 satisfying the property as in Theorem 7.5.1 is either of the form $\begin{pmatrix} 1 & 0 \\ \pm 1 & 0 \end{pmatrix}$ or $\begin{pmatrix} 0 & 1 \\ 0 & \pm 1 \end{pmatrix}$ or their scalar multiples.

In the following theorem we state the characterization theorem for right symmetric operators on ℓ_1^n, we omit the proof here, it is more of computation and details can be seen in [37, Theorem 2.2].

Theorem 7.5.3 ([37, Theorem 2.2]) *Suppose T is a linear operator on ℓ_1^n. Then T is right symmetric if and only if T attains norm at all extreme points and images of the extreme points are scalar multiples of extreme points.*

We next study left symmetric operators on ℓ_∞^n. This will once again highlight the difference between the geometry of two-dimensional and three-dimensional Banach spaces. Note that ℓ_∞^2 is isometrically isomorphic to ℓ_1^2. Therefore, the characterization of left symmetric operators on ℓ_∞^2 follows from Theorem 7.5.1. An alternate direct proof of the same is provided below.

Theorem 7.5.4 ([38, Theorem 2.3]) *Let $(0 \neq)T$ be a linear operator on ℓ_∞^2. Then T is left symmetric if and only if T attains norm at only one extreme point, say x_1, Tx_1 is a left symmetric point and image of the other extreme point is zero.*

Proof Let $Ext(\ell_\infty^2) = \{\pm x_1, \pm x_2\}$. Suppose $x_1 \in M_T$ and $Tx_2 = 0$. Let A be a linear operator such that $T \perp_B A$. Then by Theorem 4.3.3, $Tx_1 \perp_B Ax_1$. Since Tx_1 is left symmetric, we have $Ax_1 \perp_B Tx_1$. Moreover, $Ax_2 \perp_B Tx_2 = 0$. Clearly, either x_1 or x_2 is in M_A and $Ax_j \perp_B Tx_j$ for $j = 1, 2$. Hence, $A \perp_B T$.

Conversely, let T be left symmetric. Without loss of generality, suppose that $x_1 \in M_T$. We claim that $Tx_2 = 0$. If possible, suppose that $Tx_2 \neq 0$. Define a linear operator A on ℓ_∞^2 as $Ax_1 = 0, \ Ax_2 = Tx_2$. Then clearly, $M_A = \{\pm x_2\}$. Moreover,

$T \perp_B A$, as $Tx_1 \perp_B Ax_1$ and $\|Tx_1\| = \|T\|$. But since $Ax_2 \not\perp_B Tx_2$, we have $A \not\perp_B T$. Thus, $Tx_2 = 0$.
Next, we claim that Tx_1 is left symmetric. If possible, suppose that Tx_1 is not left symmetric, i.e., there exists w such that $Tx_1 \perp_B w$ but $w \not\perp_B Tx_1$. Define a linear operator A on ℓ_∞^2 by $Ax_1 = w,\ Ax_2 = 0$. Note that $M_A = \{\pm x_1\}$. Moreover, $T \perp_B A$, since $Tx_1 \perp_B Ax_1$ and $x_1 \in M_T$. But $A \not\perp_B T$, since $Ax_1 \not\perp_B Tx_1$. Thus, we have $T \perp_B A$ but $A \not\perp_B T$, a contradiction to our hypothesis. This completes the proof. □

Example 7.5.5 Note that, for the operator $T \in \mathbb{L}(\ell_\infty^2)$ defined as

$$T(1,1) = (1,0),\ T(1,-1) = (0,0),$$

$M_T = \{\pm(1,1)\}$ and $T(1,1)$ is a non-zero left symmetric point of ℓ_∞^2. Therefore, T is a left symmetric operator on ℓ_∞^2.

In the following theorem, we characterize the left symmetric linear operators on ℓ_∞^n for $n \geq 3$.

Theorem 7.5.6 ([38, Theorem 2.5]) *Suppose T is a linear operator on ℓ_∞^n, $n \geq 3$. Then T is left symmetric if and only if T is the zero operator.*

Proof The sufficient part of the proof is trivial. We only prove the necessary part. Let T be a left symmetric operator on ℓ_∞^n. Suppose that $x_1 \in M_T \cap Ext(\ell_\infty^n)$. We claim that $Te = 0$ for all $e \in Ext(\ell_\infty^n) \setminus \{\pm x_1\}$. If possible, assume that there exists $x_2 \in Ext(\ell_\infty^n) \setminus \{\pm x_1\}$ such that $Tx_2 \neq 0$. Since $x_2 \perp_{SB} x_1$, there exists a hyperplane H such that $x_1 \in H$ and $x_2 \perp_{SB} H \setminus \{0\}$. Let $\{x_1, x_3, x_4, ..., x_n\}$ be a basis of H. Then $\{x_1, x_2, x_3, x_4, ..., x_n\}$ is a basis of $\mathbb{R}^n$. Define a linear operator A on ℓ_∞^n as follows:

$$Ax_2 = Tx_2,\quad Ax_i = 0,\quad i \neq 2.$$

Verify that $M_A = \{\pm x_2\}$. Moreover, $T \perp_B A$, as $Tx_1 \perp_B Ax_1$ and $x_1 \in M_T$. But $A \not\perp_B T$, since $Ax_2 \not\perp_B Tx_2$. Therefore, $Te = 0$ for all $e \in Ext(\ell_\infty^n) \setminus \{\pm x_1\}$. Clearly, $Ext(\ell_\infty^n) \setminus \{\pm x_1\}$ contains a basis of ℓ_∞^n. Therefore, T is the zero operator. This completes the proof. □

Finally, in this section, we find a necessary and sufficient condition for an operator T on ℓ_∞^n to be right symmetric. We omit the computational details of the proof, which can be found in [38, Theorem 2.1].

Theorem 7.5.7 ([38, Theorem 2.1]) *Suppose $T = (t_{ij})$ is a non-zero linear operator on ℓ_∞^n. Then T is right symmetric if and only if for each $i \in \{1, 2, \ldots, n\}$, exactly one term of $t_{i1}, t_{i2}, \ldots, t_{in}$ is non-zero and of the same magnitude.*

We record the following observation which is now obvious.

Remark 7.5.8 The norm one right symmetric linear operators on ℓ_∞^n attains norm at all extreme points and images of the extreme points are also extreme points.

7.6 Left Symmetric and Right Symmetric Elements in $\mathbb{L}(\mathbb{X}, \mathbb{Y})$

From the previous sections, it is clear that the characterization of left and right symmetric operators in the setting of Banach spaces is more complicated than that in Hilbert spaces, due to the rich geometric structure present in the later case. In [135, Theorem 2.8], it was proved that there is no non-zero left symmetric operator on $\ell_p^2(\mathbb{R})$, where $p \geq 2,\ p \neq \infty$. The result was further generalized in [118]. The authors in [118] proved that if T is a bounded linear operator on a finite-dimensional strictly convex and smooth Banach space, then T is left symmetric if and only if T is the zero operator. One of the main tools that is being used to study left symmetric operators is the connection between orthogonality in the space of linear operators and that in the ground space. This connection was partly explored for compact linear operators in Chap. 4. The following theorem is a continuation in the same line.

Theorem 7.6.1 ([102, Theorem 2.2]) *(i) Let $\mathbb{X}$ be a reflexive, Kadets–Klee Banach space and let $\mathbb{Y}$ be any Banach space. Suppose $T \in \mathbb{K}(\mathbb{X}, \mathbb{Y})$. Then for any $A \in \mathbb{L}(\mathbb{X}, \mathbb{Y})$, $T \perp_B A$ if and only if there exist $x, y \in M_T$ such that $Ax \in (Tx)^+$ and $Ay \in (Ty)^-$.*
(ii) In addition, if $M_T = D \cup (-D)$, where D is a compact, connected subset of $S_{\mathbb{X}}$, then there exists $x \in M_T$ such that $Tx \perp_B Ax$.

Proof (i) The sufficient part is trivial. For the proof of the necessary part, assume that $T \perp_B A$. We first claim that if $\{x_n\}$ is an arbitrary norming sequence for T, then there exists $x \in M_T$ such that $x_{n_k} \to x$ for some subsequence $\{x_{n_k}\}$. Note that, $B_{\mathbb{X}}$ is weakly compact, since $\mathbb{X}$ is reflexive. Hence, there is a subsequence $\{x_{n_k}\}$ which weakly converges to some element $x \in B_{\mathbb{X}}$. Since T is compact, we have $Tx_{n_k} \to Tx$. Again, $\{x_n\}$ being a norming sequence for T, we get $\|Tx\| = \|T\|$, i.e., $x \in M_T$. Now, $\|x_{n_k}\| = \|x\| = 1$. Since $x_{n_k} \rightharpoonup x,\ \|x_{n_k}\| \to \|x\|$ and $\mathbb{X}$ is Kadets–Klee, we have $x_{n_k} \to x$. This proves our claim. Since $T \perp_B A$, from Theorem 4.4.14, it follows that either (a) or (b) is true:
(a) There exists a norming sequence $\{x_n\}$ for T such that $\|Ax_n\| \to 0$ as $n \to \infty$.
(b) There exist two norming sequences $\{x_n\},\ \{y_n\}$ for T and two sequences of positive real numbers $\{\epsilon_n\},\ \{\delta_n\}$ such that $\epsilon_n \to 0,\ \delta_n \to 0$ as $n \to \infty$ and $Ax_n \in (Tx_n)^{+\epsilon_n}$ and $Ay_n \in (Ty_n)^{-\delta_n}$ for all $n \in \mathbb{N}$.
First assume that (a) holds. Then by the above claim, we have $x \in M_T$ such that $Ax = 0$ and hence $Tx \perp_B Ax$. Now, assume that (b) holds. Then $\|Tx_n + \lambda Ax_n\| \geq \sqrt{1-\epsilon^2}\|Tx_n\|$ for all $\lambda \geq 0$ and for all $n \in \mathbb{N}$. Again we can find $x \in M_T$ such that $\|Tx + \lambda Ax\| \geq \|Tx\|$ for all $\lambda \geq 0$, i.e., $Ax \in (Tx)^+$. Similarly, we get $y \in M_T$ such that $Ay \in (Ty)^-$. This completes the proof.

(ii) Suppose there is no $x \in M_T$ such that $Tx \perp_B Ax$. Then using Lemma 4.3.5, we get $\lambda_0 \neq 0$ such that $\|Tx + \lambda_0 Ax\| < \|Tx\|\ \forall\, x \in M_T$. Let $\lambda_0 > 0$. Then $Ax \notin (Tx)^+$ for all $x \in M_T$, which contradicts (i). Similarly, $\lambda_0 < 0$ implies that $Ax \notin (Tx)^-$ for all $x \in M_T$, which again contradicts (i). Thus, there exists $x \in M_T$ such that $Tx \perp_B Ax$, completing the proof. □

The next theorem gives a necessary condition for a compact operator to be left symmetric. Such a condition was first provided in [135, Theorem 2.5] for operators on a finite-dimensional strictly convex smooth Banach space.

Theorem 7.6.2 ([102, Theorem 2.3]) *Let $\mathbb{X}$, $\mathbb{Y}$ be Banach spaces. Suppose that $\mathbb{X}$ is strictly convex and reflexive. Suppose that $T \in \mathbb{K}(\mathbb{X}, \mathbb{Y})$ is left symmetric with $\|T\| = 1$. If $x_1 \in M_T$, $y \in S_{\mathbb{X}}$, $y \perp_B x_1$, then $Ty = 0$.*

Proof If possible, let $Ty \neq 0$. Since $y \perp_B x_1$, there exists $f \in \mathbb{X}^*$ such that $\|f\| = 1$, $f(y) = 1$, $f(x_1) = 0$. Define $A \in \mathbb{K}(\mathbb{X}, \mathbb{Y})$ by $Ax := f(x)Ty$. Since $Tx_1 \perp_B Ax_1$ and $x_1 \in M_T$, we get $T \perp_B A$. Thus, $A \perp_B T$. Since $\mathbb{X}$ is strictly convex, $M_A = \{\pm y\}$. Therefore, from Theorem 4.3.10 we get that $Ay \perp_B Ty$, i.e., $Ty \perp_B Ty$. Hence $Ty = 0$, a contradiction. This completes the proof. □

Similarly, using Theorem 7.6.1, we easily get the following result.

Theorem 7.6.3 ([102, Theorem 2.4]) *Let $\mathbb{X}$ be a Banach space which is reflexive, Kadets–Klee, strictly convex and let $\mathbb{Y}$ be any Banach space. Suppose $T \in \mathbb{L}(\mathbb{X}, \mathbb{Y})$ is left symmetric with $M_T \neq \emptyset$. For $x \in M_T$ if $y \perp_B x$, then $Ty = 0$.*

Next, we prove a lemma which is needed in the characterization of left symmetric operators.

Lemma 7.6.4 ([102, Lemma 2.1]) *Let $\mathbb{X}$ be a strictly convex Banach space. Suppose $u, v \in \mathbb{X}$ and $v \perp_B u$. Let $w = (1-t)u + tv$, where $t \in (0, 1)$. Then $au \notin (av + bw)^-$, if $ab > 0$.*

Proof Let $a > 0, b > 0$. Then $av + bw = av + b(1-t)u + btv = (a + bt)v + b(1-t)u$. Note that, $a + bt > 0$ and $b(1-t) > 0$. Since $\mathbb{X}$ is strictly convex and $v \perp_B u$, we have $\|(a + bt)v\| < \|(a + bt)v + b(1-t)u\|$. Therefore,

$$\|(av + bw) - b(1-t)u\| = \|(a + bt)v\| < \|(a + bt)v + b(1-t)u\| = \|av + bw\|,$$

which implies that $u \notin (av + bw)^-$. Hence, from Proposition 3.2.2, it follows that $au \notin (av + bw)^-$, since $a > 0$.
Next, consider $a < 0, b < 0$. Then as before,

$$\|(av + bw) - b(1-t)u\| = \|(a + bt)v\| < \|(a + bt)v + b(1-t)u\| = \|av + bw\|.$$

Here $-b(1-t) > 0$. This shows that $u \notin (av + bw)^+$. Again, from Proposition 3.2.2, it follows that $au \notin (av + bw)^-$, since $a < 0$. This completes the proof. □

The following theorem on left symmetric operators is based on the idea from [118, Theorem 2.2].

Theorem 7.6.5 ([102, Theorem 2.5]) *Let $\mathbb{X}$ be a Banach space which is reflexive, Kadets–Klee, strictly convex and let $\mathbb{Y}$ be a strictly convex Banach space. Then $T \in \mathbb{L}(\mathbb{X}, \mathbb{Y})$ with $M_T \neq \emptyset$ is left symmetric if and only if T is the zero operator.*

Proof The sufficient part follows trivially. We prove the necessary part in three steps by the method of contradiction. If possible, suppose that T is a non-zero left symmetric operator.

Step 1: We prove that for each $x \in M_T$ there is a hyperspace H such that $x \perp_B H$ and $T(H) = 0$. Let $x \in M_T$. Then by Theorem 7.6.3, we get $Ty = 0$ for all y with $y \perp_B x$. Since $x \in M_T$, using Proposition 3.2.10, we get a hyperspace H such that $x \perp_B H$ and $Tx \perp_B T(H)$. We claim that $T(H) = 0$. Let $y \in H \cap S_{\mathbb{X}}$. Then there is a hyperspace H_y such that $y \perp_B H_y$. Define $A : \mathbb{X} \to \mathbb{Y}$ by $A(cy + h) = cTy$, where c is a scalar and $h \in H_y$. Note that, A is compact and $M_A = \{\pm y\}$, since $\mathbb{X}$ is strictly convex. Let $x = by + h$. Then $Ax = bTy$. Since $x \perp_B y$, we have $Tx \perp_B Ty$. Thus, $Tx \perp_B Ax$ and so $T \perp_B A$. Since T is left symmetric, we have $A \perp_B T$. Therefore, from Theorem 7.6.1 (ii), it follows that $Ay \perp_B Ty$. Thus, $Ty \perp_B Ty$, which ensures that $Ty = 0$. This justifies our claim $T(H) = 0$.

Step 2: We prove that $H \perp_B x$ and Tx is left symmetric. Let $h \in H$. Then $Th = 0$. There exists $d \in \mathbb{R}$ such that $dx + h \perp_B x$. Hence $T(dx + h) = 0$, which implies that $d = 0$. Thus $h \perp_B x$ and hence $H \perp_B x$. Now, we show that Tx is left symmetric. Let $Tx \perp_B u$. As before, define $A : \mathbb{X} \to \mathbb{Y}$ by: $A(ax + h) = au$, where a is a scalar and $h \in H$. Note that $M_A = \{\pm x\}$. Since $Tx \perp_B Ax$ and $x \in M_T$, we have $T \perp_B A$. Thus, $A \perp_B T$, since T is left symmetric. Therefore, from Theorem 7.6.1 (ii), it follows that $Ax \perp_B Tx$, i.e., $u \perp_B Tx$. This proves that Tx is left symmetric.

Step 3: We now construct an operator A such that $T \perp_B A$ but $A \not\perp_B T$. Choose $y \in H \cap S_{\mathbb{X}}$. Clearly, $x \perp_B y$ and $y \perp_B x$. Let $y \perp_B H_y$, where H_y is a subspace of H of codimension one in H. Suppose $v \in S_{\mathbb{X}}$ is such that $Tx \perp_B v$. Since Tx is left symmetric, we have $v \perp_B Tx$. Let $\|x + y\| = r$. Then from orthogonality and strict convexity it follows that $1 < r < 2$. Choose $0 < t < 1$ such that $(1 - t)(1 + \|T\|) < \frac{2-r}{1+2r}$. Choose $\epsilon \in (0, 1)$ such that $(1 - t)(1 + \|T\|) < \epsilon < \frac{2-r}{1+2r}$. Since $1 < r < 2$, we have $\frac{2-r}{1+2r} < 1$. Let $w = (1 - t)Tx + tv$. Then $\|w - v\| = (1 - t)\|Tx - v\| \leq (1 - t)(1 + \|T\|) < \epsilon$. Now, we can write each element $z \in \mathbb{X}$ as $z = ax + by + h$ where a, b are scalars and $h \in H_y$. Consider $A : \mathbb{X} \to \mathbb{Y}$ defined as $Az = av + bw$. Clearly, A is compact and $T \perp_B A$ since $x \in M_T$ and $Tx \perp_B Ax$. We now show that $A \not\perp_B T$. Note that,

$$\begin{aligned}
\|A(\frac{x+y}{\|x+y\|})\| &= \frac{\|v+w\|}{r} \\
&= \frac{\|2v + w - v\|}{r} \\
&\geq \frac{2-\epsilon}{r} \\
&> 1 + 2\epsilon.
\end{aligned}$$

Therefore, $\|A\| > 1 + 2\epsilon$. It is easy to see that $x, y \notin M_A$. Moreover, for each $h \in H_y$ we have $h \notin M_A$. Next, we show that $z \notin M_A$ if $ab < 0$. Suppose $z = -ax + by + h \in S_{\mathbb{X}}$ where $a > 0, b > 0, b - a > 0$. Using orthogonality, we have $1 = \|z\| =$

$\| -ax + by + h\| >| a |, 1 = \|z\| = \|by + h - ax\| \geq| b | - | a |= b - a$. Moreover, $| b |= \|by\| \leq \|by + h\| = \|z + ax\| \leq 2$. Therefore,

$$
\begin{aligned}
\|Az\| &= \| - av + bw\| \\
&= \|(b-a)v + b(w-v)\| \\
&\leq | b - a | + | b | \|w - v\| \\
&< | b - a | + | b | \epsilon \\
&= b - a + b\epsilon \\
&\leq 1 + 2\epsilon \\
&< \|A(\frac{x+y}{\|x+y\|})\|.
\end{aligned}
$$

Now, we consider $z = -ax + by + h \in S_{\mathbb{X}}$ where $a > 0, b > 0, b - a \leq 0$. Then also $\|Az\| <| b - a | + | b | \epsilon < a < 1$. This proves that if $z = -ax + by + h \in S_{\mathbb{X}}$, where $a > 0, b > 0$, then $z \notin M_A$. Similarly, taking $z = ax - by + h \in S_{\mathbb{X}}$, where $a > 0, b > 0$ we can prove that $z \notin M_A$. Thus, if $z = ax + by + h \in M_A$, then we must have $ab > 0$. Our next claim is that $Tz \notin (Az)^-$ for all $z \in M_A$. Let $z = ax + by + h \in M_A$. Then $ab > 0$, $Az = av + bw$ and $Tz = aTx$. Therefore, from Lemma 7.6.4, we have $Tz \notin (Az)^-$. Now, from Theorem 7.6.1, it follows that $A \not\perp_B T$. Thus, T is not left symmetric, a contradiction. This completes the proof. □

As easy consequences, we get the following two theorems.

Theorem 7.6.6 ([102, Theorem 2.6]) *Let $\mathbb{X}$ be reflexive, strictly convex Banach space and let $\mathbb{Y}$ be a strictly convex Banach space. Then $T \in \mathbb{K}(\mathbb{X}, \mathbb{Y})$ is left symmetric if and only if T is the zero operator.*

Theorem 7.6.7 ([118, Theorem 2.5]) *Let $\mathbb{X}$ be an $n-$dimensional strictly convex and smooth Banach space. $T \in \mathbb{L}(\mathbb{X})$ is left symmetric if and only if T is the zero operator.*

Remark 7.6.8 The characterization of left symmetric operators between arbitrary Banach spaces $\mathbb{X}$ and $\mathbb{Y}$ is still open.

Now, the natural question is whether there exist non-trivial left symmetric operators defined between Banach spaces. The answer is affirmative. Let us start with the following theorem.

Theorem 7.6.9 ([102, Theorems 2.7, 2.8]) *Let $\mathbb{X}, \mathbb{Y}$ be Banach spaces. Suppose that $\mathbb{X}$ is strictly convex and reflexive. Suppose that $T \in \mathbb{K}(\mathbb{X}, \mathbb{Y})$ is left symmetric with $\|T\| = 1$. Then $\dim(T(\mathbb{X})) = 1$. Also there exist $w \in S_{\mathbb{Y}}$ and $f \in S_{\mathbb{X}^*}$ such that $T(\cdot) = f(\cdot)w$.*

Proof Note that $M_T \neq \emptyset$, since T is compact and defined on a reflexive Banach space. Fix $x_1 \in M_T$. If possible, assume that $\dim(T(\mathbb{X})) > 1$. Then there exists $z \in S_{\mathbb{X}}$ such that $\dim \operatorname{span}\{Tz, Tx_1\} = 2$. Therefore, there is $y \in \operatorname{span}\{z, x_1\}$ such that $y \perp_B x_1, y \neq 0$. Hence, $Ty \neq 0$, a contradiction to Theorem 7.6.2. The next part of theorem is now obvious. □

We can strengthen Theorem 7.6.9 as follows.

Theorem 7.6.10 ([102, Theorem 2.9]) *Let $\mathbb{X}$, $\mathbb{Y}$ be Banach spaces. Suppose that $\mathbb{X}$ is strictly convex and reflexive. Let $T \in \mathbb{K}(\mathbb{X}, \mathbb{Y})$ be left symmetric with $\|T\| = 1$. Assume that $x_1 \in M_T$. Then there are $w \in S_{\mathbb{Y}}$ and $f \in S_{\mathbb{X}^*}$ such that $T(\cdot) = f(\cdot)w$ (moreover, $|M_T| = 2$ by strictly convexity of $\mathbb{X}$, so we may assume $M_T = \{\pm x_1\}$ for some $x_1 \in S_{\mathbb{X}}$). Then:*

(a) *x_1 is right symmetric, w is left symmetric,*

(b) *x_1 is left symmetric $\Leftrightarrow$ x_1 is smooth.*

Proof (a) Choose $y \in \mathbb{X} \setminus \{0\}$ such that $y \perp_B x_1$. From Theorem 7.6.2, it follows that $Ty = 0$. Thus $f(y) = 0$. Clearly, $|f(x_1)| = 1$. Therefore, from Theorem 2.1.14, we get $x_1 \perp_B y$. Hence, x_1 is right symmetric.

Now, choose $z \in \mathbb{Y} \setminus \{0\}$ such that $w \perp_B z$. Define $A \in \mathbb{K}(\mathbb{X}, \mathbb{Y})$ by $A(\cdot) := f(\cdot)z$. Clearly, $M_T = M_A$. It is easy to verify that $Tx_1 \perp_B Ax_1$, i.e., $T \perp_B A$. Therefore, $A \perp_B T$ and so by Theorem 4.3.10, $Ax_1 \perp_B Tx_1$. Hence $z \perp_B w$. This proves that w is left symmetric.

(b) If x_1 is left symmetric, then by Proposition 7.3.1 (iii), we get x_1 is smooth. On the other hand, if x_1 is smooth, then by Proposition 7.3.1 (i), we get x_1 is left symmetric. □

As a consequence of Theorem 7.6.10 and properties of the spaces $\mathbb{Y}$, $\mathbb{Y}^*$, we will prove the next theorem. The following lemma is required.

Lemma 7.6.11 ([102, Lemma 2.2]) *Let $\mathbb{X}$, $\mathbb{Y}$ be Banach space. Let $f \in \mathbb{X}^*$, $\|f\| = 1$. The spaces $\mathbb{Y}$ and $K_f(\mathbb{X}, \mathbb{Y}) := \{A_y \in \mathbb{K}(\mathbb{X}, \mathbb{Y}) : A_y(\cdot) := f(\cdot)y,\ y \in \mathbb{Y}\}$ are isometrically isomorphic.*

The proof is simple: define $\gamma\colon \mathbb{Y} \to K_f(\mathbb{X}, \mathbb{Y})$ by the formula $\gamma(y) := A_y$. The rest is clear. It follows from Lemma 7.6.11 that

$$A_z \perp_B A_w \quad \Leftrightarrow \quad z \perp_B w. \tag{7.6.1}$$

The next theorem gives a necessary condition for left symmetric operator.

Theorem 7.6.12 ([102, Theorem 2.10]) *Let $\mathbb{X}$, $\mathbb{Y}$ be reflexive Banach spaces. Suppose that $\mathbb{Y}$ is smooth. Let $T \in \mathbb{K}(\mathbb{X}, \mathbb{Y})$ be left symmetric with $\|T\| = 1$. Then there are $w \in S_{\mathbb{Y}}$ and $f \in S_{\mathbb{X}^*}$ such that $T(\cdot) = f(\cdot)w$ and w is left symmetric.*

Proof Clearly, $\mathbb{Y}^*$ is strictly convex and $T^* \in \mathbb{K}(\mathbb{Y}^*, \mathbb{X}^*)$ is left symmetric. Since $\mathbb{Y}^*$ is reflexive and T^* is compact, we get $M_{T^*} \neq \emptyset$. Now, from Theorem 7.6.9, we get that $\dim(T^*(\mathbb{Y}^*)) = 1$ and hence $\dim(T(\mathbb{X})) = 1$ by the reflexivity of $\mathbb{X}$, $\mathbb{Y}$. Therefore, there exist $w \in S_{\mathbb{Y}}$ and $f \in S_{\mathbb{X}^*}$ such that $T(\cdot) = f(\cdot)w$.

Now, we show that w is left symmetric. Choose $z \in \mathbb{Y} \setminus \{0\}$ such that $w \perp_B z$. Define $A_z \in \mathbb{K}(\mathbb{X}, \mathbb{Y})$ by $A_z(\cdot) := f(\cdot)z$. Similarly, by the reflexivity of $\mathbb{X}$ and by the compactness of T, we have $M_T \neq \emptyset$. For some $x_1 \in M_T$, we get $Tx_1 \perp_B A_z x_1$ and so $T \perp_B A_z$. Since T is left symmetric, we have $A_z \perp_B T$. It is easy to observe that $T = A_w$. Applying (7.6.1) and Lemma 7.6.11, we get $z \perp_B w$. This proves that w is left symmetric. □

Now, we are in a position to exhibit non-zero left symmetric operators. Let $\mathbb{X} \oplus_1 \mathbb{Z}$ denote the space $\mathbb{X} \times \mathbb{Z}$ with $\|(x, z)\|_1 := \|x\| + \|z\|$. We characterize left symmetric linear operators from $\mathbb{X} \oplus_1 \mathbb{R}$ into a reflexive smooth space $\mathbb{Y}$.

Theorem 7.6.13 ([102, Theorem 2.11]) *Let $\mathbb{X}$ be reflexive Banach space. Suppose that $\mathbb{Y}$ is reflexive and smooth Banach space. Let $T \in \mathbb{K}(\mathbb{X} \oplus_1 \mathbb{R}, \mathbb{Y})$ be a non-zero operator with $\|T\| = 1 = \|T(0, 1)\|$. Then the following conditions are equivalent:*
(a) *T is left symmetric,*
(b) *there are $w \in S_{\mathbb{Y}}$, $f \in S_{(\mathbb{X}\oplus_1\mathbb{R})^*}$ such that $T(\cdot) = f(\cdot)w$, $\mathbb{X} = ker(f)$ and w is left symmetric.*

Proof (a)$\Rightarrow$(b). Using Theorem 7.6.12, we get $w \in S_{\mathbb{Y}}$ and $f \in S_{(\mathbb{X}\oplus_1\mathbb{R})^*}$ such that $T(\cdot) = f(\cdot)w$, and w is left symmetric.

In order to prove $\mathbb{X} = \ker(f)$, we observe that $\mathbb{X} \perp_B (0, 1)$. From Theorem 7.6.2, it follows that $T(\mathbb{X}) = \{0\}$. Therefore $\mathbb{X} \subset \ker(f)$. Since $\operatorname{co}\dim(\mathbb{X}) = 1$, we have $\mathbb{X} = \ker(f)$.

(b)$\Rightarrow$(a). Suppose that there are $w \in S_{\mathbb{Y}}$, $f \in S_{(\mathbb{X}\oplus_1\mathbb{R})^*}$ such that $T(\cdot) = f(\cdot)w$, $\mathbb{X} = \ker(f)$ and w is left symmetric. Let $T \perp_B A$ and $A \neq 0$. It is easy to check that there is a connected closed subset $D \subset S_{\mathbb{X}}$ such that $M_T = D \cup -D$. Hence, there is $x_1 \in M_T$ such that $Tx_1 \perp_B Ax_1$ and $Tx_1 = w$, i.e., $w \perp_B Ax_1$. Since w is left symmetric, we have $Ax_1 \perp_B w$ and so we get $Ax_1 \perp_B Tx_1$. Let $x_2 \in M_A$. If $x_2 \in \{\pm x_1\}$, then $Ax_1 \perp_B Tx_1$ and so $A \perp_B T$. But, if $x_2 \notin \{\pm x_1\}$ (more precisely, $x_2 \notin \{\pm(0, 1)\}$), then there is an extreme point $(e, 0) \in Ext(B_{\mathbb{X}\oplus_1\mathbb{R}})$ such that $\|A(e, 0)\| = \|A\|$. Since $\mathbb{X} = \ker(f)$, we have $T(e, 0) = 0$. It implies that $A(e, 0) \perp_B T(e, 0)$. Thus $A \perp_B T$, completing the proof. □

Using theorem 7.6.13, the left symmetric operators from ℓ_1^n to a reflexive, smooth Banach space $\mathbb{Y}$ were characterized in [102, Theorem 2.12]. The result was further generalized in [66, Theorem 3.6] by removing the additional assumption that $\mathbb{Y}$ needs to be smooth and reflexive Banach space.

Theorem 7.6.14 ([66, Theorem 3.6]) *Let $\mathbb{Y}$ be any Banach space and let $T \in \mathbb{L}(\ell_1^n, \mathbb{Y})$ be such that $\|T\| = 1$, where $n \geq 2$. Then the following are equivalent:*
(i) T is left symmetric.
(ii) There exists i, $1 \leq i \leq n$, such that Te_i is a left symmetric point in $\mathbb{Y}$ and $Te_j = 0$ for all $j \neq i$.

Proof $(i) \Rightarrow (ii)$. Since T must attain norm at an extreme point of $B_{\ell_1^n}$, there exists some i, $1 \leq i \leq n$, such that $e_i \in M_T$. If possible, let Te_i be not a left symmetric point in $\mathbb{Y}$. Then there exists an element $y \in S_{\ell_1^n}$ such that $Te_i \perp_B y$ but $y \not\perp_B Te_i$. Define a linear operator $A : \ell_1^n \to \mathbb{Y}$ as follows:

$$Ae_i = y, \ Ae_j = 0 \text{ for all } j \neq i.$$

Note that, $M_A = \{\pm e_i\}$. Now, $Te_i \perp_B y$ implies that $Te_i \perp_B Ae_i$ and so $T \perp_B A$. From our assumption, we get $Ae_i \not\perp_B Te_i$. Therefore, using Theorem 4.3.10, we have

$A \not\perp_B T$. This contradicts that T is left symmetric. Thus, Te_i is a left symmetric point in $\mathbb{Y}$. Now, we prove that $Te_j = 0$ for all $j \neq i$. Assume to the contrary that $Te_j \neq 0$ for some $j \neq i$. Define a linear operator $A : \ell_1^n \to \mathbb{Y}$ as follows:

$$Ae_j = Te_j, \ Ae_k = 0 \text{ for all } k \neq j.$$

Then $M_A = \{\pm e_j\}$ and $Te_i \perp_B Ae_i$. Thus $T \perp_B A$. From our assumption, we get $Ae_j \not\perp_B Te_j$. Therefore, using Theorem 4.3.10, we can conclude that $A \not\perp_B T$. Therefore, $Te_j = 0$ for all $j \neq i$.

$(ii) \Rightarrow (i)$. Suppose Te_i is a left symmetric point in $\mathbb{Y}$ and $Te_j = 0$ for all $j \neq i$. Then clearly $M_T = \{\pm e_i\}$. Let $A \in \mathbb{L}(\ell_1^n, \mathbb{Y})$ and let $T \perp_B A$. From Theorem 4.3.10, it follows that $Te_i \perp_B Ae_i$. Since $T \in \mathbb{L}(\ell_1^n, \mathbb{Y})$ is left symmetric, we have $Ae_i \perp_B Te_i$. Note that $e_j \in M_A$ for at least one j, $1 \leq j \leq n$. If $j = i$, then $A \perp_B T$, since $Ae_i \perp_B Te_i$. If $j \neq i$, then also $Ae_j \perp_B Te_j$, since $Te_j = 0$ for all $i \neq j$. This proves that $A \perp_B T$. Consequently, T is left symmetric, completing the proof of the theorem. □

We now show that if T is a right symmetric operator from $\ell_1^n, n \geq 2$, to any Banach space $\mathbb{Y}$, then T must attain norm at all of the extreme points of $B_{\ell_1^n}$.

Theorem 7.6.15 ([66, Theorem 3.7]) *Let $\mathbb{Y}$ be any Banach space and let $T \in \mathbb{L}(\ell_1^n, \mathbb{Y})$ be a non-zero right symmetric operator with $\|T\| = 1$, $n \geq 2$. Then T attains norm at each $e_i \in Ext(B_{\ell_1^n})$, $1 \leq i \leq n$.*

Proof If possible, let there exist some i, $1 \leq i \leq n$, such that T does not attain norm at e_i. Without any loss of generality, assume that $\{e_1, \ldots, e_k\} \subseteq M_T$ and none of $e_i \in \{e_{k+1}, \ldots, e_n\}$ lies in M_T, where $1 \leq k < n$. Suppose $u \in S_{\mathbb{Y}}$ is such that $u \perp_B Te_{k+1}$. Define a linear operator $A : \ell_1^n \to \mathbb{Y}$ as follows:

$$Ae_i = \frac{1}{2}Te_i \text{ for } 1 \leq i \leq k, \ Ae_{k+1} = u, \text{ and } Ae_i = 0 \text{ for } k+2 \leq i \leq n.$$

There exists i, $1 \leq i \leq n$, such that $e_i \in M_A$. Clearly, $e_{k+1} \in M_A$ and $Ae_{k+1} \perp_B Te_{k+1}$. Thus $A \perp_B T$. We claim that $Ax = \frac{1}{2}Tx$ for all $x \in M_T$. To prove our claim, we first show that if $x = \sum_{i=1}^n a_i e_i \in S_{\ell_1^n}$ and $a_i \neq 0$ for at least one $i \in \{k+1, \ldots, n\}$, then $x \notin M_T$. Clearly, for such x, we have

$$\|Tx\| \leq \sum_{i=1}^{n} |a_i| \|Te_i\| < \sum_{i=1}^{n} |a_i| < 1.$$

Thus $M_T \subseteq \{x \in S_{\ell_1^n} : x = \sum_{i=1}^k a_i e_i\}$, proving our claim. Now, using Theorem 4.3.7, we get $T \not\perp_B A$. This contradicts that T is right symmetric. Thus $e_i \in M_T$ for all $1 \leq i \leq n$, completing the proof of the theorem. □

As an application of Theorems 7.6.14 and 7.6.15, we can obtain the following characterization of symmetric operators in $\mathbb{L}(\ell_1^n, \mathbb{Y})$, $n \geq 2$.

Corollary 7.6.16 ([66, Corollary 3.7.1]) *Let $\mathbb{Y}$ be any Banach space and let $T \in \mathbb{L}(\ell_1^n, \mathbb{Y})$, $n \geq 2$. Then T is symmetric if and only if T is the zero operator.*

The study of right symmetric operators between Banach spaces is evidently more challenging than that of left symmetric operators. So far we do not have any concrete characterizations of right symmetric operators, in the general scenario. However, we do have some interesting observations related to them. We begin with the following result which states that a smooth operator is not right symmetric if the domain space is reflexive, strictly convex and the range space is strictly convex.

Theorem 7.6.17 ([102, Theorem 3.1]) *Let $\mathbb{X}$ be reflexive, strictly convex Banach space and let $\mathbb{Y}$ be a strictly convex Banach space. If $T \in \mathbb{K}(\mathbb{X}, \mathbb{Y})$ is smooth then T cannot be right symmetric.*

Proof If possible, suppose that T is right symmetric. Then using Proposition 7.3.1, we get T is left symmetric. Thus T must be zero operator, which is impossible since T is smooth. Hence, T cannot be right symmetric. □

Remark 7.6.18 We would like to remark that the previous theorem was first proved in [118, Theorem 2.3] using another approach, for finite-dimensional space $\mathbb{X}$. However, Theorem 7.6.17 gives an elegant and simple proof of the same.

In the following proposition, we study the properties of Tx, where T is a right symmetric operator and $M_T = \{\pm x\}$.

Proposition 7.6.19 ([102, Proposition 3.2]) *Let T be a compact linear operator on a reflexive Banach space $\mathbb{X}$ such that $M_T = \{\pm x\}$. If T is right symmetric then Tx is right symmetric. Moreover, if Tx is smooth, then Tx is also left symmetric.*

Proof If possible, suppose there exists $y \in S_{\mathbb{X}}$ such that $y \perp_B Tx$ but $Tx \not\perp_B y$. Since $x \in M_T$, by Proposition 3.2.10, we get a hyperspace H_x such that $x \perp_B H_x$ and $Tx \perp_B T(H_x)$. Define $A : \mathbb{X} \to \mathbb{X}$ by $A(ax + h) = ay$ where a is a scalar and $h \in H_x$. Clearly, A is compact, $x \in M_A$ and $Ax \perp_B Tx$. Thus, $A \perp_B T$. Since $M_T = \{\pm x\}$ and $Tx \not\perp_B Ax$, using Theorem 4.3.10, we get $T \not\perp_B A$. This contradicts that T is right symmetric. Therefore, Tx must be right symmetric.

If Tx is smooth, then from Proposition 7.3.1 (i), we get that Tx is also left symmetric. □

Now, we are ready to prove that under some assumptions, every norm one right symmetric operator must be an extreme point of the unit ball.

Theorem 7.6.20 ([102, Theorem 3.3]) *Let $\mathbb{X}, \mathbb{Y}$ be strictly convex Banach spaces. Suppose that* $\dim(\mathbb{X}) = 2$. *Assume that $\mathbb{Y}$ is smooth. Let $T \in \mathbb{K}(\mathbb{X}, \mathbb{Y})$, $\|T\| = 1$. If T is right symmetric, then T is an extreme point of the closed unit ball of $\mathbb{K}(\mathbb{X}, \mathbb{Y})$.*

Proof Suppose $T \in \mathbb{K}(\mathbb{X}, \mathbb{Y})$ is right symmetric and $\|T\| = 1$. We claim that $|M_T| \geq 4$. If possible, suppose that $|M_T| = 2$. Then from Theorem 6.3.2, we have T is smooth. Now, from Theorem 7.6.17, it follows that T is not right symmetric, a contradiction. This proves our claim that $|M_T| \geq 4$. Choose $x, y \in M_T$ such that x, y are linearly independent. Next, we prove that $T \in Ext(B_{\mathbb{K}(\mathbb{X},\mathbb{Y})})$. Let $T = \lambda U + (1-\lambda)W$ for some $U, W \in B_{\mathbb{K}(\mathbb{X},\mathbb{Y})}$ and for some $\lambda \in (0, 1)$. Then $Tx = \lambda Ux + (1-\lambda)Wx$ and $Ty = \lambda Uy + (1-\lambda)Wy$. In particular, $Ux, Wx, Uy, Wy \in B_{\mathbb{Y}}$. Since $\mathbb{Y}$ is strictly convex, we get $Tx = Ux = Wx$ and $Ty = Uy = Wy$. Thus T, U and W coincide on the basis $\{x, y\}$, i.e., $T = U = W$. Therefore, $T \in Ext(B_{\mathbb{K}(\mathbb{X},\mathbb{Y})})$. □

When $\mathbb{X}$ is not necessarily strictly convex or smooth, we have the following two theorems regarding right symmetric operators.

Theorem 7.6.21 ([102, Theorem 3.4]) *Let $\mathbb{X}$ be a reflexive, Kadets–Klee Banach space and let T be a non-zero compact operator on $\mathbb{X}$. If T is smooth and $\|T\|$ is a spectral value of T, then either nullity $T = 0$ or T is not right symmetric.*

Proof Since T is compact and smooth on a reflexive Banach space, from Theorem 6.3.2, it follows that $M_T = \{\pm x\}$ for some $x \in S_{\mathbb{X}}$. Each non-zero spectral value of a compact operator being an eigenvalue of T, we have either $Tx = \|T\|x$ or $Tx = -\|T\|x$. If nullity $T = 0$, then we are done. So let nullity $T \geq 1$. Choose $u \in S_{\mathbb{X}}$ such that $Tu = 0$. Then $\|I + \lambda T\| \geq \|(I + \lambda T)u\| = \|Iu\| = \|I\|$, for each scalar λ. Thus, $I \perp_B T$. If possible, let $T \perp_B I$. Then by Theorem 7.6.1, we have $Tx \perp_B Ix$. Thus, $\pm\|T\|x \perp_B x$, a contradiction. This proves that T is not right symmetric. □

Theorem 7.6.22 ([102, Theorem 3.5]) *Suppose $\mathbb{X}, \mathbb{Y}$ are Banach spaces. Let $T \in \mathbb{L}(\mathbb{X}, \mathbb{Y})$ be such that the kernel of T contains a non-zero point. Then either $T \perp_B I$ or T is not right symmetric.*

Proof Choose $u \in S_{\mathbb{X}}$ such that $Tu = 0$. Then for each scalar λ, we have $\|I + \lambda T\| \geq \|(I + \lambda T)u\| = \|Iu\| = \|I\|$, i.e., $I \perp_B T$. If $T \perp_B I$ then we are done. Otherwise, we get $I \perp_B T$ but $T \not\perp_B I$, proving that T is not right symmetric.

□

We end this chapter with a result which characterizes symmetric operators when the domain space is reflexive, Kadets–Klee, and strictly convex. This is analogous to the characterization of left symmetric operators (see Theorem 7.6.2) when the domain space is reflexive, Kadets–Klee, and strictly convex and in addition the range space is strictly convex. The right symmetricity allows us to remove the restriction of strict convexity of the range space. We first prove the following lemma.

Lemma 7.6.23 ([66, Lemma 3.1]) *Let $\mathbb{X}$ be a two-dimensional Banach space and let $x \in S_{\mathbb{X}}$. Then there exists $v \in S_{\mathbb{X}}$ such that the following hold true:*
(i) $x \perp_B v$,
(ii) $tv + (1-t)x \in x^+ \setminus x^- \quad \forall t \in (0, 1)$.

Proof From Proposition 3.2.2, we have $\mathbb{X} = x^{\perp} \cup (x^- \setminus x^+) \cup (x^+ \setminus x^-)$. Note that, $x^{\perp}, x^- \setminus x^+$ and $x^+ \setminus x^-$ are mutually disjoint sets. We show that $x^- \setminus x^+, x^+ \setminus x^-$ are cones in $\mathbb{X}$, i.e., if $y_1, y_2 \in x^- \setminus x^+$, $z_1, z_2 \in x^+ \setminus x^-$ then $y_1 + y_2 \in x^- \setminus x^+$, $z_1 + z_2 \in x^+ \setminus x^-$ and $\alpha(x^- \setminus x^+) \subseteq x^- \setminus x^+$, $\alpha(x^+ \setminus x^-) \subseteq x^+ \setminus x^-$ for all scalars $\alpha > 0$.

Let $y, z \in x^- \setminus x^+$. Then there are $\lambda_1, \lambda_2 > 0$ such that $\|x + \lambda_1 y\| < \|x\|$ and $\|x + \lambda_2 z\| < \|x\|$. Using the convexity of the norm, we get $\|x + \alpha y\| < \|x\|$ and $\|x + \alpha z\| < \|x\|$ for all $\alpha \in (0, \lambda]$ where $\lambda = \min\{\lambda_1, \lambda_2\}$. Again using the convexity of the norm, it follows that $\|x + \alpha(y + z)\| < \|x\|$ for some $\alpha \in (0, \lambda]$. Thus, $y + z \notin x^+$. Hence, from Proposition 3.2.2, we have $y + z \in x^- \setminus x^+$. Now, from Proposition 3.2.2, it follows that $\alpha(x^- \setminus x^+) \subseteq x^- \setminus x^+$ for all $\alpha > 0$. Therefore, $x^- \setminus x^+$ is a cone in $\mathbb{X}$. Similarly we can show that $x^+ \setminus x^-$ is a cone in $\mathbb{X}$.

By taking $\epsilon = 0$ in Theorem 5.2.7, we get that $x^\perp$ is a union of normal cones, i.e., $x^\perp = K \cup (-K)$ where K is a cone and $0 \in K$. To obtain v satisfying (i) and (ii), we present a brief outline of the proof of Theorem 5.2.7. For $x \in S_{\mathbb{X}}$ there is $y \in S_{\mathbb{X}}$ such that $x \perp_B y$. Let

$$A = \{t \in [0, 1] : x \perp_B \{(1 - t)x + ty\}\},$$
$$B = \{t \in [0, 1] : x \perp_B \{-(1 - t)x + ty\}\}.$$

Then $0 \notin A \cup B$, $1 \in A \cap B$ and both A, B are non-empty, proper closed subsets of $[0, 1]$. Let $t_1 = \inf A$, $t_2 = \inf B$. Note that, both t_1 and t_2 are non-zero. Consider $u_1 = (1 - t_1)x + t_1 y$, $u_2 = -(1 - t_2)x + t_2 y$ and $v_1 = \frac{u_1}{\|u_1\|}$, $v_2 = \frac{u_2}{\|u_2\|}$. If K is the normal cone determined by v_1 and v_2, i.e., $K = \{\alpha v_1 + \beta v_2 : \alpha, \beta \geq 0\}$, then $x^\perp = K \cup (-K)$.
If we take $v = v_1$, then this v satisfies both the conditions (i) and (ii). □

Now, the promised characterization:

Theorem 7.6.24 ([66, Theorem 3.4]) *Let $\mathbb{X}$ be a reflexive, Kadets–Klee, strictly convex Banach space and let $\mathbb{Y}$ be any Banach space. Then $T \in \mathbb{L}(\mathbb{X}, \mathbb{Y})$ with $M_T \neq \emptyset$ is symmetric if and only if T is the zero operator.*

Proof The sufficient part follows trivially. We only prove the necessary part. Let $x \in M_T$. Then from Proposition 3.2.10, we get a hyperspace H such that $x \perp_B H$ and $Tx \perp_B T(H)$. Since T is left symmetric, it is easy to check that $T(H) = 0$. Following similar arguments like Step 2 of Theorem 7.6.5, we can prove that Tx is a left symmetric point in $\mathbb{Y}$. Since $\mathbb{X}$ is strictly convex, we obtain $M_T = \{\pm x\}$. Since T is right symmetric and $M_T = \{\pm x\}$, using the similar arguments for left symmetry of Tx, it is easy to verify that Tx is also a right symmetric point in $\mathbb{Y}$.

Choose $u \in S_{\mathbb{Y}}$ such that $Tx \perp_B u$. Let $Z = span\{Tx, u\}$. Then using Lemma 7.6.23, we can find $v \in S_Z$ such that $Tx \perp_B v$ and $tv + (1 - t)Tx \in Tx^+ \setminus Tx^-$ for all $t \in (0, 1)$. Let $y \in S_H$ and $y \perp_B H_y$, where H_y is a hyperspace of H. Then each element $z \in \mathbb{X}$ can be written as $z = \alpha x + \beta y + h$, where α, β are scalars and $h \in H_y$. From the strict convexity of $\mathbb{X}$, we get $\|x + y\| = 2 - \delta$ for some $0 < \delta < 1$. Choose $0 < \epsilon < \frac{\delta}{3-\delta}$ and $w \in B(v, \epsilon)$ such that $w = tv + (1 - t)Tx$, $t \in (0, 1)$. Observe that such a choice of w is always possible. We now construct an operator $A \in \mathbb{L}(\mathbb{X}, \mathbb{Y})$ such that $T \perp_B A$ but $A \not\perp_B T$.

Define a linear operator $A : \mathbb{X} \to \mathbb{Y}$ as follows:

$$A(\alpha x + \beta y + h) = \alpha v + \beta w.$$

Clearly, $Tx \perp_B Ax$, and thus $T \perp_B A$. Note that,

$$\begin{aligned}\|A\left(\frac{x+y}{\|x+y\|}\right)\| &= \frac{\|v+w\|}{\|x+y\|}\\ &> \frac{2-\epsilon}{2-\delta}\\ &> 1+\epsilon.\end{aligned}$$

Thus, $\|A\| > 1$ and $x, y \notin M_A$. Also, for $h \in H_y, h \notin M_A$.

Our next claim is that $z \notin M_A$ if $\alpha\beta < 0$. Let $z = -\alpha x + \beta y + h \in S_{\mathbb{X}}$ where $\alpha > 0, \beta > 0, \beta - \alpha > 0$. Then by using orthogonality, we have $1 = \|z\| = \| -\alpha x + \beta y + h\| >| \alpha |,\ 1 = \|z\| = \|\beta y + h - \alpha x\| \geq| \beta | - | \alpha |= \beta - \alpha$. Also $| \beta |= \|\beta y\| \leq \|\beta y + h\| = \|z + \alpha x\| \leq 2$. Now,

$$\begin{aligned}\|Az\| &= \| -\alpha v + \beta w\|\\ &= \|(\beta - \alpha)v + b(w - v)\|\\ &\leq | \beta - \alpha | + | \beta | \|w - v\|\\ &< | \beta - \alpha | + | \beta | \epsilon\\ &= \beta - \alpha + \beta\epsilon\\ &\leq 1 + 2\epsilon\\ &< \|A(\frac{x+y}{\|x+y\|})\|.\end{aligned}$$

Next, assume that $z = -\alpha x + \beta y + h \in S_{\mathbb{X}}$ where $\alpha > 0, \beta > 0, \beta - \alpha \leq 0$. Then also $\|Az\| <| \beta - \alpha | + | \beta | \epsilon < \alpha < 1$. Thus if $z = -\alpha x + \beta y + h \in S_{\mathbb{X}}$, where $\alpha > 0, \beta > 0$, then $z \notin M_A$. Similarly, considering $z = \alpha x - \beta y + h \in S_{\mathbb{X}}$, where $\alpha > 0, \beta > 0$ we can prove that $z \notin M_A$. Therefore, if $z = \alpha x + \beta y + h \in M_A$, then we must have $\alpha\beta > 0$. We now claim that $Tz \in Az^+ \setminus Az^-$ for all $z \in M_A$. Let $\alpha x + \beta y + h \in M_A$. Then $Az = \alpha v + \beta w$, $Tz = \alpha Tx$ and $\alpha\beta > 0$. Let $\alpha, \beta > 0$. Now,

$$\begin{aligned}\alpha v + \beta w &= (\alpha+\beta)\{(\frac{\alpha}{\alpha+\beta})v + (\frac{\beta}{\alpha+\beta})w\},\\ &= (\alpha+\beta)\{(\frac{\alpha}{\alpha+\beta})v + (\frac{\beta}{\alpha+\beta})(tv + (1-t)Tx)\},\\ &= (\alpha+\beta)\{(\frac{\alpha}{\alpha+\beta} + t(\frac{\beta}{\alpha+\beta}))v + (\frac{\beta}{\alpha+\beta})(1-t)Tx\},\\ &= (\alpha+\beta)\{\mu v + (1-\mu)Tx\},\end{aligned}$$

where $\mu = (\frac{\alpha}{\alpha+\beta} + t(\frac{\beta}{\alpha+\beta}))$. It is easy to verify that $\mu \in (0, 1)$. Thus, from Lemma 7.6.23, we get $\alpha v + \beta w \in Tx^+ \setminus Tx^-$. Moreover, since $Tz = \alpha Tx$, we have $Az \in Tz^+ \setminus Tz^-$. Since Tx is a symmetric point in $\mathbb{Y}$, Tz is also a symmetric point in

$\mathbb{Y}$. Therefore, using Theorems 7.3.3 and 7.3.4, we have $Tz \in Az^+ \setminus Az^-$. Similarly, assuming $\alpha, \beta < 0$, we get the same conclusion. This proves our claim. Now, from Theorem 7.6.1, it follows that $A \not\perp_B T$. This contradicts the symmetry of T. This completes the proof of the necessary part. □

The following ending remark invites the readers to further explore the right symmetric bounded linear operators in the framework of Banach spaces.

Remark 7.6.25 The characterization of right symmetric operators between Banach spaces is still open, even under the assumption that the domain space is reflexive, Kadets–Klee, strictly convex and the range space is strictly convex.

In this chapter and the previous one, we have comprehensively discussed possible applications of the concept of B-J orthogonality towards studying the local symmetries in a Banach space and that of smoothness, respectively. Indeed, in light of the results obtained in these two chapters, it may be safely claimed that B-J orthogonality techniques are particularly advantageous in describing the various geometric attributes of a given Banach space, and of linear operators on it. As another evidence of the extensive scope of B-J orthogonality techniques in studying the geometric theory of Banach spaces, we will discuss the extremal properties of bounded linear operators between Banach spaces. In view of the classical Krein–Milman theorem, extreme points of the unit ball of the space of operators are important from the perspective of convexity analysis on operators. However, as the reader will observe in the next chapter (which also happens to be the final chapter of this monograph), understanding and identifying the extreme contractions is an extraordinarily difficult problem, even in the setting of finite-dimensional Banach spaces. Once again, we will be applying B-J orthogonality to have an analytic approach to the description of extreme contractions between Banach spaces. Indeed, the contents of the last three chapters of this monograph strongly suggest that the concept of B-J orthogonality deserves far more attention than what it has received, as far as the analytic and geometric structures of linear operators between Banach spaces are concerned.

Chapter 8
Extreme Contractions

8.1 Introduction

Extremal structure of the unit ball of an operator space is certainly an important topic in the geometry of bounded linear operators. We recall that an extreme point of the unit ball of the space of bounded linear operators between Banach spaces is called an extreme contraction. In 1951, Kadison [62] initiated the study of extreme contractions on a Hilbert space, and obtained a complete analytic characterization of the same. However, as the reader can very well guess by now, the analogous story in the setting of Banach spaces is far more intriguing and complicated. Indeed, extreme contractions are yet to be fully understood, either from a geometric or an analytic perspective. In fact, a tractable characterization of the same remains elusive, even for finite-dimensional Banach spaces. This chapter deals with some particular aspects of extreme contractions between Banach spaces. In Sect. 8.2, we introduce two notions to deal with the problem of characterizing extreme contractions. With the help of these notions, we explore extreme contractions on strictly convex, smooth Banach spaces. A major highlight of this section is the characterization of extreme contractions on two-dimensional strictly convex, smooth Banach space. In Sect. 8.3, we continue this study on polyhedral Banach spaces. In Sect. 8.4, we apply the k-smoothness of operators to study extreme contractions. It turns out that this technique is particularly helpful for studying extreme contractions on polyhedral Banach spaces. Unless otherwise specified, we work only with real Banach spaces in this chapter.

A. Mal et al., *Birkhoff–James Orthogonality and Geometry of Operator Spaces*, Infosys Science Foundation Series, https://doi.org/10.1007/978-981-99-7111-4_8

8.2 Extreme Contractions on Smooth and Strictly Convex Banach Spaces

Let us begin this section with an easy sufficient condition for an operator to be an extreme contraction.

Proposition 8.2.1 ([133, Proposition 2.6]) *Let $\mathbb{X}, \mathbb{Y}$ be Banach spaces such that $\dim(\mathbb{X}) = n$. Let $T \in S_{\mathbb{L}(\mathbb{X},\mathbb{Y})}$. Suppose that T attains its norm at n linearly independent vectors $x_1, x_2, \ldots, x_n$ and each Tx_i is an extreme point of $B_{\mathbb{Y}}$. Then T is an extreme contraction in $\mathbb{L}(\mathbb{X}, \mathbb{Y})$.*

Proof If possible, let T be not an extreme contraction. Then there exists a scalar $t \in (0, 1)$ and operators $T_1, T_2 \in \mathbb{L}(\mathbb{X}, \mathbb{Y})$ such that $T_1, T_2 \neq T$, $\|T_1\| = \|T_2\| = 1$ and $T = tT_1 + (1-t)T_2$. Therefore, $Tx_i = tT_1x_i + (1-t)T_2x_i$ for each $i \in \{1, 2, \ldots, n\}$. Moreover, $T_1x_i, T_2x_i \in B_{\mathbb{Y}}$, as $\|T_1\| = \|T_2\| = 1$. Since for each $i \in \{1, 2, \ldots, n\}$, $Tx_i \in Ext(B_{\mathbb{Y}})$, we have $T_1x_i = T_2x_i = Tx_i$. However, this implies that T_1, T_2 agree with T on a basis of $\mathbb{X}$, and therefore $T_1 = T_2 = T$. This contradicts our initial assumption that $T_1, T_2 \neq T$ and completes the proof of the proposition. □

Note that in a strictly convex Banach space, each unit vector is an extreme point of the unit ball. Therefore, using the previous proposition, we immediately observe the following.

Proposition 8.2.2 *Let $\mathbb{X}, \mathbb{Y}$ be Banach spaces such that $\dim(\mathbb{X}) = n$ and $\mathbb{Y}$ is strictly convex. Let $T \in S_{\mathbb{L}(\mathbb{X},\mathbb{Y})}$. Suppose that T attains its norm at n linearly independent vectors. Then T is an extreme contraction in $\mathbb{L}(\mathbb{X}, \mathbb{Y})$.*

In 1951, Kadison [62] and later on Halmos [49] characterized every extreme contraction on complex Hilbert spaces as an isometry or a coisometry. Recall that an operator is called a coisometry if its adjoint is an isometry. In 1982, Grząślewicz [45] proved that the same characterization of extreme contraction is still valid if the operators are defined between real Hilbert spaces.

Theorem 8.2.3 ([62, Theorem 1], [49, Sol. 136], [45]) *Suppose $\mathbb{H}_1, \mathbb{H}_2$ are real or complex Hilbert spaces. Then $T \in \mathbb{L}(\mathbb{H}_1, \mathbb{H}_2)$ is an extreme contraction if and only if T is either an isometry or a coisometry.*

If $\mathbb{H}_1, \mathbb{H}_2$ are finite-dimensional real or complex Hilbert spaces, then an elementary proof of Theorem 8.2.3 is as follows. The proof is motivated by [46].

Theorem 8.2.4 *Suppose $\mathbb{H}_1, \mathbb{H}_2$ are real or complex finite-dimensional Hilbert spaces. Then $T \in \mathbb{L}(\mathbb{H}_1, \mathbb{H}_2)$ is an extreme contraction if and only if T is either an isometry or a coisometry.*

Proof First observe that if T is an isometry, then $M_T = S_{\mathbb{H}_1}$. Therefore, from Proposition 8.2.2 it follows that T is an extreme contraction. Similarly, if T^* is an isometry, then T^* is an extreme contraction and thus T is an extreme contraction. Conversely, suppose that T is an extreme contraction. If possible, suppose that neither T nor T^*

is an isometry, i.e., $M_T \neq S_{\mathbb{H}_1}$ and $M_{T^*} \neq S_{\mathbb{H}_2}$. Thus, using Theorem 3.3.5, we get $M_T = S_U$, where U is a proper subspace of $\mathbb{H}_1$ and $T^*Tu = u$ for all $u \in U$. Similarly, $M_{T^*} = S_V$, where V is a proper subspace of $\mathbb{H}_2$ and $TT^*v = v$ for all $v \in V$. Observe that $T(U) = V$. Indeed for $u \in U$, $\|Tu\| = \|u\|$ and $T^*Tu = u$ imply that $\|Tu\| = \|T^*Tu\|$, i.e., $Tu \in V$. On the other hand, if $v \in V$, then $\|T^*v\| = \|v\|$ and $TT^*v = v$. Let $T^*v = u$. Then $Tu = v$, i.e., $T^*Tu = T^*v = u$. Now, from

$$\|u\| = \|T^*v\| = \|v\| = \|Tu\|,$$

it follows that $u \in U$, i.e., $v = Tu \in T(U)$. Thus, $T(U) = V$. Similarly, $T(U^\perp) \subseteq V^\perp$. Clearly, $U^\perp \neq \{0\} \neq V^\perp$. Since $\mathbb{H}_1$ is finite dimensional, $\|T\|_{U^\perp} < \|T\| = 1$. Choose $1 > \epsilon > 0$ such that $(1+\epsilon)\|T\|_{U^\perp} < 1$. Now, there are two cases.
Case I: $\|T\|_{U^\perp} > 0$. Let P be the orthogonal projection on U, i.e., $P(x + y) = y$, for all $x \in U$, $y \in U^\perp$. Observe that, in this case, $\epsilon TP \neq 0$ and

$$\|(T \pm \epsilon TP)(x+y)\|^2 = \|Tx + (1 \pm \epsilon)Ty\|^2 = \|Tx\|^2 + (1 \pm \epsilon)^2\|Ty\|^2 \leq \|x\|^2 + \|y\|^2 = \|x+y\|^2.$$

Thus, $\|T \pm \epsilon TP\| \leq 1$. Now, $T = \frac{1}{2}(T + \epsilon TP) + \frac{1}{2}(T - \epsilon TP)$ contradicts that T is an extreme contraction.
Case II: $\|T\|_{U^\perp} = 0$. Choose $u_0 \in S_{U^\perp}$ and $v_0 \in S_{V^\perp}$. Define $R : \mathbb{H}_1 \to V^\perp \subseteq \mathbb{H}_2$ by $R(au_0 + u) = av_0$, for all scalars a and $u \in u_0^\perp$. Now,

$$\|R(au_0 + u)\|^2 = \|av_0\|^2 \leq \|au_0\|^2 + \|u\|^2 = \|au_0 + u\|^2$$

implies that $\|R\| \leq 1$. Observe that, if $x \in U$ and $y \in U^\perp$, then $x \in u_0^\perp$, i.e., $Rx = 0$, $Ry \in V^\perp$ and $Tx \in V$. Thus,

$$\|(T \pm R)(x+y)\|^2 = \|Tx \pm Ry\|^2 = \|Tx\|^2 + \|Ry\|^2 \leq \|x\|^2 + \|y\|^2 = \|x+y\|^2.$$

Hence, $\|T \pm R\| \leq 1$. Now, $T = \frac{1}{2}(T + R) + \frac{1}{2}(T - R)$ contradicts that T is an extreme contraction. Thus, if T is an extreme contraction, then T must be an isometry or a coisometry. This completes the proof. □

The most prominent class of strictly convex Banach spaces that comes to our mind immediately after the Hilbert spaces is surely the ℓ_p spaces $(1 < p < \infty, p \neq 2)$. In 1981, Grząślewicz provided an explicit description of extreme contractions on two-dimensional ℓ_p spaces, where $1 < p < \infty, p \neq 2$. Over the years, several researchers have explored extreme contractions in $\mathbb{L}(\ell_p^2, \ell_q^2)$, where $1 < p, q < \infty$. Here we list a few known results in this direction. In the next theorem, $e_1 = (1, 0)$, $e_2 = (0, 1)$, $x = (x_1, x_2) \in \ell_p^2$ with $\|x\| = 1$, $y = (y_1, y_2)$, $x^o = (-x_2, x_1)$ and $x^{p-1} = (sgn(x_1)|x_1|^{p-1}, sgn(x_2)|x_2|^{p-1})$. Observe that x^{p-1} is the unique norm one norming linear functional of x. Here $x \otimes y$ denote rank one operator such that $(x \otimes y)z = \langle z, x\rangle y$.

Theorem 8.2.5 ([15]) *Let $1 < p, q < \infty$ and $T \in S_{\mathbb{L}(\ell_p^2, \ell_q^2)}$. Then T is an extreme contraction*
(i) *[45] for $p = q = 2$, if and only if T is an isometry.*
(ii) *[44] for $p = 2 \neq q$, if and only if T satisfies one of the following:*
 (a) T attains its norm at two linearly independent vectors.
 (b) T is of the form

$$T = \begin{cases} x \otimes e_i, & \text{if } q < 2 \\ x \otimes y + sx^o \otimes (y^o)^{q-1}, & \text{if } q > 2, \end{cases}$$

where x is any unit vector, $|y_i|^q = \frac{1}{2}$, and $s = \pm\frac{1}{\sqrt{(q-1)}}2^{\frac{q-2}{q}}$.
(iii) *[44] for $p \neq 2 = q$, if and only if T satisfies one of the following:*
 (a) T attains its norm at two linearly independent vectors.
 (b) T is of the form

$$T = \begin{cases} e_i \otimes y, & \text{if } p > 2 \\ x^{p-1} \otimes y + sx^o \otimes y^o, & \text{if } p < 2, \end{cases}$$

where y is any unit vector, $|x_i|^p = \frac{1}{2}$ and $s = \pm\sqrt{(p-1)}2^{\frac{2-p}{p}}$.
(iv) *[46] for $p = q \neq 2$, if and only if T satisfies one of the following:*
 (a) T attains its norm at two linearly independent vectors.
 (b) T is of the form

$$T = \begin{cases} e_i \otimes y, & \text{if } p > 2,\ y_1 y_2 \neq 0 \\ x^{p-1} \otimes e_j, & \text{if } p < 2,\ x_1 x_2 \neq 0. \end{cases}$$

(v) *for $1 < q < 2 < p < \infty$, if and only if T satisfies one of the following:*
 (a) T attains its norm at two linearly independent vectors.
 (b) $T = x^{p-1} \otimes y$ with x, y unit vectors and $x_1 x_2 y_1 y_2 = 0$.

Notice that even in $\mathbb{L}(\ell_p^2, \ell_q^2)$ spaces, there are remaining cases to be answered definitively in the context of extreme contraction. We refer the readers to [15, 45–47, 64] to get a flavor of the complicacy of the problem. In the next section, we will further discuss the progress in this area over the years. While the characterization problem for extreme contractions has a definite solution for various particular pairs of Banach spaces [15, 22, 45, 46, 56, 62, 67, 129, 140], to the best of our knowledge, the problem remains unsolved in its most general form, even in the two-dimensional case. This is somewhat surprising, since which norm one operators are extreme contractions, should be, at least in principle, entirely governed by the geometries of the domain and the range spaces. Indeed, this is the motivation behind our quest for a geometric property in the framework of Banach spaces, that would completely characterize the class of extreme contractions between Banach spaces. In [124], we have introduced two different geometric concepts, whose importance in studying extreme contractions between Banach spaces would be self-explanatory in due course of time. Let us recall that for $x \in S_{\mathbb{X}},\ x^{\perp} = \{y \in \mathbb{X} : x \perp_B y\} = \{y \in \mathbb{X} : f(y) = 0,$ for some $f \in J(x)\}$, where $J(x) = \{f \in S_{\mathbb{X}^*} : f(x) = 1\}$.

Definition 8.2.6 ([124]) Let $\mathbb{X},\ \mathbb{Y}$ be two Banach spaces and $x \in S_{\mathbb{X}},\ y \in S_{\mathbb{Y}}$. We say that $(x,\ y)$ is a weakly compatible point pair if there exist $z \in x^{\perp} \cap S_{\mathbb{X}},\ w \in y^{\perp} \cap S_{\mathbb{Y}}$ and two constants $r > 0,\ \mu > 0$ such that

$$ax + bz \in B(x,\ r) \cap S_{\mathbb{X}} \Rightarrow \|ay + b\mu w\| \leq 1. \tag{8.1}$$

In short, $(x,\ y)$ is said to be a weak CPP. Any pair of constants $(r,\ \mu)$, that works in the above sense, is said to be a weak CPP constant for the pair $(x,\ y)$.
In particular, if $z_0 \in x^{\perp} \cap S_{\mathbb{X}},\ w_0 \in y^{\perp} \cap S_{\mathbb{Y}}$ satisfy (8.1) then we say that $(x,\ y)$ is $\mu-$CPP with respect to $(z_0,\ w_0)$.

Definition 8.2.7 ([124]) Let $\mathbb{X},\ \mathbb{Y}$ be two Banach spaces and $x \in S_{\mathbb{X}},\ y \in S_{\mathbb{Y}}$. We say that $(x,\ y)$ is a compatible point pair if there exist two constants $r > 0,\ \mu > 0$ such that for any $z \in x^{\perp} \cap S_{\mathbb{X}}$ and for any $w \in y^{\perp} \cap S_{\mathbb{Y}},\ \ ax + bz \in B(x,\ r) \cap S_{\mathbb{X}} \Rightarrow \|ay + b\mu w\| \leq 1$. In this case we write $(x,\ y)$ is a CPP. Any pair of constants $(r,\ \mu)$, that works in the above sense, is said to be a CPP constant for the pair $(x,\ y)$.

In light of these newly introduced geometric notions, we study extreme contractions between Banach spaces. Let us first state a simple proposition, the proof of which follows from the convexity of norm.

Proposition 8.2.8 ([124, Proposition 2.1]) *Let $\mathbb{X}$ be a Banach space, let $u,\ v \in \mathbb{X}$, and let $0 < \lambda_0 \in \mathbb{R}$. If $\|u\| < 1$ and $\|u + \lambda_0 v\| \leq 1$ then $\|u + \lambda v\| \leq 1$ for all $0 \leq \lambda \leq \lambda_0$.*

In the next proposition, we explore some basic properties of (weak) CPP.

Proposition 8.2.9 ([124, Proposition 2.2]) *Let $\mathbb{X},\ \mathbb{Y}$ be Banach spaces and let $x \in S_{\mathbb{X}},\ y \in S_{\mathbb{Y}}$. Then the following are true:*
(i) *$(x,\ y)$ is a CPP implies that $(x,\ y)$ is a weak CPP, but the converse may not be true.*
(ii) *In a Hilbert space $\mathbb{H}$, $(x,\ y)$ is a CPP for any $x,\ y \in S_{\mathbb{H}}$. In this case the CPP constant may be chosen as $(r,\ \mu)$, with any $r > 0$ and $\mu = 1$.*
(iii) *$(x,\ x)$ is always a weak CPP but $(x,\ x)$ may not be a CPP.*
(iv) *$(x,\ y)$ may or may not be a weak CPP.*
(v) *If $(r_0,\ \mu_0)$ is a (weak) CPP constant for the pair $(x,\ y)$, then $(r,\ \mu)$ is a (weak) CPP constant for the pair $(x,\ y)$, for any $0 < r \leq r_0$ and for any $0 < \mu \leq \mu_0$.*
(vi) *If $(x,\ y)$ is a CPP and $x \notin Ext(B_{\mathbb{X}})$ then $y \notin Ext(B_{\mathbb{Y}})$. But the converse is not true.*

Proof The proofs of (i)–(v) are trivial. We only prove (vi). Suppose $(x,\ y)$ is a CPP with CPP constant $(r,\ \mu)$. Since $x \notin Ext(B_{\mathbb{X}})$, there exist $x_1,\ x_2 \in B_{\mathbb{X}}$ such that $x \neq x_1,\ x \neq x_2$ and $x = \frac{1}{2}(x_1 + x_2)$. It is easy to observe that $x \perp_B (x_1 - x)$. Clearly,

there is a scalar $t \in (0, 1)$ such that $x + t(x_1 - x) = (1 - t)x + tx_1 \in B(x, r) \cap S_{\mathbb{X}}$ and $x - t(x_1 - x) = (1 - t)x + tx_2 \in B(x, r) \cap S_{\mathbb{X}}$. Since (x, y) is a CPP, for each $w \in y^{\perp} \cap S_{\mathbb{Y}}$, we have $\|y \pm t\mu w\| \leq 1$. Moreover, from $y \perp_B w$ we get $\|y \pm t\mu w\| \geq 1$ and so $\|y \pm t\mu w\| = 1$. Clearly $y = \frac{1}{2}(y + t\mu w) + \frac{1}{2}(y - t\mu w)$. Hence $y \notin Ext(B_{\mathbb{Y}})$.
But in ℓ^2_{∞}, $((1, 1), (1, 0))$ is a CPP where $(1, 0) \notin Ext(B_{\ell^2_{\infty}})$ and $(1, 1) \in Ext(B_{\ell^2_{\infty}})$. □

We next establish a connection between operator norm attainment and weak CPP for bounded linear operators. For this we require the following lemma which can be proved easily.

Lemma 8.2.10 ([124, Lemma 2.1]) *Let $\mathbb{X}$, $\mathbb{Y}$ be Banach spaces and let $(x, y) \in S_{\mathbb{X}} \times S_{\mathbb{Y}}$. Suppose that there exist $z \in x^{\perp} \setminus \{0\}$ and $w \in y^{\perp} \setminus \{0\}$ such that the (unique) linear operator $T : span\{x, z\} \to span\{y, w\}$ with $Tx = y$ and $Tz = w$ satisfies that $\|T\| = 1$. Then (x, y) is a weak CPP.*

Theorem 8.2.11 ([124, Theorem 2.3]) *Let $\mathbb{X}$, $\mathbb{Y}$ be Banach spaces. Suppose $T \in \mathbb{L}(\mathbb{X}, \mathbb{Y})$ is such that $rank\ T > 1$ and $\|T\| = 1$. Let $M_T \neq \emptyset$ and $x \in M_T$. Then (x, Tx) is a weak CPP.*

Proof Let $J(Tx) = \{f \in S_{\mathbb{Y}^*} : f(Tx) = \|Tx\| = 1\}$ and $g \in J(Tx)$. It is easy to observe that $T^*(g) \in J(x)$. Since $rank\ T > 1$, there exists $z \in \ker(T^*g) \subseteq x^{\perp}$ such that $Tz \neq 0$. From $g(Tz) = T^*g(z) = 0$, it follows that $w = Tz \in (Tx)^{\perp} \setminus \{0\}$. Now, the proof follows from Lemma 8.2.10. □

From Proposition 8.2.9, we know that given any $x \in S_{\mathbb{X}}$, (x, x) is a weak CPP. In the next theorem, we establish the existence of nontrivial weak CPP in any finite-dimensional Banach space.

Theorem 8.2.12 ([124, Theorem 2.4]) *Let $\mathbb{X}$ be a finite-dimensional Banach space. Then there exist $x, y \in S_{\mathbb{X}}$, such that $y \neq \pm x$ and (x, y) is a weak CPP.*

Proof Suppose $\{x_1, x_2, x_3, \ldots, x_n\}$ is a basis of $\mathbb{X}$. Define $S \in \mathbb{L}(\mathbb{X})$ as follows: $Sx_1 = x_2$, $Sx_2 = -x_1$, $Sx_i = 0$, $\forall\ 3 \leq i \leq n$. Let $T = \frac{1}{\|S\|}S$. Then clearly, $rank\ T > 1$, $M_T \neq \emptyset$ and 0 the only real eigenvalue of T. From Theorem 8.2.11, it follows that for each $x \in M_T$, we have (x, Tx) is a weak CPP and $Tx \neq \pm x$, since otherwise, ± 1 will be an eigenvalue of T. This completes the proof of the theorem. □

Recall from Theorem 8.2.3 that for a Hilbert space $\mathbb{H}$, a bounded linear operator in $\mathbb{L}(\mathbb{H})$ is an extreme contraction if and only if it is either an isometry or a coisometry. In particular, there is no rank one extreme contraction in $\mathbb{L}(\mathbb{H})$. However, the scenario is dramatically different in case of bounded linear operators on arbitrary Banach spaces, which are not necessarily Hilbert spaces. Indeed, it follows from the works of Grząślewicz [46] (see Theorem 8.2.5) that there are rank one extreme contractions in $\mathbb{L}(\ell^2_p)$.

The next theorem provides a sufficient condition for a rank one bounded linear operator to be an extreme contraction.

Theorem 8.2.13 ([124, Theorem 2.5]) *Suppose $\mathbb{X}$, $\mathbb{Y}$ are Banach spaces. Let $T \in S_{\mathbb{L}(\mathbb{X},\mathbb{Y})}$ be a rank one operator. If $(x,\ Tx)$ is not a weak CPP for some $x \in M_T$ such that x is a smooth point and $Tx \in Ext(B_{\mathbb{Y}})$, then T must be an extreme contraction.*

Proof If possible, let T be not an extreme contraction. Then there exist $T_1,\ T_2 \in B_{\mathbb{L}(\mathbb{X},\mathbb{Y})}$ such that $T = \frac{1}{2}T_1 + \frac{1}{2}T_2$, where $T_1 \neq T,\ T_2 \neq T$. Since $Tx \in Ext(B_{\mathbb{Y}})$, from $Tx = \frac{1}{2}T_1x + \frac{1}{2}T_2x$, it follows that $T_1x = T_2x = Tx$. If $rank\ T_1 = 1$, by using that x is a smooth point we deduce that $T_1 = T$. Therefore, $rank\ T_1 > 1$ and by Theorem 8.2.11, we have $(x, T_1x) = (x, Tx)$ is a weak CPP, a contradiction. Thus, T must be an extreme contraction. □

In particular, if $\mathbb{X}$, $\mathbb{Y}$ are two-dimensional Banach spaces, the above theorem can be substantially strengthened as follows:

Theorem 8.2.14 ([124, Theorem 2.6]) *Let $\mathbb{X}$, $\mathbb{Y}$ be two-dimensional Banach spaces. Suppose $T \in S_{\mathbb{L}(\mathbb{X},\mathbb{Y})}$ is a rank one operator. If $(x,\ Tx)$ is not a CPP for some $x \in M_T$ such that x is a smooth point of $S_{\mathbb{X}}$ and Tx is a smooth, extreme point of $B_{\mathbb{Y}}$, then T is an extreme contraction.*

Proof Let T be not an extreme contraction. Then there exist $T_1,\ T_2 \in S_{\mathbb{L}(\mathbb{X},\mathbb{Y})}$ such that $T = \frac{1}{2}(T_1 + T_2)$, $T \neq T_1,\ T \neq T_2$. Clearly, $Tx = \frac{1}{2}(T_1x + T_2x)$ and so $Tx = T_1x = T_2x$, since $Tx \in Ext(B_{\mathbb{Y}})$. Thus, $x \in M_{T_1} \cap M_{T_2}$. Let $x \perp_B y$. Then from Proposition 3.2.6, it follows that $Tx \perp_B Ty$ and $T_1x \perp_B T_1y$. Since $rank\ T = 1$, so we must have $Ty = 0$ and as $T_1 \neq T$, we must have $T_1y \neq 0$. Let $T_1y = \mu w,\ T_2y = -\mu w$ for some $\mu \in \mathbb{R} \setminus \{0\}$ and $w \in (Tx)^{\perp} \cap S_{\mathbb{Y}}$. Without loss of generality, assume that $\mu > 0$. Since $(x,\ Tx)$ is not a CPP, for $r > 0,\ \mu > 0$, we get $a,\ b \in \mathbb{R}$ such that $ax + by \in B(x,\ r) \cap S_{\mathbb{X}}$ but either $\|aTx + b\mu w\| > 1$ or $\|aTx - b\mu w\| > 1$. The first case implies that $\|T_1(ax + by)\| > 1$ and the second case implies that $\|T_2(ax + by)\| > 1$. Now, $\|ax + by\| = 1$ shows that either $\|T_1\| > 1$ or $\|T_2\| > 1$, which is a contradiction to our initial assumption that $\|T_1\| = \|T_2\| = 1$. Therefore, T must be an extreme contraction. This completes the proof. □

We can relax the restriction on the dimension of the domain space $\mathbb{X}$ (codomain space $\mathbb{Y}$) if the corresponding norm on $\mathbb{X}$ (codomain space $\mathbb{Y}$) is induced by an inner product. Note that in a Hilbert space $\mathbb{H}$, if $x,\ y,\ u,\ v \in S_{\mathbb{H}}$ are such that $x \perp y$ and $u \perp v$ then $\|ax + by\|^2 = \|au + bv\|^2 = a^2 + b^2$. We now state the following theorem which can be proved following the same line of arguments of Theorem 8.2.14.

Theorem 8.2.15 ([124, Theorem 2.7]) *Let $\mathbb{H}$ be any Hilbert space and let $\mathbb{X}$, $\mathbb{Y}$ be two-dimensional strictly convex smooth Banach spaces. Suppose T is a rank one operator from $\mathbb{H}$ to $\mathbb{Y}$ (from $\mathbb{X}$ to $\mathbb{H}$) with $\|T\| = 1$. If $(x,\ Tx)$ is not a CPP for $x \in M_T$, then T is an extreme contraction.*

Let us recall that a Banach space $\mathbb{X}$ is said to have Kadets–Klee property if for any sequence $\{x_n\}$ in $\mathbb{X}$, $x_n \rightharpoonup x$ and $\|x_n\| \to \|x\|$ implies that $x_n \to x$. Note that in a Banach space $\mathbb{X}$, a sequence $\{x_n\}$ is said to be weakly convergent to x ($x_n \rightharpoonup x$), if

$f(x_n) \to f(x)$ for each $f \in \mathbb{X}^*$ and $\{x_n\}$ is said to be strongly convergent to x ($x_n \to x$), if $\|x_n - x\| \to 0$. In the following theorem, we obtain a sufficient condition for a rank one bounded linear operator T such that it is not an extreme contraction.

Theorem 8.2.16 ([124, Theorem 2.8]) *Let $\mathbb{X}$ be a smooth reflexive Banach space with Kadets–Klee property and let $\mathbb{Y}$ be a smooth Banach space. Suppose $T \in S_{\mathbb{L}(\mathbb{X},\mathbb{Y})}$ is a rank one operator. If $M_T = \{\pm x\}$ and (x, Tx) is a CPP, then T is not an extreme contraction.*

Proof Since $\mathbb{X}$ is smooth, $rank\ T = 1$ and $x \in M_T$, using Proposition 3.2.6, we can easily observe that $Th = 0$ for all $h \in x^{\perp}$. Moreover, there is a unique hyperspace H_1 of codimension 1 such that $x \perp_B H_1$. Choose $y \in H_1 \cap S_{\mathbb{X}}$. Then there is a unique hyperspace H_2 in H_1 such that $y \perp_B H_2$. Hence, we can write each $z \in \mathbb{X}$ uniquely as $ax + h_1$, where $a \in \mathbb{R},\ h_1 \in H_1$. Again, we can write h_1 uniquely as $by + h_2$, where $b \in \mathbb{R},\ h_2 \in H_2$. Therefore, for each $z \in \mathbb{X}$, there are unique scalars $a,\ b \in \mathbb{R}$ and unique $h_2 \in H_2$ such that $z = ax + by + h_2$. Choose $w \in S_{\mathbb{Y}}$ such that $Tx \perp_B w$. For each $n \in \mathbb{N}$, define T_n, S_n as follows:

$$T_n(z) = aTx + \frac{b}{n}w,\ S_n(z) = aTx - \frac{b}{n}w.$$

It is easy to observe that $T_n, S_n \in \mathbb{K}(\mathbb{X}, \mathbb{Y})$. Clearly, for all $n \in \mathbb{N}$, $T = \frac{1}{2}T_n + \frac{1}{2}S_n$ and $T \neq T_n, T \neq S_n$. If possible, assume that T is an extreme contraction. Then for each $n \in \mathbb{N}$, either $\|T_n\| > 1$ or $\|S_n\| > 1$. Without loss of generality (passing on to a subsequence, if necessary), assume that for all $n \in \mathbb{N}$, $\|T_n\| > 1$. Since each T_n is a compact operator on a reflexive Banach space, $M_{T_n} \neq \emptyset$. Let $y_n \in M_{T_n}$ for each $n \in \mathbb{N}$.

Now, since (x, Tx) is a CPP, there exist $r > 0,\ \mu > 0$ such that $ax + bh \in B(x, r) \cap S_{\mathbb{X}} \Rightarrow \|aTx + b\mu u\| \leq 1$ for all $h \in H_1 \cap S_{\mathbb{X}}$ and for all $u \in (Tx)^{\perp} \cap S_{\mathbb{Y}}$. Choose $n_0 \in \mathbb{N}$ satisfying $0 < \frac{1}{n} < \mu$ for all $n \geq n_0$. We claim that $\|T_n z\| \leq 1$ for all $n \geq n_0$. Let $z = (ax + by + h_2) \in B(x, r) \cap S_{\mathbb{X}}$, where, $h_2 \in H_2$. If $by + h_2 = 0$ then we clearly have $\|T_n z\| \leq 1$. Let $by + h_2 \neq 0$. Then $z = ax + ch$, where $c = \|by + h_2\|$ and $h = \frac{1}{c}(by + h_2)$. Since $y \perp_B h_2$, we get $|b| \leq \|by + h_2\| = c$. Hence, $z = ax + ch \in B(x, r) \cap S_{\mathbb{X}} \Rightarrow \|aTx \pm c\mu w\| \leq 1$. Now $|b| \leq c$ implies that $-c\mu < \frac{b}{n} < c\mu$ for all $n \geq n_0$. Therefore, from Proposition 8.2.8, it follows that for all $n \geq n_0$,

$$\|T_n z\| = \|T_n(ax + by + h_2)\| = \|aTx + \frac{b}{n}w\| \leq 1.$$

Note that, $y_n \notin B(x, r)$, since $y_n \in M_{T_n}$ and $\|T_n\| > 1$. Similarly, we have $y_n \notin B(-x, r)$. Now $\mathbb{X}$ being reflexive, $B_{\mathbb{X}}$ is weakly compact. Therefore, $\{y_n\}$ has a weakly convergent subsequence in $B_{\mathbb{X}}$. Without loss of generality, assume that $y_n \rightharpoonup y_0$, where $y_0 \in B_{\mathbb{X}}$.

Let $z = (ax + by + h_2) \in S_{\mathbb{X}}$. Then $|a| \leq \|(ax + by + h_2)\| = 1$ and $|b| \leq \|(by + h_2)\| = \|z - ax\| \leq 1 + |a| \leq 2$. Therefore, $\|(T_n - T)z\| = \|(T_n - T)(ax + by + h_2)\| = \|\frac{b}{n}w\| = \frac{|b|}{n} \leq \frac{2}{n}$. Thus, $\|(T_n - T)z\| \to 0$ as $n \to \infty$. Hence, $T_n \to T$.

Since T is a compact operator and $y_n \rightharpoonup y_0$, we have $Ty_n \to Ty_0$. Therefore, $\|T_n y_n - Ty_0\| \leq \|T_n y_n - Ty_n\| + \|Ty_n - Ty_0\| \leq \|T_n - T\| + \|Ty_n - Ty_0\| \to 0$ as $n \to \infty$. Hence, $T_n y_n \to Ty_0 \Rightarrow \|T_n y_n\| \to \|Ty_0\| \Rightarrow \|Ty_0\| \geq 1$, since $\|T_n y_n\| > 1$ for all $n \in \mathbb{N}$. This proves that $\|Ty_0\| = 1$, since $\|T\| = 1$ and $y_0 \in B_{\mathbb{X}}$. Therefore, $y_0 = \pm x$. Again, $y_n \rightharpoonup y_0$ and $\|y_n\| \to \|y_0\|$. Hence, by Kadets–Klee property of $\mathbb{X}$, we get $y_n \to y_0$. This implies that $y_0 \notin B(\pm x, r)$, a contradiction. Therefore, T is not an extreme contraction. This establishes the theorem. □

Note that combining Theorems 8.2.14 and 8.2.16, we get a complete characterization of rank one extreme contractions from a two-dimensional smooth Banach space to a two-dimensional strictly convex and smooth Banach space, in terms of CPP. We state the characterization in the next theorem:

Theorem 8.2.17 ([124, Theorem 2.9]) *Let $\mathbb{X}$, $\mathbb{Y}$ be two-dimensional smooth Banach spaces and in addition, let $\mathbb{Y}$ be strictly convex. Suppose $T \in S_{\mathbb{L}(\mathbb{X},\mathbb{Y})}$ is a rank one operator. Let $M_T = \{\pm x\}$. Then T is an extreme contraction if and only if (x, Tx) is not a CPP .*

Remark 8.2.18 ([124, Remark 2.1]) If we additionally assume that $\mathbb{X}$ is strictly convex, then *rank* $T = 1$ will imply that $M_T = \{\pm x\}$ and so the assumption that $M_T = \{\pm x\}$ can be relaxed from the hypothesis of Theorems 8.2.16 and 8.2.17.

As before, if the domain space $\mathbb{X}$ (codomain space $\mathbb{Y}$) is a Hilbert space then combining Theorems 8.2.15 and 8.2.16, we get the next two characterizations:

Theorem 8.2.19 ([124, Theorem 2.10]) *Let $\mathbb{H}$ be a Hilbert space and let $\mathbb{Y}$ be a two-dimensional smooth and strictly convex Banach space. Suppose $T \in S_{\mathbb{L}(\mathbb{H},\mathbb{Y})}$ is a rank one operator. Then $M_T = \{\pm x\}$. Moreover, T is an extreme contraction if and only if (x, Tx) is not a CPP.*

Theorem 8.2.20 ([124, Theorem 2.11]) *Let $\mathbb{X}$ be a two-dimensional smooth Banach space and let $\mathbb{H}$ be a Hilbert space. Suppose $T \in S_{\mathbb{L}(\mathbb{X},\mathbb{H})}$ is a rank one operator. Let $M_T = \{\pm x\}$. Then T is an extreme contraction if and only if (x, Tx) is not a CPP.*

As concrete applications of Theorem 8.2.17, we would completely describe rank one extreme contractions between two-dimensional strictly convex smooth Banach spaces, in three special cases. First, we completely describe rank one extreme contractions in $\mathbb{L}(\ell_4^2)$. In view of Theorem 8.2.17, our strategy is to determine when a pair of points on the unit sphere of ℓ_4^2 is not a CPP. To this end, we have the following theorem:

Theorem 8.2.21 ([124, Theorem 2.12]) *Let $\times = \ell_4^2$. Then $((x, y), (x_1, y_1)) \in S_{\mathbb{X}} \times S_{\mathbb{X}}$ is not a CPP if and only if $xy = 0$ and $x_1 y_1 \neq 0$.*

Proof **Sufficient part**: Suppose that $xy = 0$ and $x_1 y_1 \neq 0$. If possible, assume that $((x, y), (x_1, y_1))$ is a CPP with CPP constant (r, μ). Note that in ℓ_4^2, $(x, y) \perp_B \lambda(y^3, -x^3)$, where $\lambda \in \mathbb{R}$. Let $k = \frac{1}{\|(y^3, -x^3)\|}$ and $k_1 = \frac{1}{\|(y_1^3, -x_1^3)\|}$. Since $xy = 0$, we have

$(x,\ y) = (\pm 1, 0)$ or $(0, \pm 1)$. Let $(x,\ y) = (1, 0)$. Then $(a, b) \in B((1, 0),\ r) \cap S_{\mathbb{X}} \Rightarrow \|a(x_1,\ y_1) + b\mu k_1(y_1^3, -x_1^3)\|^4 \leq 1$. Therefore,

$$a^4 + 6a^2b^2k_1^2x_1^2y_1^2\mu^2 + 4ab^3k_1^3x_1y_1(y_1^4 - x_1^4)\mu^3 + b^4\mu^4 \leq a^4 + b^4$$

$$\Rightarrow 6a^2k_1^2x_1^2y_1^2\mu^2 + 4abk_1^3x_1y_1(y_1^4 - x_1^4)\mu^3 + b^2(\mu^4 - 1) \leq 0. \quad (8.2)$$

Now, we can choose sufficiently small b such that $(a, b) \in B((1, 0),\ r) \cap S_{\mathbb{X}}$ but (8.2) does not hold. Therefore, $((1, 0),\ (x_1,\ y_1))$ is not a CPP. Similarly, we can show that $((-1,\ 0),\ (x_1,\ y_1))$ and $((0, \pm 1),\ (x_1,\ y_1))$ are not CPP.
Necessary part: Assume that $((x,\ y),\ (x_1,\ y_1))$ is not a CPP.
If possible, let $xy = x_1y_1 = 0$. Then it is easy to check that $((x,\ y),\ (x_1,\ y_1))$ is a CPP, contradicting our assumption. Therefore, either $xy \neq 0$ or $x_1y_1 \neq 0$.
Next, let $xy \neq 0$. Choose $0 < r < \min\{\frac{3|xy|}{8k+6|xy|},\ 1\}$. Then we can find a scalar $\mu > 0$ for which $a(x, y) + bk(\mp y^3, \pm x^3) \in B((x, y),\ r) \cap S_{\mathbb{X}}$. This implies that $\|a(x_1, y_1) + b\mu k_1(\mp y_1^3, \pm x_1^3)\| \leq 1$. Thus, if $xy \neq 0$, then $((x,\ y),\ (x_1,\ y_1))$ is a CPP, contradicting our hypothesis. We note that the case $x_1y_1 \neq 0$ has already been taken care of. Hence, if $((x,\ y),\ (x_1,\ y_1))$ is not a CPP we must have $xy = 0$ and $x_1y_1 \neq 0$. This completes the argument. □

The following theorem provides a complete description of rank one extreme contractions in $\mathbb{L}(\ell_4^2)$.

Theorem 8.2.22 ([124, Theorem 2.13]) *In $\mathbb{L}(\ell_4^2)$, the matrix representations (with respect to the standard ordered basis) of the class of all rank one extreme contractions are given by*

$$\begin{bmatrix} x_1 & 0 \\ y_1 & 0 \end{bmatrix}, \quad \begin{bmatrix} 0 & x_1 \\ 0 & y_1 \end{bmatrix},$$

where $x_1y_1 \neq 0$ and $x_1^4 + y_1^4 = 1$.

Proof Let $T \in S_{\mathbb{L}(\ell_4^2)}$ be a rank one operator. Suppose $(x, y) \in M_T$ and $T(x, y) = (x_1, y_1)$. Then from Theorem 8.2.17, it follows that T is an extreme contraction if and only if $((x, y), (x_1, y_1))$ is not a CPP. Therefore, by Theorem 8.2.21, we have $xy = 0$ and $x_1y_1 \neq 0$. From $xy = 0$ we get either $(x, y) = \pm(1, 0)$ or $(x, y) = \pm(0, 1)$. Since T has rank one, applying Proposition 3.2.6, we can easily deduce that in the first case, $T(0, 1) = (0, 0)$, whereas in the second case, $T(1, 0) = (0, 0)$. Therefore, the matrix representation of T with respect to the standard ordered basis is necessarily of the form

$$\begin{bmatrix} x_1 & 0 \\ y_1 & 0 \end{bmatrix}, \quad \begin{bmatrix} 0 & x_1 \\ 0 & y_1 \end{bmatrix},$$

where $x_1y_1 \neq 0$ and $x_1^4 + y_1^4 = 1$. □

We would like to remark that Theorem 8.2.22 falls completely within the scope of the analysis of extreme contractions in $\mathbb{L}(\ell_p^2)$, presented in Theorem 8.2.5. However, the true strength of Theorems 8.2.17 and 8.2.19 lies in the fact that it is applicable for general two-dimensional spaces, and furthermore we are allowed to take the domain space and the range space different from one another. In the next theorem, we present a result in this direction by proving that there does not exist any rank one extreme contraction in $\mathbb{L}(\mathbb{H}, \ell_p^2)$, whenever p is even and $\mathbb{H}$ is any Hilbert space. Clearly, this generalizes Theorem 8.2.5 (ii) to some extent, and moreover this once again illustrates how drastically the nature of extreme contractions can change if we choose the domain and the range space to be different from one another.

Theorem 8.2.23 ([124, Theorem 2.14]) *Let $T \in S_{\mathbb{L}(\mathbb{H},\ell_p^2)}$ be a rank one operator, where p is even and $\mathbb{H}$ is an arbitrary Hilbert space. Then T is not an extreme contraction.*

Proof Suppose $p = 2m$. We first claim that if $\widetilde{x} \in S_{\mathbb{H}}$ and $(x_1, y_1) \in S_{\ell_p^2}$, then $(\widetilde{x}, (x_1, y_1))$ is a CPP. Choose $r \in (0, 1)$. Let $\widetilde{u} \in \widetilde{x}^{\perp} \cap S_{\mathbb{H}}$. Then from $a\widetilde{x} + b\widetilde{u} \in B(\widetilde{x}, r) \cap S_{\mathbb{H}}$, we get $a^2 + b^2 = 1$, $|a - 1| < r$ and $|b| < r$. Note that, $(x_1, y_1) \perp_B (-y_1^{p-1}, x_1^{p-1})$. Let $k_1 = \frac{1}{\|(-y_1^{p-1}, x_1^{p-1})\|}$. In order to prove our claim, we have to find $\mu > 0$ such that $\|a(x_1, y_1) + b\mu k_1(\mp y_1^{p-1}, \pm x_1^{p-1})\|^p \leq 1$. Check that, $\|a(x_1, y_1) + b\mu k_1(-y_1^{p-1}, x_1^{p-1})\|^p \leq 1$ if and only if $a^p + \binom{p}{2}(ax_1y_1)^{p-2}(b\mu k_1)^2 + \binom{p}{3}(ax_1y_1)^{p-3}(b\mu k_1)^3(x_1^p - y_1^p) + \ldots + (b\mu)^p \leq 1 = (a^2 + b^2)^m$. Choose $0 < \mu \leq 1$ satisfying

$$\mu\left[\binom{p}{2}(1+r)^{p-2}|x_1y_1|^{p-2}k_1^2 + \binom{p}{3}(1+r)^{p-3}rk_1^3|x_1y_1|^{p-3}|x_1^p - y_1^p| + \ldots + r^{p-2}\right] \leq \binom{m}{1}(1-r)^{p-2}.$$

Now it is easy to check that

$$\binom{p}{2}(ax_1y_1)^{p-2}(b\mu k_1)^2 + \binom{p}{3}(ax_1y_1)^{p-3}(b\mu k_1)^3(x_1^p - y_1^p) + \ldots\ldots + (b\mu)^p \leq \binom{m}{1}a^{p-2}b^2 + \binom{m}{2}a^{p-4}b^4 + \ldots + b^p.$$

Thus, $\|a(x_1, y_1) + b\mu k_1(-y_1^{p-1}, x_1^{p-1})\|^p \leq 1$.
Similarly, we get $\|a(x_1, y_1) + b\mu k_1(y_1^{p-1}, -x_1^{p-1})\|^p \leq 1$. This proves that $(\widetilde{x}, (x_1, y_1))$ is a CPP. Therefore, if $T \in S_{\mathbb{L}(\mathbb{H},\, \ell_p^2)}$ is a rank one operator, then using Theorem 8.2.19, we conclude that T is not an extreme contraction. □

Approaching similarly, we can characterize the class of all rank one extreme contractions in $\mathbb{L}(\ell_4^2, \mathbb{H})$, where $\mathbb{H}$ is an arbitrary Hilbert space, as follows.

Theorem 8.2.24 ([124, Theorem 2.15]) *Suppose* $\mathbb{H}$ *is a Hilbert space. Let* $T \in S_{\mathbb{L}(\ell_4^2,\mathbb{H})}$ *be a rank one operator. Then* T *is an extreme contraction if and only if* $M_T = \{\pm(1,0)\}$ *or* $M_T = \{\pm(0,1)\}$. *In particular, if* $\mathbb{H} = \ell_2^n$ *then the matrix representations (with respect to the standard ordered basis) of the class of all rank one extreme contractions are given by*

$$\begin{bmatrix} x_1 & x_2 & \dots & x_n \\ 0 & 0 & \dots & 0 \end{bmatrix}^{\top}, \quad \begin{bmatrix} 0 & 0 & \dots & 0 \\ x_1 & x_2 & \dots & x_n \end{bmatrix}^{\top},$$

where $x_1^2 + x_2^2 + \dots x_n^2 = 1$.

Remark 8.2.25 ([124, Remark 2.2]) From Theorem 8.2.24 it follows that the matrix representations (with respect to the standard ordered basis) of the class of all rank one extreme contractions in $\mathbb{L}(\ell_4^2, \ell_2^2)$ are given by

$$\begin{bmatrix} x_1 & 0 \\ y_1 & 0 \end{bmatrix}, \quad \begin{bmatrix} 0 & x_1 \\ 0 & y_1 \end{bmatrix},$$

where $x_1^2 + y_1^2 = 1$. This description of rank one extreme contractions in $\mathbb{L}(\ell_4^2, \ell_2^2)$ also follows from Theorem 8.2.5 (iii). However, Theorem 8.2.24 is applicable for an arbitrary Hilbert space as the range space, whereas Theorem 8.2.5 (iii) is applicable only for two-dimensional spaces.

Note that the significance of the domain space and the range space in the study of extreme contractions between Banach spaces is further illustrated by Theorems 8.2.23 and 8.2.24. In $\mathbb{L}(\mathbb{H}, \ell_4^2)$, there does not exist any rank one extreme contraction. However, in $\mathbb{L}(\ell_4^2, \mathbb{H})$, indeed there are rank one extreme contractions.

The following example shows that the condition *rank* $T = 1$ in Theorem 8.2.16 is not necessary for T to be such that T is not an extreme contraction.

Example 8.2.26 ([124, Example 2.15.1]) Define $T : {\ell_p}^2 \longrightarrow {\ell_p}^2$ by

$$T(1,\ 0) = (1,\ 0),\ T(0,\ 1) = \frac{1}{2}(0,\ 1).$$

Then it is easy to check that $M_T = \{\pm(1,\ 0)\}$. Again $((1,\ 0),\ (1,\ 0))$ is a CPP. From Theorem 8.2.5 (iv), it is easy to see that T is not an extreme contraction.

The above example calls for our attention towards the problem of characterizing rank two extreme contractions on a two-dimensional strictly convex and smooth Banach space. In this context, it is worthwhile to observe that if a norm one linear operator T, defined between two-dimensional strictly convex Banach spaces, attains norm at two linearly independent unit vectors then T is surely an extreme contraction. Therefore, it is sufficient to restrict our attention to rank two linear operators between two-dimensional strictly convex and smooth Banach spaces, which attain norm at only one pair of points. As we will see in the following theorem, in order to obtain a desired characterization in this case, the notion of $\mu-$CPP plays a vital role. We consider $T \in \mathbb{L}(\mathbb{X}, \mathbb{Y})$, where $\mathbb{X}$, $\mathbb{Y}$ are two-dimensional smooth Banach spaces and $\mathbb{Y}$ is strictly convex. Let $\|T\| = 1,\ rank\ T = 2,\ M_T = \{\pm x\}$. Let $y \in x^{\perp} \cap S_{\mathbb{X}}$. Then clearly $0 < \|Ty\| < 1$. As $\mathbb{X}$ and $\mathbb{Y}$ are both smooth so $Tx \perp_B Ty$. Let $w = \frac{Ty}{\|Ty\|}$. In the following theorem, we prove that T is not an extreme contraction if and only if $(x,\ Tx)$ is a μ-CPP with respect to $(y,\ w)$ for some $\mu > \|Ty\|$.

Theorem 8.2.27 ([124, Theorem 2.16]) *Let $\mathbb{X}$, $\mathbb{Y}$ be two-dimensional smooth Banach spaces and in addition, let $\mathbb{Y}$ be strictly convex. Suppose $T \in S_{\mathbb{L}(\mathbb{X},\mathbb{Y})}$ is a rank two operator and $M_T = \{\pm x\}$. Then T is not an extreme contraction if and only if $(x,\ Tx)$ is a $\mu-$CPP with respect to the pair $(y,\ w)$ for some $\mu > \|Ty\|$.*

Proof **Necessary part**: Let $\|Ty\| = k$. Since T is not an extreme contraction, there exist $T_1,\ T_2 \in S_{\mathbb{L}(\mathbb{X},\mathbb{Y})}$ such that $T = \frac{1}{2}(T_1 + T_2),\ T \neq T_1,\ T \neq T_2$. Clearly $Tx = \frac{1}{2}(T_1x + T_2x)$. Since $Tx \in Ext(B_{\mathbb{Y}})$, we have $Tx = T_1x = T_2x$, i.e., $x \in M_{T_1} \cap M_{T_2}$. On the other hand, applying Proposition 3.2.6 and the fact that $x \perp_B y$, we get $T_1x \perp_B T_1y$, i.e., $Tx \perp_B T_1y$. Therefore, there exists $\delta \in \mathbb{R} \setminus \{0\}$ such that $T_1y = (k + \delta)w$ and $T_2y = (k - \delta)w$. Let $ax + by \in S_{\mathbb{X}}$. Then $\|T_1(ax + by)\| \leq 1 \Rightarrow \|aTx + b(k + \delta)w\| \leq 1$ and $\|T_2(ax + by)\| \leq 1 \Rightarrow \|aTx + b(k - \delta)w\| \leq 1$. Therefore, $\|aTx + b(k + |\delta|)w\| \leq 1$. Thus, $(x,\ Tx)$ is a $\mu-$CPP with respect to the pair $(y,\ w)$ for some $\mu > \|Ty\|$. This establishes the proof of the necessary part.
Sufficient part: Suppose that $(x,\ Tx)$ is a $\mu-$CPP with respect to the pair $(y,\ w)$ for some $\mu > \|Ty\| = k$. Then there exists $\epsilon > 0$ such that $\mu = k + \epsilon$. Moreover, there exists $r > 0$ such that $ax + by \in B(x,\ r) \cap S_{\mathbb{X}}$ implies that $\|aTx + b(k + \epsilon)w\| \leq 1$. Since $M_T = \{\pm x\}$, there exists $\delta > 0$ such that $\sup\{\|Th\| : h \in B(x,\ r)^c \cap S_{\mathbb{X}}\} = 1 - \delta$. Choose $0 < c < min\{k, \frac{\delta}{2}, \epsilon\}$. Define $T_1,\ T_2 \in \mathbb{L}(\mathbb{X}, \mathbb{Y})$ as follows:

$$T_1x = Tx \qquad T_2x = Tx$$
$$T_1y = (k + c)w \qquad T_2y = (k - c)w$$

Clearly, $T = \frac{1}{2}(T_1 + T_2),\ T \neq T_1,\ T \neq T_2$. Let $ax + by \in B(x,\ r) \cap S_{\mathbb{X}}$. Then since $\|aTx + b(k + \epsilon)w\| \leq 1$, using $0 < k - c < k + c \leq k + \epsilon$ and Proposition 8.2.8, we obtain $\|T_1(ax + by)\| = \|aTx + b(k + c)w\| \leq 1$ and $\|T_2(ax + by)\| = \|aTx + b(k - c)w\| \leq 1$. Now, let $ax + by \in B(x,\ r)^c \cap S_{\mathbb{X}}$. Since $x \perp_B y$ and $\|ax + by\| = 1$, we have $|b| \leq \|ax + by\| + \|ax\| \leq 2$. Therefore, $\|T_1(ax + by)\| = \|aTx + b(k + c)w\| = \|T(ax + by) + bcw\| \leq \|T(ax + by)\| + |bc| \leq 1 - \delta + 2c < 1 - \delta + \delta = 1$. This proves that $\|T_1\| = 1$. Similarly we can show that $\|T_2\| = 1$. Hence, T is not an extreme contraction. □

Now we are in a position to completely characterize extreme contractions between two-dimensional strictly convex and smooth Banach spaces. We would like to remark that to the best of our knowledge, such a general characterization of extreme contractions was presented for the very first time in [124]. Indeed, we have the following theorem:

Theorem 8.2.28 ([124, Theorem 2.17]) *Let $\mathbb{X}$, $\mathbb{Y}$ be two-dimensional smooth, strictly convex Banach spaces. Let $T \in \mathbb{L}(\mathbb{X}, \mathbb{Y})$ be a norm one linear operator. Then T is an extreme contraction if and only if exactly one of the following three conditions holds true:*
(i)*: (a) Rank T is one,*
(b) $M_T = \{\pm x\}$ for some $x \in S_{\mathbb{X}}$,
(c) (x, Tx) is not a CPP.

(ii)*: (a) Rank T is two,*
(b) $M_T = \{\pm x\}$ for some $x \in S_{\mathbb{X}}$,
(c) $Ty = kw$ for some $y \in x^{\perp} \cap S_{\mathbb{X}}$, $w \in (Tx)^{\perp} \cap S_{\mathbb{X}}$ and $0 < k < 1$
(d) (x, Tx) is not a $\mu-$CPP with respect to the pair (y, w) for any $\mu > k$,

(iii)*: (a) Rank T is two,*
(b) T attains norm at at least two pairs of linearly independent unit vectors $\pm x$, $\pm y$.

So far, our analysis has been mostly confined to two-dimensional (strictly convex and smooth) Banach spaces. A more general result, without any restriction on the dimension of the space, is given in the form of the following theorem:

Theorem 8.2.29 ([124, Theorem 2.18]) *Let $\mathbb{X}$ be a smooth, reflexive Banach space with Kadets–Klee property and let $\mathbb{Y}$ be a smooth Banach space. Suppose $T \in S_{\mathbb{K}(\mathbb{X},\mathbb{Y})}$ is a smooth operator. Then $M_T = \{\pm x\}$, for some $x \in S_{\mathbb{X}}$. Moreover, if we additionally assume that (x, Tx) is a CPP with CPP constants (r, μ) and $0 < \|T\|_H < \mu$, where H is the unique hyperspace such that $x \perp_B H$, then T is not an extreme contraction.*

Proof Since T is a compact smooth operator, from Theorem 6.3.2, it follows that $M_T = \{\pm x\}$, for some $x \in S_{\mathbb{X}}$. Because of $\|T\|_H > 0$, we can choose $y \in H \cap S_{\mathbb{X}}$ such that $Ty \neq 0$. There is a unique hyperspace H_0 of H such that $y \perp_B H_0$. Hence, for each $z \in \mathbb{X}$, there are unique $a, b \in \mathbb{R}$, $h_0 \in H_0$ such that $z = ax + by + h_0$. For each $n \in \mathbb{N}$, let us define $T_n, S_n \in \mathbb{L}(\mathbb{X}, \mathbb{Y})$ as follows:

$$T_n(z) = aTx + (1 + \frac{1}{n})(bTy + Th_0),$$

$$S_n(z) = aTx + (1 - \frac{1}{n})(bTy + Th_0).$$

Using the sequential criteria of a compact operator, it is easy to check that $T_n, S_n \in \mathbb{K}(\mathbb{X}, \mathbb{Y})$, for each $n \in \mathbb{N}$. Clearly, $T = \frac{1}{2}T_n + \frac{1}{2}S_n$ and $T \neq T_n, T \neq S_n$ for all $n \in \mathbb{N}$. It is easy to verify that $\|S_n\| \leq 1$ for all $n \in \mathbb{N}$. If $\|T_n\| \leq 1$ for some n, then T is not an extreme contraction and we are done. Therefore, assume that $\|T_n\| > 1$ for each $n \in \mathbb{N}$. Since each T_n is a compact operator on a reflexive Banach space, $M_{T_n} \neq \emptyset$. Choose $y_n \in M_{T_n}$ for each $n \in \mathbb{N}$.
Now, since $0 < \|T\|_H < \mu$, there exists $n_0 \in \mathbb{N}$ such that $\frac{\mu}{\|T\|_H} > 1 + \frac{1}{n}$ for all $n \geq n_0$. Let $z = (ax + by + h_0) \in B(x, r) \cap S_{\mathbb{X}}$. We claim that $\|T_n z\| \leq 1$ for all $n \geq$

n_0. For $by + h_0 = 0$, we clearly have $\|T_n z\| \leq 1$ for each $n \in \mathbb{N}$. Suppose $by + h_0 \neq 0$. Then $z = ax + ch$, where $c = \|by + h_0\|$ and $h = \frac{1}{c}(by + h_0)$. Note that, $|a| \leq \|ax + ch\| = \|z\| = 1$ and $|b| \leq \|by + h_0\| = \|ch\| = c$. First assume $Th \neq 0$. Then from $z = ax + ch \in B(x, r) \cap S_{\mathbb{X}}$ and the fact that (x, Tx) is a CPP with CPP constant (r, μ), we have $\|aTx \pm \frac{c\mu}{\|Th\|} Th\| \leq 1$. Moreover, $\frac{\mu}{\|Th\|} \geq \frac{\mu}{\|T\|_H} > 1 + \frac{1}{n}$ for all $n \geq n_0$. Thus, using Proposition 8.2.8, we get

$$\|T_n z\| = \|aTx + (1 + \frac{1}{n})\{bTy + Th_0\}\| = \|aTx + c(1 + \frac{1}{n})Th\| \leq 1.$$

Next, assume that $Th = 0$. Then $\|T_n z\| = \|aTx + (1 + \frac{1}{n})(bTy + Th_0)\| = \|aTx + (1 + \frac{1}{n})cTh\| = \|aTx\| \leq 1$. Since $y_n \in M_{T_n}$ and $\|T_n\| > 1$, we have $y_n \notin B(x, r)$ for all $n \geq n_0$. Similarly, we can show that $y_n \notin B(-x, r)$ for all $n \geq n_0$. Now $\mathbb{X}$ being reflexive and $B_{\mathbb{X}}$ being weakly compact, the sequence $\{y_n\}$ has a weakly convergent subsequence in $B_{\mathbb{X}}$. Without loss of generality, assume that $y_n \rightharpoonup y_0$, where $y_0 \in B_{\mathbb{X}}$. Let $z = (ax + by + h_0) \in S_{\mathbb{X}}$. Then $|a| \leq \|(ax + by + h_0)\| = 1$ and $\|(by + h_0)\| = \|z - ax\| \leq 1 + |a| \leq 2$. Now, $\|(T_n - T)z\| = \|(T_n - T)(ax + by + h_0)\| = \|\frac{1}{n}(bTy + Th_0)\| \leq \frac{1}{n}\|by + h_0\| \leq \frac{2}{n}$. This proves that $\|(T_n - T)z\| \to 0$ as $n \to \infty$. Hence, $T_n \to T$. Since T is a compact operator and $y_n \rightharpoonup y_0$, we have $Ty_n \to Ty_0$. Using this, we get the following:

$$\|T_n y_n - Ty_0\| \leq \|T_n y_n - Ty_n\| + \|Ty_n - Ty_0\| \leq \|T_n - T\| + \|Ty_n - Ty_0\| \to 0 \text{ as } n \to \infty.$$

Hence, $T_n y_n \to Ty_0$, which shows that $\|T_n y_n\| \to \|Ty_0\|$, i.e., $\|Ty_0\| \geq 1$, since $\|T_n y_n\| > 1$ for all $n \in \mathbb{N}$. Since $y_0 \in B_{\mathbb{X}}$ and $\|T\| = 1$, we have $\|Ty_0\| = 1$. Therefore, $y_0 = \pm x$. Again, $y_n \rightharpoonup y_0$ and $\|y_n\| \to \|y_0\|$. Since $\mathbb{X}$ has Kadets–Klee property, we get $y_n \to y_0$. This implies that $y_0 \notin B(\pm x, r)$, a contradiction. Therefore, T is not an extreme contraction. This proves the theorem. □

We would also like to illustrate that extreme contractions and this newly introduced notion of CPP play an important role in determining the geometry of the underlying space. First we deal with strict convexity. Indeed, our next two theorems separately give a characterization of strict convexity (in the two-dimensional case, the necessary part of which is trivial) and a sufficient condition for strict convexity (without any restriction on the dimension of the underlying space), using the notions of extreme contractions and CPP.

Theorem 8.2.30 ([124, Theorem 2.19]) *Let $\mathbb{X}$ be a two-dimensional Banach space. Then $\mathbb{X}$ is strictly convex if and only if for any $T \in S_{\mathbb{L}(\mathbb{X})}$, M_T contains two linearly independent vectors implies that T is an extreme contraction.*

Proof We only prove the sufficient part. Suppose on the contrary that $\mathbb{X}$ is not strictly convex. Then there exist $x, y \in S_{\mathbb{X}}$ and $\lambda_0 > 0$ such that the line segment joining $x + \lambda_0 y$ and $x - \lambda_0 y$ is a subset of $S_{\mathbb{X}}$. It is now easy to observe that $x \perp_B y$. Define $T, T_1, T_2 \in \mathbb{L}(\mathbb{X})$ as follows:

$$Tx = x \qquad T_1x = x + \frac{1}{2}\lambda_0 y \qquad T_2x = x - \frac{1}{2}\lambda_0 y$$
$$Ty = 0 \qquad T_1y = 0 \qquad T_2y = 0.$$

Now, it is easy to see that the linearly independent vectors $x + \lambda_0 y$ and $x - \lambda_0 y$ are in M_T. Moreover, verify that $\|T\| = \|T_1\| = \|T_2\| = 1$. Indeed if $z = ax + by \in S_{\mathbb{X}}$, then $\|T_1 z\| = \|a(x + \frac{1}{2}\lambda_0 y)\| = |a| \leq \|ax + by\| = 1$ and $\|T_1 x\| = 1$. Thus $T = \frac{1}{2}(T_1 + T_2)$ and $T \neq T_1,\ T \neq T_2$, which implies that T is not an extreme contraction. This clearly contradicts the hypothesis. Therefore, $\mathbb{X}$ must be strictly convex. □

Theorem 8.2.31 ([124, Theorem 2.20]) *Let $\mathbb{X}$ be a Banach space such that for any $x \in S_{\mathbb{X}}$, $(x,\ x)$ is a CPP. Then $\mathbb{X}$ is strictly convex.*

Proof Suppose on the contrary that $\mathbb{X}$ is not strictly convex. Then there exist $y,\ z \in S_{\mathbb{X}}$ and $\lambda_0 > 0$ such that the line segment joining $y + \lambda_0 z$ and $y - \lambda_0 z$ is a subset of $S_{\mathbb{X}}$. Consider the set $S = \{\lambda > 0 : \|y + \lambda z\| > 1\}$. Note that, for $|\lambda| > 2$, we have $\|y + \lambda z\| \geq |\lambda| - 1 > 1$. Thus, $S \neq \emptyset$. Assume that $k = \inf S$. Let $x = y + kz$. If $\|y + kz\| > 1$ then we can find $\delta > 0$ such that $\|y + \lambda z\| > 1$ for all $\lambda \in (k - \delta,\ k + \delta)$, which contradicts the fact that $k = \inf S$. So $\|x\| = 1$. Clearly, $x \perp_B z$ and $\|x - \lambda z\| = 1$ for all $\lambda \in [0, k]$. We now show that $(x,\ x)$ is not a CPP which leads to a contradiction. Let $r > 0$. Then there exists $\lambda \in [0, k]$ such that $x - \lambda z \in B(x, r) \cap S_{\mathbb{X}}$ but $\|x + \mu\lambda z\| > 1$ for any $\mu > 0$. This establishes the theorem. □

We end the section with a characterization of Hilbert spaces among Banach spaces in terms of CPP. Recall that for any two elements $x, y \in \mathbb{X}$, x is said to be Isosceles orthogonal [61] to y, written as $x \perp_I y$, if $\|x + y\| = \|x - y\|$. To serve our purpose we state a theorem characterizing Hilbert spaces in terms of equivalence of B-J orthogonality and Isosceles orthogonality.

Theorem 8.2.32 ([4, Theorem 5.1]) *Let $\mathbb{X}$ be a Banach space. Then the following are equivalent:*
(i) $x, y \subset \mathbb{X},\ x \perp_I y \Rightarrow x \perp_B y$.
(ii) $x, y \in \mathbb{X},\ x \perp_B y \Rightarrow x \perp_I y$.
(iii) $x, y \in S_{\mathbb{X}},\ x \perp_I y \Rightarrow x \perp_B y$.
(iv) $x, y \in S_{\mathbb{X}},\ x \perp_B y \Rightarrow x \perp_I y$.
(v) $x, y \in S_{\mathbb{X}},\ x \perp_I y \Rightarrow x + y \perp_B x - y$.
(vi) *$\mathbb{X}$ is a Hilbert space.*

Theorem 8.2.33 ([124, Theorem 2.21]) *A Banach space $\mathbb{X}$ is a Hilbert space if and only if for any $x \in S_{\mathbb{X}}$, $(x,\ x)$ is a CPP with CPP constant $(r,\ 1)$ for any $r > 0$.*

Proof The necessary part is a consequence of Proposition 8.2.9 (ii). For the sufficient part, we prove that for $x,\ y \in S_{\mathbb{X}}$, if $x \perp_B y$ then $x \perp_I y$. Since for each $x \in S_{\mathbb{X}}$, $(x,\ x)$ is a CPP, from Theorem 8.2.31, it follows that $\mathbb{X}$ is strictly convex. Let $x,\ y \in S_{\mathbb{X}}$ and $x \perp_B y$. Then $\|x + y\| > 1$. Let $c = \frac{1}{\|x+y\|}$. Then $0 < c < 1$ and $c(x + y) \in S_{\mathbb{X}}$. Clearly, $c(x + y) \in B(x,\ 1)$. Hence, $\|cx - cy\| \leq 1$, since $(x,\ x)$ is a CPP. We claim that $\|cx - cy\| = 1$. Suppose on the contrary that $\|cx - cy\| < 1$.

Note that the function $f : \mathbb{R} \longrightarrow \mathbb{R}$ defined by $f(a) = \|cx - ay\|$ is continuous on $\mathbb{R}$ and $f(c) < 1$. Moreover, $f(c+2) = \|cx - (c+2)y\| \geq c + 2 - c = 2 > 1$. Therefore, by continuity of f, we get some $\lambda \in (c,\ c+2)$ such that $f(\lambda) = 1$, i.e., $\|cx - \lambda y\| = 1$. Since $(x,\ x)$ is a CPP, $\|cx + \lambda y\| \leq 1$. Therefore,

$$cx + cy = (1 - \frac{c}{\lambda})cx + \frac{c}{\lambda}(cx + \lambda y)$$

$$\Rightarrow \|cx + cy\| \leq (1 - \frac{c}{\lambda})\|cx\| + \frac{c}{\lambda}\|(cx + \lambda y)\|$$

$$\Rightarrow \|cx + cy\| < (1 - \frac{c}{\lambda}) + \frac{c}{\lambda} = 1, \quad \text{a contradiction.}$$

Thus, $\|cx - cy\| = 1$ and so $\|x + y\| = \|x - y\|$. Hence, $x \perp_B y \Rightarrow x \perp_I y$. It now follows from Theorem 8.2.32 that $\mathbb{X}$ is a Hilbert space and completes the proof. □

8.3 Extreme Contractions on Polyhedral Banach Spaces

A convex set C in a finite-dimensional Banach space is said to be a polyhedron or a polytope [70] if C is the intersection of a finite number of closed half-spaces. Equivalently, C is a polytope if and only if C has finitely many extreme points. A Banach space $\mathbb{X}$ is said to be polyhedral [70] if the unit ball of each finite-dimensional subspace of $\mathbb{X}$ is a polytope. Note that c_0 (the space of all sequences converging to 0) is a polyhedral Banach space (see [70, Proposition 4.7]). Indeed, c_0 is the simplest example of an infinite-dimensional polyhedral Banach space, in the sense that each infinite-dimensional polyhedral Banach space contains a isomorphic copy of c_0 (see [35]). Therefore, in this section our first objective is to describe the extreme contractions on c_0. The notion of nice operators plays an important role here.

Definition 8.3.1 ([95]) Let $\mathbb{X}$, $\mathbb{Y}$ be Banach spaces. An operator $T \in \mathbb{L}(\mathbb{X}, \mathbb{Y})$ is said to be a nice operator if $T^*(Ext(B_{\mathbb{Y}^*})) \subseteq Ext(B_{\mathbb{X}^*})$.

Note that each nice operator is an extreme contraction. However, the converse is not always true. See [21, 137, 138] for example of non-nice extreme operators.

Definition 8.3.2 ([22]) A Banach space $\mathbb{X}$ is said to be nice if for each Banach space $\mathbb{Y}$, the collection of all extreme contractions from $\mathbb{Y}$ to $\mathbb{X}$ coincides with the collection of all nice operators from $\mathbb{Y}$ to $\mathbb{X}$.

Cabrera-Serrano and Mena-Jurado [22] proved that c_0 is a nice Banach space. Recall that for an arbitrary set I, $c_0(I)$ is the space of all scalar valued functions f on I which vanishes at infinity, i.e., $c_0(I)$ is the collection of all $f : I \to \mathbb{R}$ such that for each $\epsilon > 0$, the set $\{\gamma \in I : |f(\gamma)| > \epsilon\}$ is finite.

Theorem 8.3.3 ([22, Proposition 2.1]) *Let I be any non-empty set. Then $c_0(I)$ is nice.*

Hence, from the previous theorem we can conclude the following.

Theorem 8.3.4 *Let $\mathbb{X}$ be any Banach space. Then $T \in \mathbb{L}(\mathbb{X}, c_0)$ is an extreme contraction if and only if T is a nice operator.*

In the rest of this section, our main focus is to explore extreme contractions on finite-dimensional polyhedral Banach spaces. However, before that, we would like to briefly discuss extreme contractions on some other spaces which are not necessarily polyhedral. After Hilbert spaces and $\ell_p(1 < p < \infty)$ spaces, the most explored spaces in the context of extreme contraction are $C(K)$ space (space of all continuous functions on a compact Hausdorff space K with supremum norm), the space ℓ_∞ and the space ℓ_1. Therefore, it is important to describe the extreme contraction on these spaces. Extreme contractions on many $C(K)$-type spaces actually coincide with the nice operators. We recall from [97, 145] that the nice operators between continuous function spaces are weighted composition operators. Specifically, if K_1, K_2 are compact Hausdorff spaces, then $T \in \mathbb{L}(C(K_1), C(K_2))$ is a nice operator if and only if there exists $h \in C(K_2)$ with $|h(k)| = 1$ for all $k \in K_2$ and a continuous function $\phi : K_2 \to K_1$ such that

$$Tf(k) = h(k)f(\phi(k)), \text{ for all } f \in C(K_1),\ k \in K_2. \tag{8.3}$$

Some of the known description of extreme contractions on $C(K)$-type spaces can be found in [7, 21, 28, 56, 140]. The following theorem is worth mentioning here.

Theorem 8.3.5 ([7, 21, 28, 56, 140]) *Let K_1, K_2 be compact Hausdorff spaces. Let $T \in \mathbb{L}(C(K_1), C(K_2))$ and $\|T\| = 1$. Suppose either of the following is true:*
(i) *[21] K_1 is metrizable.*
(ii) *[7] K_1 is Eberlein-compact (that means K_1 is homeomorphic to a weakly compact subset of a Banach space) and K_2 is metrizable.*
(iii) *[56, 140] K_2 is extremally disconnected.*
(iv) *[140] K_1 is a dispersed compact Hausdorff space.*
(v) *[21] T is compact.*
(vi) *[28] T is weakly compact and K_2 is separable.*
Then T is extreme if and only if T is a nice operator if and only if T is of the form (8.3). Moreover, in case of (iii) and (iv), the result is true for complex scalars too.

Al-Halees and Fleming explored extreme contraction on vector valued function spaces.

Theorem 8.3.6 ([3, Theorem 4]) *Suppose K_1 is a compact Hausdorff space and K_2 is an extremally disconnected compact Hausdorff space. Suppose T is an extreme contraction on $\mathbb{L}(C(K_1, E), C(K_2, E))$, where E is either ℓ_∞^n or c_0. Then T is a nice operator.*

Now, for any index set A, let $\ell_\infty(A)$ be the space of all bounded real valued functions on A. Then $\ell_\infty(A)$ is isometrically isomorphic to $C(\beta A)$, where βA is the Stone–Čech compactification of A. Therefore, using Theorem 8.3.5 (iii), we can characterize extreme contractions on $\mathbb{L}(\ell_\infty(A), \ell_\infty(B))$, where A, B are arbitrary index sets. In [67], Kim independently characterized extreme contractions on $\ell_\infty(= \ell_\infty(\mathbb{N}))$.

Theorem 8.3.7 ([67, Theorem 1]) *An operator $T \in \mathbb{L}(\ell_\infty, \ell_\infty)$ is an extreme contraction if and only if $T(fg) = T1TfTg$, where $f, g \in \ell_\infty$ and $|T1(i)| = 1$ for all $i \in \mathbb{N}$.*

A Banach space is said to be an L^1-space if it is isometric to a space $L(S, \Sigma, \mu)$, where S is a non-empty set, Σ is a $\sigma-$algebra on S, and μ is a $\sigma-$finite measure on Σ. In 1973, Sharir investigated the extreme contractions on L^1-spaces. Recall that the dual of an L^1 space can be identified with a $C(X)$ space for some appropriate hyperstonian space X. The following results related to extreme contractions on L^1 space are worth mentioning here.

Theorem 8.3.8 ([139, Theorem 2.2]) *Let $\mathbb{X}$, $\mathbb{Y}$ be L^1-spaces. Then every extreme contraction $T : \mathbb{X} \to \mathbb{Y}$ is nice.*

Theorem 8.3.9 ([139, Theorem 2.9]) *Let $\mathbb{X}$ be an L^1-space and $\mathbb{Y}$ be an arbitrary Banach space. Let T be an extreme contraction in $\mathbb{L}(\mathbb{X}, \mathbb{Y})$. Then $T(Ext(B_{\mathbb{X}})) \subseteq Ext(B_{\mathbb{Y}})$.*

Note that the composition of two nice operators on a Banach space is again a nice operator. Therefore, the set of all nice operators on a Banach space forms a semigroup. From Theorems 8.3.5 and 8.3.8, we get that there are a large number of Banach spaces, where the set of all extreme contractions form a semigroup under composition. Motivated by these facts, Lindenstrauss and Perles in [80] explored the spaces where the set of all extreme contractions form a semigroup. In this context, they first proved the following equivalence.

Theorem 8.3.10 ([80, Theorem 1.1]) *If $\mathbb{X}$ is a finite-dimensional Banach space, then the following statements are equivalent:*
(1) $T \in Ext(B_{\mathbb{L}(\mathbb{X},\mathbb{X})}),\ \ x \in Ext(B_{\mathbb{X}}) \Rightarrow Tx \in Ext(B_{\mathbb{X}})$.
(2) $T_1, T_2 \in Ext(B_{\mathbb{L}(\mathbb{X},\mathbb{X})}) \Rightarrow T_1 \circ T_2 \in Ext(B_{\mathbb{L}(\mathbb{X},\mathbb{X})})$.
(3) $\{T_i\}_{i=1}^m \subseteq Ext(B_{\mathbb{L}(\mathbb{X},\mathbb{X})}) \Rightarrow \|T_1 \circ \ldots \circ T_m\| = 1$, *for all* m.

Proof (1) $\Rightarrow$ (2). Note that, since $\mathbb{X}$ is finite dimensional, by the Krein–Milman theorem there are at least m linearly independent extreme points say $x_1, x_2, \ldots, x_m \in Ext(B_{\mathbb{X}})$. Now, by (1), $T_1 \circ T_2(x_i) \in Ext(B_{\mathbb{X}})$ for all $1 \leq i \leq m$. Thus, $\|T_1 \circ T_2\| \leq \|T_1\|\|T_2\| = 1$ implies that $x_i \in M_{T_1 \circ T_2}$ for all $1 \leq i \leq m$. Therefore, from Proposition 8.2.1, it follows that $T_1 \circ T_2 \in Ext(B_{\mathbb{L}(\mathbb{X},\mathbb{X})})$.
The implication (2) $\Rightarrow$ (3) is trivial. The implication (3) $\Rightarrow$ (1) is quite long and involved. So we skip this proof here. □

In the same paper, Lindenstrauss and Perles characterized those spaces $\mathbb{X}$ with $\dim(\mathbb{X}) < 5$, such that $\mathbb{X}$ satisfies properties $(1) - (3)$ of the previous theorem.

Theorem 8.3.11 ([80, Theorem 1.2]) *Let $\mathbb{X}$ be a Banach space of dimension strictly less than 5. Then $\mathbb{X}$ has the properties* (1), (2) *and* (3), *of Theorem 8.3.10, if and only if either of the following is true:*
(i) *$\mathbb{X}$ is an inner product space.*
(ii) *$B_{\mathbb{X}}$ is a polytope with the property that for every facet (i.e., maximal proper face) K of $B_{\mathbb{X}}$, $B_{\mathbb{X}}$ is the convex hull of $K \bigcup (-K)$.*

Later on, Larman proved in [74] that Theorem 8.3.11 is true even if $\dim(\mathbb{X}) \leq 6$. Motivated by these results, the following definition was introduced to study extreme contraction further.

Definition 8.3.12 ([129, Definition 1.2]) Let $\mathbb{X}$, $\mathbb{Y}$ be Banach spaces. We say that the pair $(\mathbb{X}, \mathbb{Y})$ has L-P (abbreviated form of Lindenstrauss–Perles) property if a norm one bounded linear operator $T \in \mathbb{L}(\mathbb{X}, \mathbb{Y})$ is an extreme contraction if and only if $Ext(B_{\mathbb{X}}) \subseteq M_T$

In due course of time, we will present the relevance of the notion of L-P property in the study of extreme contractions. Our next objective is to obtain a necessary condition for an operator to be an extreme contraction on a polyhedral Banach space. We need the following easy proposition to prove the desired result.

Proposition 8.3.13 ([129, Proposition 2.1]) *Let $\mathbb{X}$ be an $n-$dimensional Banach space and let $A \subseteq \mathbb{X}$ be a bounded set. Suppose $\{x_1, x_2, \ldots, x_n\}$ is a basis of $\mathbb{X}$. Then the following set $S = \{|\alpha_i| : z = \sum_{i=1}^{n} \alpha_i x_i \text{ and } z \in A\}$ is bounded.*

Next, we prove the desired theorem.

Theorem 8.3.14 ([129, Theorem 2.2]) *Suppose $\mathbb{X}$ is an $n-$dimensional polyhedral Banach space and $\mathbb{Y}$ is any Banach space. Let $T \in \mathbb{L}(\mathbb{X}, \mathbb{Y})$ be an extreme contraction. Then $span(M_T \cap Ext(B_{\mathbb{X}})) = \mathbb{X}$. Moreover, if $M_T \cap Ext(B_{\mathbb{X}})$ contains exactly $2n$ elements then $T(M_T \cap Ext(B_{\mathbb{X}})) \subset Ext(B_{\mathbb{Y}})$.*

Proof We first show that $span(M_T \cap Ext(B_{\mathbb{X}})) = \mathbb{X}$. Since $\dim(\mathbb{X}) = n$, using Krein–Milman theorem we see that $|Ext(B_{\mathbb{X}})| \geq 2n$ and at least n points of $Ext(B_{\mathbb{X}})$ are linearly independent. If $M_T \cap Ext(B_{\mathbb{X}})$ contains n linearly independent points, then we are done. If possible, suppose that $M_T \cap Ext(B_{\mathbb{X}})$ does not contain n linearly independent points. Suppose $M_T \cap Ext(B_{\mathbb{X}})$ has at most k linearly independent points, where $k < n$. Let $\{x_1, x_2, \ldots, x_k\}$ be a linearly independent subset of $M_T \cap Ext(B_{\mathbb{X}})$. We extend it to a basis $\{x_1, x_2, \ldots, x_k, x_{k+1}, \ldots, x_n\}$ of $\mathbb{X}$ such that each $x_i \in Ext(B_{\mathbb{X}})$, for $i = 1, 2, \ldots, n$. Now, choose $w(\neq 0) \in B_{\mathbb{Y}}$ such that $Tx_n \pm \frac{w}{j} \neq 0$ for every $j \in \mathbb{N}$. For each $j \in \mathbb{N}$, define linear operators $T_j,\ S_j : \mathbb{X} \to \mathbb{Y}$ as follows:

$$T_j(x_i) = Tx_i \ (1 \leq i \leq n-1) \qquad S_j(x_i) = Tx_i \ (1 \leq i \leq n-1)$$

$$T_j(x_n) = Tx_n + \frac{w}{j} \qquad S_j(x_n) = Tx_n - \frac{w}{j}.$$

Clearly, for all $j \in \mathbb{N}$, we have $T \neq T_j$, $T \neq S_j$ and $T = \frac{1}{2}(T_j + S_j)$. Since T is an extreme contraction, there exists a subsequence $\{T_{j_k}\}$ of $\{T_j\}$ such that for all k, either $\|T_{j_k}\| > 1$ or $\|S_{j_k}\| > 1$. Without loss of generality, assume that $\|T_j\| > 1$ for all $j \in \mathbb{N}$. Now, $\mathbb{X}$ being finite dimensional, $M_{T_j} \cap Ext(B_{\mathbb{X}}) \neq \emptyset$ for each j. Suppose $y_j \in M_{T_j} \cap Ext(B_{\mathbb{X}})$, i.e., $\|T_j y_j\| = \|T_j\| > 1$ for each j.

Let $z = a_1x_1 + \ldots + a_{n-1}x_{n-1} + a_nx_n \in S_{\mathbb{X}}$ be arbitrary. Then from Proposition 8.3.13, it follows that there exists $r > 0$ such that $|a_n| \leq r$. Now, from

$$\|(T_j - T)z\| = \|(T_j - T)(a_1x_1 + \ldots + a_nx_n)\| = \|\frac{a_n w}{j}\| \leq \frac{r}{j} \to 0,$$

we get that $T_j \to T$ as $j \to \infty$. As each $y_j \in S_{\mathbb{X}}$ and $S_{\mathbb{X}}$ is compact, we may assume without loss of generality that $y_j \to y_0$ for some $y_0 \in S_{\mathbb{X}}$. Noting that the set $Ext(B_{\mathbb{X}})$ is finite and each $y_j \in Ext(B_{\mathbb{X}})$, we can further assume that $y_j = y_0$ for each j. So, $T_jy_j = T_jy_0 \to Ty_0$ as $j \to \infty$. Therefore, $\|Ty_0\| \geq 1$. As $\|T\| = 1$, $\|Ty_0\| = 1$. So, $y_0 \in M_T$. Since $y_0 \in Ext(B_{\mathbb{X}})$ and M_T has at most k linearly independent extreme points, the set $\{y_0, x_1, x_2, \ldots, x_k\}$ is linearly dependent. Let $y_0 = a_1x_1 + a_2x_2 + \ldots + a_kx_k$. Then, $T_jy_j = T_jy_0 = T_j(a_1x_1 + \ldots + a_kx_k) = a_1Tx_1 + \ldots + a_kTx_k = Ty_0$. Therefore, $\|T_jy_j\| = \|Ty_0\| = 1$, which contradicts that $\|T_j\| > 1$ for all $j \in \mathbb{N}$. Thus, if T is an extreme contraction, then $span(M_T \cap Ext(B_{\mathbb{X}})) = \mathbb{X}$.

Now, we prove the next part of the theorem. Suppose $Ext(B_{\mathbb{X}}) = \{\pm x_1, \pm x_2, \ldots, \pm x_m\}$, where $m \geq n$ and $(M_T \cap Ext(B_{\mathbb{X}})) = \{\pm x_i : 1 \leq i \leq n\}$. Clearly, $\|T\| = \|Tx_i\| = 1$ for all $1 \leq i \leq n$ and $\|Tx_j\| < \|T\| = 1$ for all $n+1 \leq j \leq m$. Therefore, there exists $\epsilon > 0$ such that $\|Tx_j\| < 1 - \epsilon$ for all $n+1 \leq j \leq m$. Clearly, $\{x_1, x_2, \ldots, x_n\}$ is a basis of $\mathbb{X}$. Hence there exist scalars $a_{1j}, a_{2j}, \ldots, a_{nj}$ such that

$$x_j = a_{1j}x_1 + a_{2j}x_2 + \ldots + a_{nj}x_n, \ \forall\, n+1 \leq j \leq m.$$

Now, using Proposition 8.3.13, we get $r > 0$ such that $|a_{ij}| \leq r$ for all $1 \leq i \leq n$ and $n+1 < j < m$. Suppose $Tx_k \notin Ext(B_{\mathbb{Y}})$ for some $k \in \{1, 2, \ldots, n\}$. Then there exist $y, z \in S_{\mathbb{Y}} \cap B[Tx_k, \frac{\epsilon}{2r}]$ such that $Tx_k = (1-t)y + tz$ for some $t \in (0, 1)$. Let us define a linear operator $T_1 : \mathbb{X} \to \mathbb{Y}$ as follows:

$$T_1x_k = y \text{ and } T_1x_i = Tx_i \ \forall\, 1 \leq i(\neq k) \leq n.$$

Similarly, define $T_2 : \mathbb{X} \to \mathbb{Y}$ as follows:

$$T_2x_k = z \text{ and } T_2x_i = Tx_i \ \forall\, 1 \leq i(\neq k) \leq n.$$

Clearly, $T_1 \neq T$, $T_2 \neq T$ and $T = (1-t)T_1 + tT_2$. We claim that $\|T_1\| = \|T_2\| = 1$. Note that, $\|T_1x_i\| = \|Tx_i\| = 1$ for all $1 \leq i(\neq k) \leq n$. Moreover, since $y \in S_{\mathbb{Y}} \cap B[Tx_k, \frac{\epsilon}{2r}]$, we have $\|T_1x_k\| = \|y\| = 1$. Furthermore, for each $n+1 \leq j \leq m$, we get

$$\begin{aligned}\|T_1x_j\| &= \|a_{1j}T_1x_1 + \ldots + a_{kj}T_1x_k + \ldots + a_{nj}T_1x_n\| \\ &= \|a_{1j}Tx_1 + \ldots + a_{kj}y + \ldots + a_{nj}Tx_n\| \\ &= \|a_{1j}Tx_1 + \ldots + a_{nj}Tx_n + a_{kj}(y - Tx_k)\| \\ &\leq \|Tx_j\| + |a_{kj}|\|y - Tx_k\| \\ &\leq 1 - \epsilon + r\frac{\epsilon}{2r} \\ &= 1 - \frac{\epsilon}{2} < 1.\end{aligned}$$

Thus, for each $x_i \in Ext(B_{\mathbb{X}})$, $\|T_1x_i\| \leq 1$ and this proves that $\|T_1\| = 1$. Similarly, we can show that $\|T_2\| = 1$. Therefore, $T = (1-t)T_1 + tT_2$ and $\|T_1\| = \|T_2\| = 1$, contradicting that T is an extreme contraction. Thus, if $|M_T \cap Ext(B_{\mathbb{X}})| = 2n$, then $T(M_T \cap Ext(B_{\mathbb{X}})) \subset Ext(B_{\mathbb{Y}})$. This completes the proof. □

As an immediate application of Theorem 8.3.14, it is possible to characterize extreme contractions from a finite-dimensional polyhedral Banach space to a strictly convex Banach space. We present the characterization in the next theorem.

Theorem 8.3.15 ([129, Theorem 2.3]) *Let $\mathbb{X}$ be a finite-dimensional polyhedral Banach space and let $\mathbb{Y}$ be a strictly convex Banach space. Then $T \in \mathbb{L}(\mathbb{X}, \mathbb{Y})$, with $\|T\| = 1$, is an extreme contraction if and only if $span(M_T \cap Ext(B_{\mathbb{X}})) = \mathbb{X}$.*

Proof The necessary part follows from Theorem 8.3.14. We only prove the sufficient part. Let T be not an extreme contraction. Then there exist $T_1, T_2 \in S_{\mathbb{L}(\mathbb{X},\mathbb{Y})}$ such that $T_1, T_2 \neq T$ and $T = (1-t)T_1 + tT_2$ for some $t \in (0, 1)$. Let $\{x_1, x_2, \ldots, x_n\} \subseteq M_T \cap Ext(B_{\mathbb{X}})$ be a basis of $\mathbb{X}$. Then $Tx_i = (1-t)T_1x_i + tT_2x_i$, for each $i \in \{1, 2, \ldots, n\}$. Now, $\mathbb{Y}$ being strictly convex, $Ext(B_{\mathbb{Y}}) = S_{\mathbb{Y}}$. Therefore, $Tx_i \in Ext(B_{\mathbb{Y}})$. On the other hand, $T_1x_i, T_2x_i \in B_{\mathbb{Y}}$, since $\|T_1\| = \|T_2\| = 1$. Hence, we must have $Tx_i = T_1x_i = T_2x_i$, for each $1 \leq i \leq n$. However, this implies that T_1, T_2 agree with T on a basis of $\mathbb{X}$. Thus $T_1 = T_2 = T$, contradicting our initial assumption. Therefore T must be an extreme contraction. This establishes the theorem. □

As another useful application of Theorem 8.3.14, it is possible to characterize extreme contractions from ℓ_1^n to $\mathbb{Y}$, where $\mathbb{Y}$ is any Banach space. It is worth mentioning that the following corollary also follows from [139, Lemma 2.8].

Corollary 8.3.16 ([129, Corrollary 2.4]) *Let $\mathbb{X} = \ell_1^n$ and let $\mathbb{Y}$ be any Banach space. Let $T \in \mathbb{L}(\mathbb{X}, \mathbb{Y})$ with $\|T\| = 1$. Then T is extreme contraction if and only if $Ext(B_{\mathbb{X}}) \subseteq M_T$ and $T(Ext(B_{\mathbb{X}})) \subseteq Ext(B_{\mathbb{Y}})$.*

Proof The necessary part follows directly from Theorem 8.3.14, by noting that $Ext(B_{\ell_1^n}) = \{\pm e_i : 1 \leq i \leq n\}$, where $e_i = (\underbrace{0, \ldots, 0, 1}_{i}, 0, \ldots, 0)$. To prove the sufficient part, assume that T is not an extreme contraction. Then there exist $T_1, T_2 \in S_{\mathbb{L}(\mathbb{X},\mathbb{Y})}$ such that $T_1, T_2 \neq T$ and $T = (1-t)T_1 + tT_2$ for some $t \in (0, 1)$. So $Te_i = (1-t)T_1e_i + tT_2e_i$, for each $1 \leq i \leq n$. Clearly, $T_1e_i, T_2e_i \in B_{\mathbb{Y}}$, as $\|T_1\| = \|T_2\| = 1$. Since $Te_i \in Ext(B_{\mathbb{Y}})$, we have $Te_i = T_1e_i = T_2e_i$, for each

$1 \leq i \leq n$. However, this shows that T_1, T_2 agree with T on a basis of $\mathbb{X}$. Thus, $T_1 = T_2 = T$, contradicting our initial assumption. Therefore, T must be an extreme contraction, completing the proof. □

Remark 8.3.17 ([129, Remark 2.5]) If $\mathbb{Y}$ is a polyhedral Banach space with $|Ext(B_{\mathbb{Y}})| = 2p$, then the number of extreme contractions in $L(\ell_1^n, \mathbb{Y})$ is $(2p)^n$. Moreover, since the number of extreme contractions in $\mathbb{L}(\ell_1^n, \mathbb{Y})$ is finite, each extreme point of the unit ball of $\mathbb{L}(\ell_1^n, \mathbb{Y})$ is also an exposed point of the unit ball of $\mathbb{L}(\ell_1^n, \mathbb{Y})$. Therefore, the number of exposed points of the unit ball of $\mathbb{L}(\ell_1^n, \mathbb{Y})$ is also $(2p)^n$. In particular,
(i) The number of extreme contractions in $\mathbb{L}(\ell_1^n, \ell_1^n)$ is $2^n n^n$.
(ii) The number of extreme contractions in $\mathbb{L}(\ell_1^n, \ell_\infty^n)$ is 2^{n^2}.

Now, if we choose the domain space to be something other than polyhedral Banach space, for example, say ℓ_2^n space, and ℓ_∞^n as the codomain space, then we get the following result.

Proposition 8.3.18 ([129, Proposition 2.6]) *Let $T \in \mathbb{L}(\mathbb{X}, \mathbb{Y})$ with $\|T\| = 1$, where $\mathbb{X} = \ell_2^n$ and $\mathbb{Y} = \ell_\infty^n$. Then T is an extreme contraction if and only if corresponding matrix representation of T with respect to standard ordered basis is of the form*

$$\begin{bmatrix} a_{11} & a_{12} & \cdots & a_{1n} \\ a_{21} & a_{22} & \cdots & a_{2n} \\ \vdots & \vdots & \ddots & \vdots \\ a_{n1} & a_{n2} & \cdots & a_{nn} \end{bmatrix},$$

where $\sum_{j=1}^{n} {a_{ij}}^2 = 1$ for all $i = 1, 2, \ldots, n$.

Proof The proof directly follows from Corollary 8.3.16, by noting that T is an extreme contraction if and only if T^* is an extreme contraction, since the domain and the codomain are reflexive. □

In the following two examples, we further illustrate the variation in the norm attainment set of extreme contractions, depending on the domain and the range space.

Example 8.3.19 ([129, Example 2.7]) In $\mathbb{L}(\ell_2^2, \ell_\infty^2)$,

$$T = \begin{bmatrix} 1 & 0 \\ 1 & 0 \end{bmatrix}$$

is an extreme contraction although $M_T = \{\pm(1, 0)\}$.

Example 8.3.20 ([129, Example 2.8]) In $\mathbb{L}(\ell_2^2, \ell_\infty^2)$,

$$T = \begin{bmatrix} 1 & 0 \\ 0 & 1 \end{bmatrix}$$

is an extreme contraction. Here, $\pm(1,0), \pm(0,1) \in M_T$ but $T(1,0) = (1,0)$ and $T(0,1) = (0,1)$ are not in $Ext(B_{\ell_\infty^2})$.

Let us now focus on the notion of L-P property for a pair of Banach spaces $(\mathbb{X}, \mathbb{Y})$. It is easy to observe that ℓ_1^n is a universal L-P space, in the sense that the pair $(\ell_1^n, \mathbb{Y})$ has L-P property for every Banach space $\mathbb{Y}$. We would also like to note that it follows from the works [76, 77] of Lima that there exist finite-dimensional polyhedral Banach spaces $\mathbb{X}$, $\mathbb{Y}$, with $\mathbb{X} \neq \mathbb{Y}$, such that the pair $(\mathbb{X}, \mathbb{Y})$ has L-P property. Also, there are finite-dimensional polyhedral Banach spaces $\mathbb{X}, \mathbb{Y}$ with $\mathbb{X} \neq \mathbb{Y}$, such that the pair $(\mathbb{X}, \mathbb{Y})$ does not satisfy L-P property. As for example, it follows from [76, Theorem 2.1] that the pair $(\ell_\infty^3, \ell_1^3)$ has L-P property, whereas from [76, Lemma 3.2] it follows that the pair $(\ell_\infty^4, \ell_1^4)$ does not satisfy L-P property.

Let us observe that Proposition 8.3.18 implies, in particular, that the pair $(\ell_2^n, \ell_\infty^n)$ does not have L-P property. Now, we prove that this observation is actually a consequence of the following general result.

Theorem 8.3.21 ([129, Theorem 2.9]) *Let $\mathbb{X}$ be a strictly convex Banach space and let $\mathbb{Y}$ be a finite-dimensional polyhedral Banach space. Then the pair $(\mathbb{X}, \mathbb{Y})$ does not have L-P property.*

Proof Suppose that the pair $(\mathbb{X}, \mathbb{Y})$ satisfies L-P property. Then for any extreme contraction $T \in \mathbb{L}(\mathbb{X}, \mathbb{Y})$, we have $T(Ext(B_{\mathbb{X}})) \subseteq Ext(B_{\mathbb{Y}})$. Since $\mathbb{X}$ is strictly convex, $Ext(B_{\mathbb{X}}) = S_{\mathbb{X}}$. On the other hand, $\mathbb{Y}$ being a finite-dimensional polyhedral Banach space, $Ext(B_{\mathbb{Y}})$ is a finite set. Thus, for any extreme contraction $T \in \mathbb{L}(\mathbb{X}, \mathbb{Y})$, T cannot be one to one. Therefore, there exists $x \in S_{\mathbb{X}}$ such that $T(x) = 0$, which contradicts that $T(Ext(B_{\mathbb{X}})) = T(S_{\mathbb{X}}) \subseteq Ext(B_{\mathbb{Y}})$. This completes the proof. □

Remark 8.3.22 ([129, Remark 2.10]) From Theorem 8.3.21, it follows that the pairs (ℓ_p^n, ℓ_1^n) and $(\ell_p^n, \ell_\infty^n)$ do not have L-P property for any $1 < p < \infty$. Furthermore, we already know that the pair $(\ell_1^n, \mathbb{Y})$ has L-P property for every Banach space $\mathbb{Y}$. This illustrates that there exist Banach spaces $\mathbb{X}, \mathbb{Y}$ such that the pair $(\mathbb{X}, \mathbb{Y})$ has L-P property, whereas the pair $(\mathbb{Y}, \mathbb{X})$ does not have L-P property.

Remark 8.3.23 ([129, Remark 2.11]) Theorem 8.3.21 holds for each Banach space $\mathbb{Y}$ having countably many extreme points.

Now, we give a characterization of finite-dimensional Hilbert spaces among all finite-dimensional strictly convex Banach spaces in terms of L-P property.

Theorem 8.3.24 ([129, Theorem 2.12]) *Let $\mathbb{X}$ be a finite-dimensional strictly convex Banach space. Then $\mathbb{X}$ is a Hilbert space if and only if the pair $(\mathbb{X}, \mathbb{X})$ has L-P property.*

Proof First we claim that the pair $(\mathbb{X}, \mathbb{X})$ has L-P property if and only if every extreme contraction in $\mathbb{L}(\mathbb{X}, \mathbb{X})$ is an isometry. Clearly, if every extreme contraction in $\mathbb{L}(\mathbb{X}, \mathbb{X})$ is an isometry, then the pair $(\mathbb{X}, \mathbb{X})$ has L-P property, since $S_{\mathbb{X}} = Ext(B_{\mathbb{X}})$. Conversely, since $(\mathbb{X}, \mathbb{X})$ satisfies L-P property, $T \in \mathbb{L}(\mathbb{X}, \mathbb{X})$ is an extreme contraction if and only if $T(Ext(B_{\mathbb{X}})) \subseteq Ext(B_{\mathbb{X}})$, i.e., $T(S_{\mathbb{X}}) \subseteq S_{\mathbb{X}}$. In other words, we must have $M_T = S_{\mathbb{X}}$ and $\|T\| = 1$, i.e., $T \in \mathbb{L}(\mathbb{X}, \mathbb{X})$ is an isometry. This proves our claim. Now, the result follows from a theorem by Navarro [96], which states that a finite-dimensional Banach space $\mathbb{X}$ is a Hilbert space if and only if every extreme contraction in $\mathbb{L}(\mathbb{X}, \mathbb{X})$ is an isometry. □

Applying Theorem 8.3.24, we get a characterization of Hilbert spaces among all strictly convex Banach spaces in terms of L-P property.

Corollary 8.3.25 ([129, Corollary 2.13]) *A strictly convex Banach space $\mathbb{X}$ is a Hilbert space if and only if for every two-dimensional subspace $\mathbb{Y}$ of $\mathbb{X}$, the pair $(\mathbb{Y}, \mathbb{Y})$ has L-P property.*

Proof **Necessary part**: Note that each two-dimensional subspace $\mathbb{Y}$ of a Hilbert space $\mathbb{X}$ is also a Hilbert space. Moreover, we know that for a finite-dimensional Hilbert space $\mathbb{Y}$, isometries are the only extreme contractions in $\mathbb{L}(\mathbb{Y}, \mathbb{Y})$ and $Ext(B_{\mathbb{Y}}) = S_{\mathbb{Y}}$. Thus, for every two-dimensional subspace $\mathbb{Y}$ of $\mathbb{X}$, the pair $(\mathbb{Y}, \mathbb{Y})$ has L-P property.

Sufficient part: Note that, each two-dimensional subspace $\mathbb{Y}$ of a strictly convex space $\mathbb{X}$ is also a strictly convex space. Since the pair $(\mathbb{Y}, \mathbb{Y})$ has L-P property, it follows from Theorem 8.3.24 that $\mathbb{Y}$ is a Hilbert space. Thus every two-dimensional subspace of $\mathbb{X}$ is a Hilbert space which means that $\mathbb{X}$ itself has to be a Hilbert space. This proves the corollary. □

8.4 Extreme Contractions Through k-Smoothness

Recall that an element $x \in S_{\mathbb{X}}$ is said to be an exposed point of the unit ball $B_{\mathbb{X}}$, if there exists a supporting linear functional x^* of x such that x^* attains norm only at $\pm x$. We also observe that in a finite-dimensional polyhedral Banach space, a point is an extreme point of the unit ball if and only if it is an exposed point of the same. In this section, we first characterize exposed points of the unit ball of a finite-dimensional polyhedral Banach space, which clearly follows from Theorems 6.5.1 and 6.5.2. We write these observations in the following proposition.

Proposition 8.4.1 ([86, Proposition 2.1]) *Let $\mathbb{X}$ be a polyhedral Banach space of dimension n. Let $x \in S_{\mathbb{X}}$. Then the following are equivalent:*
(a) x is an exposed point of $B_{\mathbb{X}}$.
(b) x is an extreme point of $B_{\mathbb{X}}$.
(c) x is $n-$smooth.

Note that by [71, Theorem 2.11] a finite-dimensional Banach space is polyhedral if and only if its dual is polyhedral. In the next theorem, we prove a characterization of extreme contractions defined between finite-dimensional polyhedral Banach spaces in terms of $k-$smoothness.

Theorem 8.4.2 ([86, Theorem 2.2]) *Let* $\mathbb{X}, \mathbb{Y}$ *be polyhedral Banach spaces such that* $\dim(\mathbb{X}) = n$ *and* $\dim(\mathbb{Y}) = m$. *Then* $T \in S_{\mathbb{L}(\mathbb{X},\mathbb{Y})}$ *is an extreme contraction if and only if* T *is* $mn-$*smooth.*

Proof The spaces $\mathbb{X}, \mathbb{Y}$ being finite dimensional, from [78, Theorem 1], it follows that

$$Ext(B_{\mathbb{L}(\mathbb{X},\mathbb{Y})^*}) = Ext(B_{\mathbb{Y}^*}) \otimes Ext(B_{\mathbb{X}}).$$

Since $Ext(B_{\mathbb{Y}^*})$ and $Ext(B_{\mathbb{X}})$ are finite sets, $Ext(B_{\mathbb{L}(\mathbb{X},\mathbb{Y})^*})$ is also a finite set. Therefore, $\mathbb{L}(\mathbb{X}, \mathbb{Y})^*$ is a polyhedral Banach space. Moreover, $\mathbb{L}(\mathbb{X}, \mathbb{Y})^*$ is a finite-dimensional Banach space. From these two facts we get that $\mathbb{L}(\mathbb{X}, \mathbb{Y})$ is a finite-dimensional polyhedral Banach space. Now, $\dim(\mathbb{L}(\mathbb{X}, \mathbb{Y})) = mn$. Hence, using Proposition 8.4.1, we get that $T \in S_{\mathbb{L}(\mathbb{X},\mathbb{Y})}$ is an extreme contraction if and only if T is $mn-$smooth. □

Next, we characterize the extreme contractions between two-dimensional polygonal Banach spaces using Theorem 8.4.2. Recall that a two-dimensional polyhedral Banach space is known as a polygonal Banach space.

Theorem 8.4.3 ([86, Theorem 2.3]) *Suppose* $\mathbb{X}, \mathbb{Y}$ *are polygonal Banach spaces such that* $\dim(\mathbb{X}) = \dim(\mathbb{Y}) = 2$. *Let* $T \in S_{\mathbb{L}(\mathbb{X},\mathbb{Y})}$. *Then* T *is an extreme contraction if and only if either of the following holds:*
(i) $M_T \cap Ext(B_{\mathbb{X}}) = \{\pm x_1, \pm x_2\}$ *and* $Tx_1, Tx_2 \in Ext(B_{\mathbb{Y}})$.
(ii) $M_T \cap Ext(B_{\mathbb{X}}) = \{\pm x_1, \pm x_2, \pm x_3\}$ *and* $|\{x_i : Tx_i \in Ext(B_{\mathbb{Y}}), 1 \leq i \leq 3\}| \geq 2$.
(iii) $M_T \cap Ext(B_{\mathbb{X}}) = \{\pm x_1, \pm x_2, \pm x_3\}$, $Tx_1 \in Ext(B_{\mathbb{Y}})$, $Tx_2, Tx_3 \notin Ext(B_{\mathbb{Y}})$ *and there exist edges* F, G *of* $B_{\mathbb{Y}}$ *such that* $Tx_2 \in F$, $Tx_3 \in G$ *and* $F \neq \pm G$.
(iv) $|M_T \cap Ext(B_{\mathbb{X}})| \geq 8$ *and there exists* $x \in M_T \cap Ext(B_{\mathbb{X}})$ *such that* $Tx \in Ext(B_{\mathbb{Y}})$.
(v) $|M_T \cap Ext(B_{\mathbb{X}})| \geq 8$ *and for each* $x \in M_T \cap Ext(B_{\mathbb{X}})$, $Tx \notin Ext(B_{\mathbb{Y}})$. *Moreover, there exist* $x_i \in M_T \cap Ext(B_{\mathbb{X}})$ *and* $y_i^* \in Ext\ J(Tx_i)$ *for* $1 \leq i \leq 4$ *such that* $x_2 = ax_1 + bx_3, x_4 = cx_1 + dx_3, y_2^* = \alpha_1 y_1^* + \alpha_2 y_3^*, y_4^* = \beta_1 y_1^* + \beta_2 y_3^*$ *and* $\beta_1\alpha_2 ad - \beta_2\alpha_1 bc \neq 0$.

Proof Using Theorem 8.4.2, we get that T is an extreme contraction if and only if T is $4-$smooth. If T is an extreme contraction, then from Theorem 8.3.14, it follows that $span(M_T \cap Ext(B_{\mathbb{X}})) = \mathbb{X}$, i.e., $|M_T \cap Ext(B_{\mathbb{X}})| \geq 4$. Therefore, we only assume that $|M_T \cap Ext(B_{\mathbb{X}})| \geq 4$.

First consider the case $|M_T \cap Ext(B_{\mathbb{X}})| = 4$. In this case, we show that T is an extreme contraction if and only if (i) holds. Suppose $M_T \cap Ext(B_{\mathbb{X}}) = \{\pm x_1, \pm x_2\}$

for some $x_1, x_2 \in S_{\mathbb{X}}$. Clearly, $\{x_1, x_2\}$ is linearly independent. Therefore, applying Corollary 6.6.7, we can say that T is extreme contraction, i.e., T is 4−smooth if and only if Tx_1 and Tx_2 are 2−smooth. Thus, by Proposition 8.4.1, $Tx_1, Tx_2 \in Ext(B_{\mathbb{Y}})$. Therefore, if $|M_T \cap Ext(B_{\mathbb{X}})| = 4$, then T is an extreme contraction if and only if (i) holds.

Now, assume that $|M_T \cap Ext(B_{\mathbb{X}})| = 6$. In this case, we show that T is an extreme contraction if and only if either (ii) or (iii) holds. Let $M_T \cap Ext(B_{\mathbb{X}}) = \{\pm x_1, \pm x_2, \pm x_3\}$ for some $x_1, x_2, x_3 \in S_{\mathbb{X}}$. In this case, from Theorem 6.7.1, renaming vectors x_1, x_2 and x_3 if necessary, we get T is an extreme contraction, i.e., T is 4−smooth if and only if Tx_1 is non-smooth, Tx_2, Tx_3 are not interior point of same line segment of $S_{\mathbb{Y}}$ and $Tx_2, -Tx_3$ are not interior point of same line segment of $S_{\mathbb{Y}}$. Now, Tx_1 is non-smooth implies that Tx_1 is 2−smooth. Hence, $Tx_1 \in Ext(B_{\mathbb{Y}})$. Observe that, if at least one of $Tx_2, Tx_3 \in Ext(B_{\mathbb{Y}})$, then (ii) holds. Suppose $Tx_2, Tx_3 \notin Ext(B_{\mathbb{Y}})$. Then Tx_2, Tx_3 are interior points of line segments of $S_{\mathbb{Y}}$. Therefore, there are edges F, G of $B_{\mathbb{Y}}$ such that $Tx_2 \in F, Tx_3 \in G$. Since Tx_2, Tx_3 are not interior point of same line segment of $S_{\mathbb{Y}}$ and $Tx_2, -Tx_3$ are not interior point of same line segment of $S_{\mathbb{Y}}$, we get that $F \neq \pm G$. Thus, (iii) holds. Therefore, if $|M_T \cap Ext(B_{\mathbb{X}})| = 6$, then T is an extreme contraction if and only if either (ii) or (iii) holds.

Next, assume that $|M_T \cap Ext(B_{\mathbb{X}})| \geq 8$. Then from Theorem 6.7.3, we can easily conclude that T is an extreme contraction if and only if either (iv) or (v) holds. □

In the next theorem, we characterize the extreme contractions of $\mathbb{L}(\ell_\infty^3, \mathbb{Y})$, where $\mathbb{Y}$ is a two-dimensional polygonal Banach space.

Theorem 8.4.4 ([83, Theorem 3.12]) *Suppose $\mathbb{X} = \ell_\infty^3$ and $\mathbb{Y}$ is a two-dimensional polygonal Banach space. Let $T \in S_{\mathbb{L}(\mathbb{X},\mathbb{Y})}$. Then T is an extreme contraction if and only if $|M_T \cap Ext(B_{\mathbb{X}})| \geq 6$ and $T(M_T \cap Ext(B_{\mathbb{X}})) \subseteq Ext(B_{\mathbb{Y}})$.*

Proof First assume that T is an extreme contraction. Then from Theorem 8.4.2, it follows that T is 6−smooth. Using Theorem 8.3.14, we get $span(M_T \cap Ext(B_{\mathbb{X}})) = \mathbb{X}$, i.e., $|M_T \cap Ext(B_{\mathbb{X}})| \geq 6$. Let $|M_T \cap Ext(B_{\mathbb{X}})| = 6$. Then $M_T \cap Ext(B_{\mathbb{X}})$ is of the form $\{\pm x_1, \pm x_2, \pm x_3\}$, for a linearly independent set $\{x_1, x_2, x_3\}$. Now, from Corollary 6.6.7 it is clear that Tx_i is 2−smooth for each $1 \leq i \leq 3$. Therefore, from Proposition 8.4.1, it follows that $Tx_i \in Ext(B_{\mathbb{Y}})$ for all $1 \leq i \leq 3$. Now, assume that $|M_T \cap Ext(B_{\mathbb{X}})| = 8$. Then from Theorem 6.7.8, we can conclude that for all $x \in M_T \cap Ext(B_{\mathbb{X}})$, Tx is 2−smooth, i.e., $Tx \in Ext(B_{\mathbb{Y}})$.
Conversely, let $|M_T \cap Ext(B_{\mathbb{X}})| \geq 6$ and $T(M_T \cap Ext(B_{\mathbb{X}})) \subseteq Ext(B_{\mathbb{Y}})$. If $|M_T \cap Ext(B_{\mathbb{X}})| = 6$, then from Corollary 6.6.7, it follows that T is 6−smooth. Hence, by Theorem 8.4.2, we get T is an extreme contraction. If $|M_T \cap Ext(B_{\mathbb{X}})| = 8$, then from Theorem 6.7.8, we have T is 6−smooth. Thus, again from Theorem 8.4.2, it follows that T is an extreme contraction. This establishes the theorem. □

Note that Theorem 8.4.4 once again emphasizes the significance of Theorem 8.4.2 in the study of extreme contraction between finite-dimensional polyhedral Banach spaces. For further continuation of the study of extreme contraction on finite-dimensional Banach spaces, Ray et al. introduced the notion of weak L-P property in [111].

Definition 8.4.5 ([111, Definition 1.2]) Let $\mathbb{X}, \mathbb{Y}$ be Banach spaces. We say that the pair $(\mathbb{X}, \mathbb{Y})$ has weak L-P property if for each extreme contraction $T \in S_{\mathbb{L}(\mathbb{X},\mathbb{Y})}$, $T(Ext(B_{\mathbb{X}})) \cap Ext(B_{\mathbb{Y}}) \neq \emptyset$.

Note that if $\mathbb{X}$ is a reflexive Banach space, then each functional $f \in \mathbb{X}^*$ attains its norm at an extreme point of $B_{\mathbb{X}}$. As a consequence, the pair $(\mathbb{X}, \mathbb{R})$ satisfies weak L-P property. Here we obtain some non-trivial examples of pair of polyhedral Banach spaces satisfying weak L-P property. The following lemma will be used in the next theorem.

Lemma 8.4.6 ([86, Lemma 2.4]) *Let $\mathbb{X}, \mathbb{Y}$ be Banach spaces. Suppose $\emptyset \neq E_1 \subseteq \mathbb{X}$, $\emptyset \neq E_2 \subseteq \mathbb{Y}^*$ and $B_i \subseteq E_i$ is a basis of span E_i for $i = 1, 2$. Then $\{y^* \otimes x : x \in B_1, y^* \in B_2\}$ is a basis of span $\{y^* \otimes x : x \in E_1, y^* \in E_2\}$.*

Proof We first prove the linear independence of the set $\{y^* \otimes x : x \in B_1, y^* \in B_2\}$. Consider a finite subset $\{y_i^* \otimes x_j : x_j \in B_1, y_i^* \in B_2, 1 \leq i \leq n, 1 \leq j \leq m\}$ of $\{y^* \otimes x : x \in B_1, y^* \in B_2\}$. Suppose c_{ij} are scalars such that

$$\sum_{1\leq i\leq n, 1\leq j\leq m} c_{ij} y_i^* \otimes x_j = 0. \tag{8.4}$$

Choose $y \in \mathbb{Y}, \phi \in \mathbb{X}^*$. Let us define $S \in \mathbb{L}(\mathbb{X}, \mathbb{Y})$ by $Sx = \phi(x)y$ for all $x \in \mathbb{X}$. Now, using (8.4), we get

$$\begin{aligned}
& \sum_{1\leq i\leq n, 1\leq j\leq m} c_{ij} y_i^* \otimes x_j(S) = 0 \\
\Rightarrow & \sum_{1\leq i\leq n, 1\leq j\leq m} c_{ij} y_i^* S(x_j) = 0 \\
\Rightarrow & \sum_{1\leq i\leq n, 1\leq j\leq m} c_{ij} \phi(x_j) y_i^*(y) = 0 \\
\Rightarrow & \phi\Bigg(\sum_{1\leq i\leq n, 1\leq j\leq m} c_{ij} x_j y_i^*(y) \Bigg) = 0 \\
\Rightarrow & \sum_{1\leq i\leq n, 1\leq j\leq m} c_{ij} x_j y_i^*(y) = 0, \text{ (since } \phi \in \mathbb{X}^* \text{ is arbitrary)} \\
\Rightarrow & \sum_{1\leq i\leq n} c_{ij} y_i^*(y) = 0 \text{ for all } 1 \leq j \leq m \\
\Rightarrow & \sum_{1\leq i\leq n} c_{ij} y_i^* = 0 \text{ (since } y \in \mathbb{Y} \text{ is arbitrary)} \\
\Rightarrow & c_{ij} = 0 \text{ for all } 1 \leq j \leq m, 1 \leq i \leq n.
\end{aligned}$$

This proves the linear independence of the set $\{y_i^* \otimes x_j : 1 \leq i \leq n, 1 \leq j \leq m\}$ and consequently the linear independence of the set $\{y^* \otimes x : x \in B_1, y^* \in B_2\}$.
Now, let $y^* \in E_2$ and $x \in E_1$. Then there are scalars a_i, b_j and $x_j \in B_1, y_i^* \in B_2$ for $1 \leq i \leq n, 1 \leq j \leq m$ such that $y^* = \sum_{i=1}^n a_i y_i^*, x = \sum_{j=1}^m b_j x_j$. Thus,

$$y^* \otimes x = \sum_{1 \leq i \leq n, 1 \leq j \leq m} a_i b_j y_i^* \otimes x_j.$$

This proves that $y^* \otimes x \in span\ \{y^* \otimes x : x \in B_1, y^* \in B_2\}$. Therefore, $\{y^* \otimes x : x \in B_1, y^* \in B_2\}$ is a basis of $span\ \{y^* \otimes x : x \in E_1, y^* \in E_2\}$. This establishes the lemma. □

In the next theorem, we obtain a sufficient condition for a pair of finite-dimensional polyhedral Banach spaces to satisfy weak L-P property.

Theorem 8.4.7 ([86, Theorem 2.5]) *Let $\mathbb{X}, \mathbb{Y}$ be polyhedral Banach spaces such that $\dim(\mathbb{X}) = n, \dim(\mathbb{Y}) = m$. Suppose $|Ext(B_{\mathbb{X}})| = 2(n+p)$. If $mp < n + p$, then the pair $(\mathbb{X}, \mathbb{Y})$ satisfies weak L-P property.*

Proof Consider an extreme contraction $T \in S_{\mathbb{L}(\mathbb{X},\mathbb{Y})}$. We show that there exists $x \in M_T \cap Ext(B_{\mathbb{X}})$ such that $Tx \in Ext(B_{\mathbb{Y}})$. From Theorem 8.3.14, we get $span(M_T \cap Ext(B_{\mathbb{X}})) = \mathbb{X}$. Therefore, there are at least n linearly independent elements in $M_T \cap Ext(B_{\mathbb{X}})$. Let $M_T \cap Ext(B_{\mathbb{X}}) = \{\pm x_i : 1 \leq i \leq r\}$ such that $\{x_1, x_2, \ldots, x_n\}$ is linearly independent. Then $r \leq n + p$. If possible, suppose that $Tx_i \notin Ext(B_{\mathbb{Y}})$, for any i, $1 \leq i \leq r$. Then using Proposition 8.4.1, we can say that Tx_i is not $m-$smooth for each $1 \leq i \leq r$. Let Tx_i be k_i-smooth for all $1 \leq i \leq r$. Then $k_i \leq (m-1)$ for all $1 \leq i \leq r$. Clearly, $k_i = \dim(span\ Ext\ J(Tx_i))$. Let $\{y_{ij}^* \in Ext\ J(Tx_i) : 1 \leq j \leq k_i\}$ be a basis of $span\ Ext\ J(Tx_i)$ for each $1 \leq i \leq r$. Consider the space

$$W_i = span\ \{y^* \otimes x_i : y^* \in Ext\ J(Tx_i)\} \text{ for each } 1 \leq i \leq r.$$

From Lemma 8.4.6 it follows that $B_i = \{y_{ij}^* \otimes x_i : 1 \leq j \leq k_i\}$ is a basis of W_i and so for each $1 \leq i \leq r$, $\dim(W_i) = k_i$. Now, let T be $k-$smooth. Then

$$\begin{aligned} k &= \dim(span\ J(T)) \\ &= \dim(span\ Ext\ J(T)) \\ &= \dim(span\{y^* \otimes x_i : y^* \in Ext\ J(Tx_i), 1 \leq i \leq r\}) \\ &= \dim(W), \text{ where,} \\ W &= span\ \{y^* \otimes x_i : y^* \in Ext\ J(Tx_i), 1 \leq i \leq r\}. \end{aligned}$$

Clearly, $W \subseteq W_1 + W_2 + \ldots + W_r$. Therefore,

$$k = \dim(W) \leq \dim(\sum_{i=1}^r W_i) \leq \sum_{i=1}^r \dim(W_i) = \sum_{i=1}^r k_i \leq (m-1)r \leq (m-1)(n+p).$$

Now, $mp < n + p$ implies that $(m - 1)(n + p) < mn$, and thus $k < mn$. Therefore, from Theorem 8.4.2, we conclude that T is not an extreme contraction, a contradiction. This proves that there exists $1 \leq i \leq r$ such that $Tx_i \in Ext(B_{\mathbb{Y}})$. Hence, the pair $(\mathbb{X}, \mathbb{Y})$ satisfies weak L-P property. This establishes the theorem. □

In [111, Theorem 2.1], Ray et al. proved that if $\mathbb{X}$ is an n-dimensional polyhedral Banach space with exactly $(2n + 2)$ extreme points and $m \leq n$, then the pair $(\mathbb{X}, \ell_\infty^n)$ satisfies weak L-P property. As a simple consequence of Theorem 8.4.7, we get a more general result.

Corollary 8.4.8 ([86, Corollary 2.6]) *Suppose* $\mathbb{X}, \mathbb{Y}$ *are polyhedral Banach spaces such that* $\dim(\mathbb{X}) = n$, $\dim(\mathbb{Y}) = m$. *Let* $|Ext(B_{\mathbb{X}})| = 2n + 2$ *and* $m \leq n$. *Then the pair* $(\mathbb{X}, \mathbb{Y})$ *satisfies weak L-P property.*

The natural question that arises now is whether the sufficient condition given in Theorem 8.4.7 for a pair of Banach spaces to satisfy weak L-P property is also a necessary condition or not. In the next theorem, we show that the condition is both necessary and sufficient for the pair $(\mathbb{X}, \ell_\infty^2)$, where $\mathbb{X}$ is a two-dimensional polygonal Banach space. Note that, for $x_1, x_2 \in \mathbb{X}$, $L[x_1, x_2]$, $L(x_1, x_2)$ and $L[x_1, x_2[$ represents the following sets:

$$L[x_1, x_2] = \{tx_1 + (1 - t)x_2 : 0 \leq t \leq 1\},$$

$$L(x_1, x_2) = \{tx_1 + (1 - t)x_2 : 0 < t < 1\} \text{ and}$$

$$L[x_1, x_2[= L[x_1, x_2] \cup \{tx_2 - (t - 1)x_1 : t > 1\}.$$

Theorem 8.4.9 ([86, Theorem 2.8]) *Let* $\mathbb{X}$ *be a two-dimensional polygonal Banach space. Then the pair* $(\mathbb{X}, \ell_\infty^2)$ *satisfies weak L-P property if and only if* $|Ext(B_{\mathbb{X}})| \leq 6$.

Proof Let $|Ext(B_{\mathbb{X}})| \leq 6$. Then from Theorem 8.4.7, we conclude that the pair $(\mathbb{X}, \ell_\infty^2)$ satisfies weak L-P property. Conversely, let $|Ext(B_{\mathbb{X}})| \geq 8$. We show that the pair $(\mathbb{X}, \ell_\infty^2)$ does not satisfy weak L-P property. Clearly, $\mathbb{X}^*$ is a two-dimensional polygonal Banach space such that $|Ext(B_{\mathbb{X}^*})| \geq 8$. Choose $\{x_1, x_2, x_3, x_4\} \subseteq Ext(B_{\mathbb{X}^*})$ such that $L[x_1, x_3]$ and $L[x_1, -x_3]$ are not edges of $B_{\mathbb{X}^*}$. Let us define $T \in \mathbb{L}(\ell_1^2, \mathbb{X}^*)$ by $Te_1 = x_1$ and $Te_2 = x_3$, where $e_1 = (1, 0)$ and $e_2 = (0, 1)$. Note that $M_T = \{\pm e_1, \pm e_2\}$. Since $x_1, x_3 \in Ext(B_{\mathbb{X}^*})$ and $\mathbb{X}^*$ is polygonal Banach space, by Proposition 8.4.1, we say that x_1, x_3 are $2-$smooth points. Thus, from Corollary 6.6.7, it follows that T is $4-$smooth. Hence, by Theorem 8.4.2, T is an extreme contraction. Clearly, $T^* : \mathbb{X} \to \ell_\infty^2$ is also an extreme contraction. We claim that $T^*(Ext(B_{\mathbb{X}})) \cap Ext(B_{\ell_\infty^2}) = \emptyset$. If possible, suppose that there exists $u \in Ext(B_{\mathbb{X}})$ such that $T^*u \in Ext(B_{\ell_\infty^2})$. Then T^*u is $2-$smooth. Clearly, $Ext\ J(T^*u) \subset Ext(B_{\ell_1^2}) = \{\pm e_1, \pm e_2\}$ and so without loss of generality, we may assume that $Ext\ J(T^*u) = \{e_1, e_2\}$. Thus,

$$\begin{aligned} & e_1(T^*u) = e_2(T^*u) = 1 \\ \Rightarrow\ & u(Te_1) = u(Te_2) = 1 \\ \Rightarrow\ & u(x_1) = u(x_3) = 1 \\ \Rightarrow\ & u(tx_1 + (1-t)x_3) = 1 \text{ for all } t \in [0,1] \\ \Rightarrow\ & \|tx_1 + (1-t)x_3\| = 1 \text{ for all } t \in [0,1]. \end{aligned}$$

Hence, $L[x_1, x_3]$ is an edge of $B_{\mathbb{X}^*}$, a contradiction. Therefore, $T^*(Ext(B_{\mathbb{X}})) \cap Ext(B_{\ell_\infty^2}) = \emptyset$. Thus, if $|Ext(B_{\mathbb{X}})| \geq 8$, then the pair $(\mathbb{X}, \ell_\infty^2)$ does not satisfy weak L-P property. This establishes the theorem. □

Remark 8.4.10 ([86, Remark 2.9]) If $(\mathbb{X}, \mathbb{Y})$ and $(\mathbb{X}_0, \mathbb{Y}_0)$ are pair of Banach spaces such that $\mathbb{X}$ is isometrically isomorphic to $\mathbb{X}_0$ and $\mathbb{Y}$ is isometrically isomorphic to $\mathbb{Y}_0$, it is clear that $(\mathbb{X}, \mathbb{Y})$ satisfies L-P property (or weak L-P property) if and only if the same happens with $(\mathbb{X}_0, \mathbb{Y}_0)$. Thus, Theorem 8.4.9 is also true if ℓ_∞^2 is replaced by a two-dimensional Banach space $\mathbb{Y}$ such that $|Ext(B_{\mathbb{Y}})| = 4$.

In the next theorem, we show that given a two-dimensional polygonal Banach space $\mathbb{X}$ with $|Ext(B_{\mathbb{X}})| \geq 8$, there exists a two-dimensional polygonal Banach space $\mathbb{Y}$ with $|Ext(B_{\mathbb{Y}})| \geq 6$, such that the pair $(\mathbb{X}, \mathbb{Y})$ does not satisfy weak L-P property.

Theorem 8.4.11 ([86, Theorem 2.10]) *Let $\mathbb{X}$ be a two-dimensional polygonal Banach space such that $|Ext(B_{\mathbb{X}})| \geq 8$. Then for each $n \in \mathbb{N} \setminus \{1, 2\}$, there exists a two-dimensional polygonal Banach space $\mathbb{Y}$ such that $|Ext(B_{\mathbb{Y}})| = 2n$ and the pair $(\mathbb{X}, \mathbb{Y})$ does not satisfy weak L-P property.*

Proof Suppose $\{x_1, x_2, x_3, x_4\} \subseteq Ext(B_{\mathbb{X}})$ such that $x_i \neq \pm x_j$ for all $1 \leq i, j \leq 4$ with $i \neq j$ and $L[x_i, x_{i+1}]$ is an edge of $B_{\mathbb{X}}$ for each $1 \leq i \leq 3$. Assume that $L[x_1, x_2[\ \cap L[x_4, x_3[= \{y_1\}$ and $L[x_2, x_1[\ \cap L[-x_3, -x_4[= \{y_2\}$. Suppose $z_1 = \frac{y_1 + x_2}{2}$ and $z_{n-1} = \frac{y_1 + x_3}{2}$. Now, for each $2 \leq i \leq n-2$, we can easily choose vectors $z_i = a_i z_1 + b_i z_{n-1}$, where $a_i, b_i > 0$ such that the convex hull of $\{\pm y_2, \pm z_j : 1 \leq j \leq n-1\}$ is a symmetric convex set having $2n$ extreme points $\{\pm y_2, \pm z_j : 1 \leq j \leq n-1\}$. Let $\mathbb{Y}$ be the Banach space such that $B_{\mathbb{Y}}$ is the convex hull of $\{\pm y_2, \pm z_j : 1 \leq j < n-1\}$. Then $|Ext(B_{\mathbb{Y}})| = 2n$. Clearly, x_i $(1 \leq i \leq 4)$ are smooth points of $\mathbb{Y}$. Now, consider the operator $I : \mathbb{X} \to \mathbb{Y}$ defined by $Ix = x$ for all $x \in \mathbb{X}$. Note that $M_I \cap Ext(B_{\mathbb{X}}) = \{\pm x_1, \pm x_2, \pm x_3, \pm x_4\}$. Moreover, $x_1, x_2 \in L[z_1, y_2]$ and $x_3, x_4 \in L[-y_2, z_{n-1}]$. So there exist $f, g \in S_{\mathbb{Y}^*}$ such that $J(x_1) = J(x_2) = \{f\}$ and $J(x_3) = J(x_4) = \{g\}$. Suppose I is $k-$smooth, where $1 \leq k \leq 4$. Then

$$\begin{aligned} k &= \dim(span\ J(I)) \\ &= \dim(span\ Ext\ J(I)) \\ &= \dim(span\ \{f \otimes x_1, f \otimes x_2, g \otimes x_3, g \otimes x_4\}) \\ &= 4, \end{aligned}$$

which implies that I is $4-$smooth. Hence, using Theorem 8.4.2, we conclude that I is an extreme contraction. Now, observe that if $x \in Ext(B_{\mathbb{X}}) \setminus M_I$, then $\|Ix\| <$

$\|I\| = 1$, i.e., $x \notin Ext(B_{\mathbb{Y}})$. If $x \in M_I \cap Ext(B_{\mathbb{X}})$, then Ix is smooth point of $B_{\mathbb{Y}}$. Therefore, $I(Ext(B_{\mathbb{X}})) \cap Ext(B_{\mathbb{Y}}) = \emptyset$. This proves that the pair $(\mathbb{X}, \mathbb{Y})$ does not satisfy weak L-P property, establishing the theorem. □

Although the above theorem indicates that the condition stated in Theorem 8.4.7 may be necessary for a pair $(\mathbb{X}, \mathbb{Y})$ to satisfy weak L-P property, where $\mathbb{X}, \mathbb{Y}$ are arbitrary two-dimensional polygonal Banach spaces, the answer is still not known in its full generality. However, if $\dim(\mathbb{X}) > 2$, then the condition is not necessary. We exhibit polyhedral Banach spaces $\mathbb{X}, \mathbb{Y}$ such that $(\mathbb{X}, \mathbb{Y})$ satisfies weak L-P property but $\dim(\mathbb{X}) = n$, $\dim(\mathbb{Y}) = m$, $|Ext(B_{\mathbb{X}})| = 2(n + p)$ and $mp \geq n + p$ hold. To prove our claim, we first obtain a bound of the order of smoothness of a class of bounded linear operators defined between ℓ_∞^4 and a two-dimensional Banach space.

Theorem 8.4.12 ([86, Theorem 2.14]) *Let $\mathbb{X} = \ell_\infty^4$ and $\mathbb{Y}$ be a two-dimensional Banach space. Suppose $T \in S_{\mathbb{L}(\mathbb{X},\mathbb{Y})}$ is such that $Ext(B_{\mathbb{X}}) \subseteq M_T$ and Tx is smooth for all $x \in Ext(B_{\mathbb{X}})$. Then T is $k-$smooth where $k \leq 6$.*

Proof Let us write $Ext(B_{\mathbb{X}}) = \{\pm x_1, \pm x_2, \ldots, \pm x_8\}$, where $\{x_1, x_2, x_3, x_4\}$ is linearly independent. Suppose $S = \{x_1, x_2, \ldots, x_8\}$.
First consider that $Rank(T) = 1$. Then there exists $y^* \in S_{\mathbb{Y}^*}$ such that for any $i \in \{1, 2, \ldots, 8\}$, $J(Tx_i)$ is either $\{y^*\}$ or $\{-y^*\}$. Let T be $k-$smooth. From Lemma 6.6.1, it follows that $\{y^* \otimes x_i : 1 \leq i \leq 4\}$ is linearly independent. Therefore, we have

$$\begin{aligned} k &= \dim(span\ J(T)) \\ &= \dim(span\ Ext\ J(T)) \\ &= \dim(span\ \{y^* \otimes x_i : 1 \leq i \leq 8\}) \\ &= \dim(span\ \{y^* \otimes x_i : 1 \leq i \leq 4\}) \\ &= 4. \end{aligned}$$

Hence T is $4-$smooth.
Let $Rank(T) = 2$. Then from Lemma 6.7.6, it follows that $T(B_{\mathbb{X}})$ is a convex set with four extreme points. Without loss of generality, assume that $\pm Tx_1, \pm Tx_2$ are four distinct extreme points of $T(B_{\mathbb{X}})$. Suppose $Tx_i \in L(Tx_1, Tx_2)$ for some $3 \leq i \leq 8$. We claim that for each $3 \leq j \leq 8$, $Tx_j \in L[Tx_1, Tx_2] \cup L[-Tx_1, -Tx_2]$. Using $\|Tx_i\| = 1$, it is easy to check that $L[Tx_1, Tx_2] \subseteq S_{\mathbb{Y}}$. Let $J(Tx_i) = \{y^*\}$. Then for each $y \in L[Tx_1, Tx_2]$, we have $y^*(y) = 1$. If possible, suppose there exists $3 \leq j(\neq i) \leq 8$ such that $Tx_j \in L(Tx_1, -Tx_2)$. Let $J(Tx_j) = \{z^*\}$. Then $y^* \neq \pm z^*$. Now, for all $y \in L[Tx_1, -Tx_2]$, we have $z^*(y) = 1$. Thus $y^*, z^* \in J(Tx_1)$, which contradicts that Tx_1 is smooth. Therefore, $Tx_j \notin L(Tx_1, -Tx_2)$. Similarly, we can show that $Tx_j \notin L(-Tx_1, Tx_2)$. Thus, for each $i \in \{1, 2, \ldots, 8\}$, $J(Tx_i) = \{y^*\}$ or $\{-y^*\}$. Now, it is easy to verify that T is $4-$smooth.

Now, suppose that $Tx_i \notin L(\pm Tx_1, \pm Tx_2)$ for any $3 \leq i \leq 8$. Then $Tx_i \in \{\pm Tx_1, \pm Tx_2\}$ for all $3 \leq i \leq 8$. Let $J(Tx_1) = \{y_1^*\}$ and $J(Tx_2) = \{y_2^*\}$. Then for each $i \in \{1, 2, \ldots, 8\}$,

$$J(Tx_i) = \{y_1^*\} \text{ or } \{-y_1^*\} \text{ or } \{y_2^*\} \text{ or } \{-y_2^*\}.$$

Thus, there exist two subsets S_1 and S_2 ($S_1 \cap S_2 = \emptyset$, $S_1 \cup S_2 = S$) of S such that $T(S_1) = \pm Tx_1$ and $T(S_2) = \pm Tx_2$. Therefore, for $i \in \{1, 2, \ldots, 8\}$,

$$\begin{aligned} J(Tx_i) &= \{y_1^*\} \text{ or } \{-y_1^*\}, \text{ if } x_i \in S_1 \\ &= \{y_2^*\} \text{ or } \{-y_2^*\}, \text{ if } x_i \in S_2. \end{aligned}$$

Clearly, any 4 elements of S_1 as well as S_2 are linearly dependent. Otherwise, if $\{x_{11}, x_{12}, x_{13}, x_{14}\}$ is a linearly independent subset of S_1, then for any $x \in \mathbb{X}$,

$$\begin{aligned} x &= \sum_{i=1}^{4} \lambda_i x_{1i}, \text{ where } \lambda_i \text{ are scalars,} \\ \Rightarrow Tx &= \sum_{i=1}^{4} \lambda_i Tx_{1i} \in span\{Tx_1\}. \end{aligned}$$

Hence, $Rank(T) = 1$, a contradiction. Thus, maximal linearly independent subsets of S_1 and S_2 contain at most 3 elements. Let us write those linearly independent subsets of S_1 and S_2, respectively, by $A_1 = \{x_{1i} : 1 \leq i \leq n_1\}$ and $A_2 = \{x_{2i} : 1 \leq i \leq n_2\}$, where $n_1, n_2 \leq 3$. Now, if T is $k-$smooth, then

$$\begin{aligned} k &= \dim(span\ J(T)) \\ &= \dim(span\ Ext\ J(T)) \\ &= \dim(span\ \{y_1^* \otimes x, y_2^* \otimes z : x \in S_1, z \in S_2\}) \\ &= \dim(span\ \{y_1^* \otimes x_{1i}, y_2^* \otimes x_{2i} : 1 \leq i \leq n_1, 1 \leq j \leq n_2\}) \\ &\leq n_1 + n_2 \leq 6. \end{aligned}$$

Consequently, T is $k-$smooth, where $k \leq 6$. This establishes the theorem. □

Now, we are in a position to show that although the pair $(\ell_\infty^4, \mathbb{Y})$ does not satisfy the condition given in Theorem 8.4.7, the pair $(\ell_\infty^4, \mathbb{Y})$ satisfies weak L-P property, where $\mathbb{Y}$ is a two-dimensional polygonal Banach space.

Theorem 8.4.13 ([86, Theorem 2.15]) *Let $\mathbb{Y}$ be a two-dimensional polygonal Banach space. Then the pair $(\ell_\infty^4, \mathbb{Y})$ satisfies weak L-P property.*

Proof Let $\mathbb{X} = \ell_\infty^4$. Then $|Ext(B_{\mathbb{X}})| = 16 = 2(4+4)$. Note that, comparing with Theorem 8.4.7, here we get $m = 2$, $n = 4$ and $p = 4$. Thus, $mp < n + p$ is not satisfied. However, we show that the pair $(\mathbb{X}, \mathbb{Y})$ satisfies weak L-P property. For this, consider an extreme contraction $T \in S_{\mathbb{L}(\mathbb{X},\mathbb{Y})}$. First assume that $|M_T \cap Ext(B_{\mathbb{X}})| = 16$.

If $Tx \in Ext(B_{\mathbb{Y}})$ for some $x \in M_T \cap Ext(B_{\mathbb{X}})$, then we are done. If possible, let $Tx \notin Ext(B_{\mathbb{Y}})$ for any $x \in M_T \cap Ext(B_{\mathbb{X}})$. Then from Proposition 8.4.1, it follows that Tx is smooth for all $x \in M_T \cap Ext(B_{\mathbb{X}})$. Now, using Theorem 8.4.12, we see that T is $k-$smooth, where $k \leq 6$. Consequently, from Theorem 8.4.2, we get T is not an extreme contraction. This contradiction proves that $Tx \in Ext(B_{\mathbb{Y}})$ for some $x \in M_T \cap Ext(B_{\mathbb{X}})$. Now, let $|M_T \cap Ext(B_{\mathbb{X}})| = 2q$, where $1 \leq q \leq 7$. Suppose $Tx \notin Ext(B_{\mathbb{Y}})$ for any $x \in M_T \cap Ext(B_{\mathbb{X}})$. Clearly, Tx is smooth for all $x \in M_T \cap Ext(B_{\mathbb{X}})$. Let $M_T \cap Ext(B_{\mathbb{X}}) = \{\pm x_1, \pm x_2, \ldots, \pm x_q\}$ and $J(Tx_i) = \{y_i^*\}$, $1 \leq i \leq q$. Now, if T is $k-$smooth, then

$$\begin{aligned}
k &= \dim(span\ J(T)) \\
&= \dim(span\ Ext\ J(T)) \\
&= \dim(span\ \{y_i^* \otimes x_i : 1 \leq i \leq q\}) \\
&\leq q \leq 7.
\end{aligned}$$

This shows that T cannot be $8-$smooth and hence not an extreme contraction. Again this is a contradiction. Therefore, $Tx \in Ext(B_{\mathbb{Y}})$ for some $x \in M_T \cap Ext(B_{\mathbb{X}})$. Thus, the pair $(\mathbb{X}, \mathbb{Y})$ satisfies weak L-P property. This completes the proof. □

We end this chapter with an application of Theorem 8.4.3 (or Theorem 8.4.7), by computing the exact number of extreme contractions on $\mathbb{X}$, where $S_{\mathbb{X}}$ is a regular hexagon.

Theorem 8.4.14 ([86, Theorem 2.16]) *Let $\mathbb{X}$ be a two-dimensional polygonal Banach space such that $S_{\mathbb{X}}$ is a regular hexagon. Then* $|Ext(B_{\mathbb{L}(\mathbb{X},\mathbb{X})})| = 30$.

Proof Without loss of generality, assume that the vertices of $S_{\mathbb{X}}$ are $\pm x_1 = \pm(1, 0)$, $\pm x_2 = \pm(\frac{1}{2}, \frac{\sqrt{3}}{2})$, $\pm x_3 = \pm(-\frac{1}{2}, \frac{\sqrt{3}}{2})$. Let $T \in Ext(B_{\mathbb{L}(\mathbb{X},\mathbb{X})})$. Then using Theorem 8.4.3, we get that either (i) $|M_T \cap Ext(B_{\mathbb{X}})| = 4$ or (ii) $|M_T \cap Ext(B_{\mathbb{X}})| = 6$.

(i) First consider the case $|M_T \cap Ext(B_{\mathbb{X}})| = 4$. Suppose $M_T \cap Ext(B_{\mathbb{X}}) = \{\pm x_1, \pm x_2\}$. Then by Theorem 8.4.3, we get $Tx_1, Tx_2 \in Ext(B_{\mathbb{X}})$. Observe that, if x, y are two distinct extreme points of $B_{\mathbb{X}}$, then $\|x - y\| \geq 1$. Now, $x_3 = x_2 - x_1$ and $x_3 \notin M_T$ implies that Tx_1 and Tx_2 cannot be distinct extreme points of $B_{\mathbb{X}}$. Therefore, $Tx_1 = Tx_2$. Clearly, there are 6 possibilities for Tx_1. Hence, there are 6 extreme contractions $T \in \mathbb{L}(\mathbb{X}, \mathbb{X})$ such that $M_T \cap Ext(B_{\mathbb{X}}) = \{\pm x_1, \pm x_2\}$. Similar arguments show that there are 6 extreme contractions $T \in \mathbb{L}(\mathbb{X}, \mathbb{X})$ such that $M_T \cap Ext(B_{\mathbb{X}}) = \{\pm x_2, \pm x_3\}$. Now, assume that $M_T \cap Ext(B_{\mathbb{X}}) = \{\pm x_1, \pm x_3\}$. Since $x_2 = x_1 + x_3$ and $\|T\| = 1$, $Tx_1 \neq Tx_3$. Observe that, if x, y are two linearly independent points of $Ext(B_{\mathbb{X}})$, then $\|x + y\| \geq 1$. Therefore, from the fact that $Tx_1, Tx_3 \in Ext(B_{\mathbb{X}})$ and $x_2 \notin M_T$, we conclude that Tx_1, Tx_3 are not linearly independent. Hence, $Tx_1 = -Tx_3$. In this case, there are 6 possibilities for Tx_1. Therefore, there are 6 extreme contractions $T \in \mathbb{L}(\mathbb{X}, \mathbb{X})$ such that $M_T \cap Ext(B_{\mathbb{X}}) = \{\pm x_1, \pm x_3\}$. So we get 18 extreme contractions T such that $|M_T \cap Ext(B_{\mathbb{X}})| = 4$.

(ii) Now, consider the case $|M_T \cap Ext(B_{\mathbb{X}})| = 6$, i.e., $M_T \cap Ext(B_{\mathbb{X}}) = \{\pm x_1, \pm x_2, \pm x_3\}$. In this case, we show that T is an isometry. By Theorem 8.4.3

(or Theorem 8.4.7), we can say that $Tx_i \in Ext(B_{\mathbb{X}})$ for some $1 \leq i \leq 3$. Without loss of generality, assume that $Tx_1 \in Ext(B_{\mathbb{X}})$ and $Tx_1 = x_1$. Now, there are the following 6 possibilities for Tx_3 :

$$(1)\ Tx_3 \in L[x_1, x_2], \quad (2)\ Tx_3 \in L[x_2, x_3], \quad (3)\ Tx_3 \in L[x_3, -x_1],$$
$$(4)\ Tx_3 \in L[-x_1, -x_2], \quad (5)\ Tx_3 \in L[-x_2, -x_3], \quad (6)\ Tx_3 \in L[-x_3, x_1].$$

We take into account of each case separately.
(1) Suppose $Tx_3 \in L[x_1, x_2]$. Then for some $t \in [0, 1]$, $Tx_3 = tx_1 + (1-t)x_2$. Therefore, $Tx_2 = Tx_1 + Tx_3 = x_1 + tx_1 + (1-t)x_2 = (1+t)x_1 + (1-t)x_2 = \left(\frac{3}{2} + \frac{t}{2}, (1-t)\frac{\sqrt{3}}{2}\right)$. Thus, $\|Tx_2\| > 1$, a contradiction.
(2) Suppose $Tx_3 \in L[x_2, x_3]$. Then there exists $t \in [0, 1]$ such that $Tx_3 = tx_2 + (1-t)x_3$. Thus, $Tx_2 = tx_1 + x_2 = (t + \frac{1}{2}, \frac{\sqrt{3}}{2})$. Since $\|Tx_2\| = 1$, we must have $t = 0$. So in this case $Tx_3 = x_3$.
(3) Suppose $Tx_3 \in L[x_3, -x_1]$. Then there exists $t \in [0, 1]$ such that $Tx_3 = tx_3 - (1-t)x_1$. Thus, $Tx_2 = tx_2$. Since $\|Tx_2\| = 1$, we have $t = 1$. So in this case $Tx_3 = x_3$.
(4) Suppose $Tx_3 \in L[-x_1, -x_2]$. Then there exists $t \in [0, 1]$ such that $Tx_3 = -tx_1 - (1-t)x_2$. Thus, $Tx_2 = -(1-t)x_3$. Since $\|Tx_2\| = 1$, we have $t = 0$. So in this case $Tx_3 = -x_2$.
(5) Suppose $Tx_3 \in L[-x_2, -x_3]$. Then there exists $t \in [0, 1]$ such that $Tx_3 = -tx_2 - (1-t)x_3$. Thus, $Tx_2 = (1-t)x_1 - x_3 = \left(\frac{3}{2} - t, -\frac{\sqrt{3}}{2}\right)$. Since $\|Tx_2\| = 1$, we have $t = 1$. So in this case $Tx_3 = -x_2$.
(6) Suppose $Tx_3 \in L[-x_3, x_1]$. Then there exists $t \in [0, 1]$ such that $Tx_3 = tx_1 - (1-t)x_3$. Proceeding similarly as case (1), we can show that $\|Tx_2\| > 1$, a contradiction.

Therefore, if $Tx_1 = x_1$, then considering all possibilities for Tx_3, we have either $Tx_3 = x_3$ or $Tx_3 = -x_2$. In each case, T is an isometry. Clearly, an isometry is an extreme contraction. Now, it is easy to verify that there are 12 isometries on $\mathbb{X}$. Therefore, there are 12 extreme contractions T such that $|M_T \cap Ext(B_{\mathbb{X}})| = 6$. Combining (i) and (ii), we can conclude that there are total $18 + 12 = 30$ extreme contractions on $\mathbb{X}$. □

Research Problems

We would like to end this monograph with a list of concrete problems which might be of some interest to the inquisitive reader. Each of the problems is intricately related to the subject matter developed in this monograph. Moreover, it is strongly hoped that a thorough reading of the relevant results would serve as a motivating factor behind exploring these queries in further detail. We have consciously tried to present the problems using a mix-and-match approach: some are relatively easier while some involve considerable computational difficulties. We believe that a detailed study of the problems, along with the relevant related results which have already been proved, might pave a way towards fruitful exploration, leading to further development of the state of the art in the geometry of bounded linear operators between Banach spaces.

1. Let $\mathbb{X}$ be an n-dimensional ($n > 2$) real Banach space and let $T \in \mathbb{L}(\mathbb{X})$. Then prove or disprove the statement "T satisfies the B Š Property implies M_T has at most two components". (For $\dim(\mathbb{X}) = 2$, see Theorem 4.3.13.)
2. Let $\mathbb{X}$, $\mathbb{Y}$ be Banach spaces and let $T \in \mathbb{L}(\mathbb{X}, \mathbb{Y})$ be such that
   (i) $M_T = \{\alpha x_0 : |\alpha| = 1\}$ for some $x_0 \in S_{\mathbb{X}}$.
   (ii) Tx_0 is smooth point in $\mathbb{Y}$.
   (iii) $dist(T, \mathbb{K}(\mathbb{X}, \mathbb{Y})) < \|T\|$.
   Is T necessarily smooth?
3. Suppose that $\mathbb{X}$ and $\mathbb{Y}$ are Banach spaces and that $T \in S_{\mathbb{L}(\mathbb{X},\mathbb{Y})}$. What are the necessary and sufficient conditions for T to be a multi-smooth point of finite order? Specifically, one can consider the case $\mathbb{X} = \ell_\infty^n$, $\mathbb{Y} = \ell_1^n$, where $n \geq 3$.
4. Let $\mathbb{X}$, $\mathbb{Y}$ be strictly convex Banach spaces and let $\mathbb{Y}$ be smooth. Suppose that dim $\mathbb{X} > 2$. If $T \in \mathbb{K}(\mathbb{X}, \mathbb{Y})$ is right symmetric then whether T is an extreme point of the closed unit ball of $\mathbb{K}(\mathbb{X}, \mathbb{Y})$. (For $\dim(\mathbb{X}) = 2$, see Theorem 7.6.20.)
5. Let $\mathbb{X}$ be a reflexive, Kadets–Klee, strictly convex Banach space and let $\mathbb{Y}$ be strictly convex. Suppose $T \in \mathbb{L}(\mathbb{X}, \mathbb{Y})$ with $M_T \neq \emptyset$. Then what are the necessary and sufficient conditions for T to be right symmetric? (For left symmetric operators, see Theorem 7.6.5.)

A. Mal et al., *Birkhoff–James Orthogonality and Geometry of Operator Spaces*, Infosys Science Foundation Series, https://doi.org/10.1007/978-981-99-7111-4

6. Characterize extreme contractions in $\mathbb{L}(\mathbb{X}, \mathbb{Y})$, where
(i) $\mathbb{X} = \ell_p^n$ and $\mathbb{Y} = \ell_q^n$ with $1 < p, q < \infty$.
(ii) $\mathbb{X} = \ell_p$ and $\mathbb{Y} = \ell_q$ with $1 < p, q < \infty$.
(iii) $\mathbb{X} = \ell_\infty^n$ and $\mathbb{Y} = \ell_1^n$,
where $\mathbb{X}$ and $\mathbb{Y}$ are considered over $\mathbb{R}$ as well as $\mathbb{C}$.

References

1. T.J. Abatzoglou, Norm derivatives on spaces of operators. Math. Ann. **239**, 129–135 (1979)
2. Y.A. Abramovich, C.D. Aliprantis, O. Burkinshaw, The daugavet equation in uniformly convex banach spaces. J. Funct. Anal. **97**, 215–230 (1991)
3. H. Al-Halees, R. Fleming, Extreme contractions on continuous vector-valued function spaces. Proc. Am. Math. Soc. **134**, 2661–2666 (2006)
4. J. Alonso, H. Martini, S. Wu, On Birkhoff orthogonality and isosceles orthogonality in normed linear spaces. Aequationes Math. **83**, 153–189 (2012)
5. J. Alonso, Some properties of Birkhoff and isosceles orthogonality in normed linear spaces, inner product spaces and applications, in *Pitman Research Notes in Mathematics Series*, vol. 376, Longman, Harlow, pp. 1–11 (1997)
6. C. Alsina, J. Sikorska, M.S. Tomás, *Norm Derivatives and Characterizations of Inner Product Spaces* (World Scientific, Hackensack, NJ, 2010)
7. D. Amir, J. Lindenstrauss, The structure of weakly compact sets in Banach Spaces. Ann. Math. **88**, 35–46 (1968)
8. T. Ando, *Distance to the Set of Thin Operators*, unpublished report (1972)
9. L. Arambašíc, R. Rajíc, The Birkhoff-James orthogonality in Hilbert C*-modules. Linear Algebra Appl. **437**, 1913–1929 (2012)
10. L. Arambašíc, R. Rajíc, A strong version of the Birkhoff-James orthogonality in Hilbert C^*-modules. Ann. Funct. Anal. **5**(1), 109–120 (2014)
11. L. Arambašíc, R. Rajíc, On three concepts of orthogonality in Hilbert C^*-modules. Linear Multilinear Algebra **63**(7), 1485–1500 (2015)
12. L. Arambašíc, R. Rajíc, On symmetry of the (strong) Birkhoff-James orthogonality in Hilbert C^*-modules. Ann. Funct. Anal. **7**(1), 17–23 (2016)
13. L. Arambašíc, R. Rajíc, On Birkhoff-James and Roberts orthogonality. Spec. Matrices **6**, 229–236 (2018)
14. V. Balestro, H. Martini, R. Teixeira, A new construction of Radon curves and related topics. Aequationes Math. **90**(5), 1013–1024 (2016)
15. P. Bandyopadhyay, A.K. Roy, Extreme contractions in $\mathbb{L}(\ell_p^2, \ell_q^2)$ and the mazur intersection property in $\ell_p^2 \otimes_\pi \ell_q^2$. Real Anal. Exchange **20**(2), 681–698 (1994/95)
16. C. Benítez, M. Fernández, M.L. Soriano, Orthogonality of matrices. Linear Algebra Appl. **422**, 155–163 (2007)
17. R. Bhatia, P. Šemrl, Orthogonality of matrices and some distance problem. Linear Algebra Appl. **287**, 77–85 (1999)
18. R. Bhatia, R. Sharma, Some inequalities for positive linear maps. Linear Algebra Appl. **436**, 1562–1571 (2012)
19. T. Bhattacharyya, P. Grover, Characterization of Birkhoff-James orthogonality. J. Math. Anal. Appl. **407**, 350–358 (2013)
20. G. Birkhoff, Orthogonality in linear metric spaces. Duke Math. J. **1**, 169–172 (1935)
21. R.M. Blumenthal, J. Lindenstrauss, R.R. Phelps, Extreme operators into $C(K)$. Pacific J. Math. **15**, 747–756 (1965)
22. A.M. Cabrera-Serrano, J.F. Mena-Jurado, On extreme operators whose adjoints preserve extreme points. J. Convex Anal. **22**(1), 247–258 (2015)

23. A. Chattopadhyay, D. Sain, T. Senapati, Characterization of symmetric points in ℓ_p^n-spaces. Linear Multilinear Algebra **69**(16), 2998–3009 (2021)
24. J. Chmieliński, On an ϵ—Birkhoff orthogonality. J. Inequal. Pure Appl. Math. **6**(3), Article 79 (2005)
25. J. Chmieliński, T. Stypula, P. Wójcik, Approximate orthogonality in normed spaces and its applications. Linear Algebra Appl. **531**, 305–317 (2017)
26. J. Chmieliński, P. Wójcik, ρ-orthogonality and its preservation-revisited, recent developments in functional equations and inequalities. Banach Center Publ. **99**, 17–30 (2013)
27. J.B. Conway, *A Course in Functional Analysis* (Springer, New York, 1990)
28. H. Corson, J. Lindenstrauss, Continuous selections with nonmetrizable range. Trans. Amer. Math. Soc. **121**, 492–504 (1966)
29. M.M. Day, *Normed Linear Spaces* (Academic Press Inc., Springer, New York, Berlin-Göttingen-Heidelberg, 1962)
30. M.M. Day, Some characterizations of inner-product spaces. Trans. Amer. Math. Soc. **62**, 320–337 (1947)
31. W. Deeb, R. Khalil, Exposed and smooth points of some classes of operation in $L(\ell_p)$. J. Funct. Anal. **103**, 217–228 (1992)
32. S.S. Dragomir, On approximation of continuous linear functionals in normed linear spaces. An. Univ. Timişoar a Ser. Ştiinţ. Mat. **29**(1), 51–58 (1991)
33. S.S. Dragomir, J.J. Koliha, Two mappings related to semi-inner products and their applications in geometry of normed linear spaces. Appl. Math. **45**, 337–355 (2000)
34. M. Fabian, P. Habala, P. Hajek, V.M. Santalucia, J. Pelant, V. Zizler, *Functional Analysis and Infinite-dimensional Geometry* (Springer, New York). ISBN 0-387-95219-5
35. V.P. Fnof, Polyhedral Banach spaces. Math. Notes **30**, 809–813 (1981)
36. M. Fujii, R. Nakamoto, An estimation of the transcendental radius of an operator. Math. Japonica **27**, 637–638 (1982)
37. P. Ghosh, D. Sain, K. Paul, Symmetric properties of orthogonality of linear operators on $(R^n, \|.\|_1)$. Novi Sad J. Math. **47**(2), pp. 41–46 (2017)
38. P. Ghosh, D. Sain, K. Paul, On symmetry of Birkhoff-James orthogonality of linear operators. Adv. Oper. Theory **2**(4), 428–434 (2017)
39. P. Ghosh, D. Sain, K. Paul, Orthogonality of bounded linear operators. Linear Algebra Appl **500**, 43–51 (2016)
40. J.R. Giles, Classes of semi-inner-product spaces. Trans. Amer. Math. Soc. **129**, 436–446 (1967)
41. P. Grover, Orthogonality to matrix subspaces, and a distance formula. Linear Algebra Appl. **445**, 280–288 (2014)
42. P. Grover, S. Singla, Birkhoff-James orthogonality and applications: a survey, A Survey. in *Operator Theory, Functional Analysis and Applications. Operator Theory: Advances and Applications*, ed. by Bastos, M.A., Castro, L., Karlovich, A.Y., vol. 282. Birkhäuser, Cham
43. R. Grząślewicz, Smoothness in $\mathcal{L}(C(X), C(Y))$. Acta Univ. Carolin. Math. Phys. **37**, 25–28 (1996)
44. R. Grząślewicz, Extremal structure of $\mathcal{L}(\ell_m^2, \ell_n^p)$. Linear Multilinear Algebra **24**, 117–125 (1989)
45. R. Grząślewicz, Extreme contractions on real Hilbert spaces. Math. Ann. **261**, 463–466 (1982)
46. R. Grząślewicz, Extreme operators on 2-dimensional ℓ_p-spaces. Colloq. Math. **44**, 309–315 (1981)
47. R. Grząślewicz, A note on extreme contractions on ℓ_p-spaces. Portugal. Math. **40**, 413–419 (1981)
48. R. Grząślewicz, R. Younis, Smooth points and M-ideals. J. Math. Anal. Appl. **175**, 91–95 (1993)
49. P.R. Halmos, A Hilbert space problem book. *Graduate Texts in Mathematics* (Springer, 1967)
50. P. Harmand, D. Werner, W. Werner, *M-Ideals in Banach Spaces and Banach Algebras*. Lecture Notes in Mathematics, vol. 1547 (Springer-Verlag, Berlin, 1993) viii+387pp
51. E. Heil, Abschätzungen für einige Affininvarianten konvexer Kurven. Monatsh. Math. **71**, 405–423 (1967)

52. S. Heinrich, The differentiability of the norm in the spaces of operators. Funkcional Anal. i Priložen, **9**, 93–94 (1975); English transl.: Funct. Anal. Appl. **9**, 360–362 (1975)
53. J. Hennefield, Smooth, compact operators. Proc. Am. Math. Soc. **77**, 87–90 (1979)
54. R.B. Holmes, Geometric functional analysis and its applications. *Graduate Texts Mathematics*, vol. 24 (Springer, New York (NY), 1975), x+246pp
55. J.R. Holub, On the metric geometry of ideals of operators on hilbert space. Math. Ann. **201**, 157–163 (1973)
56. A. Iwanik, Extreme contractions on certain function spaces. Colloq. Math. **40**(1), 147–153 (1978)
57. R.C. James, Reflexivity and the sup of linear functionals. Israel J. Math. **13**(1972), 289–300 (1973)
58. R.C. James, Characterizations of reflexivity. Studia Math. **23**, 205–216 (1964)
59. R.C. James, Orthogonality and linear functionals in normed linear spaces. Trans. Am. Math. Soc. **61**, 265–292 (1947)
60. R.C. James, Inner product in normed linear spaces. Bull. Amer. Math. Soc. **53**(6), 559–566 (1947)
61. R.C. James, Orthogonality in normed linear spaces. Duke Math. J. **12**, 291–302 (1945)
62. R.V. Kadison, Isometries of operator algebras. Ann. of Math. **54**(2), 325–338 (1951)
63. S. Kakutani, Some characterizations of Euclidean space. Jpn. J. Math. **16**, 93–97 (1939)
64. C.H. Kan, A class of extreme L_p contractions, $p \neq 1, 2, \infty$ and real 2×2 extreme matrices. Illinois J. Math. **30**, 612–635 (1986)
65. R. Khalil, A. Saleh, Multi-smooth points of finite order, Missouri. J. Math. Sci. **17**, 76–87 (2005)
66. D. Khurana, S. Roy, D. Sain, Symmetric points in spaces of linear operators between Banach spaces. Acta Sci. Math. (Szeged) **86**, 617–634 (2020)
67. C.W. Kim, Extreme contraction operators on l_∞. Math. Z. **151**(2), 101–110 (1976)
68. F. Kittaneh, On zero-trace matrices. Linear Algebra Appl. **151**, 119–124 (1991)
69. F. Kittaneh, R. Younis, Smooth points of certain operator spaces. Integral Equations Operator Theory **13**, 849–855 (1990)
70. V. Klee, Polyhedral sections of convex bodies. Acta Math. **103**, 243–267 (1960)
71. V. Klee, Some characterizations of convex polyhedra. Acta Math. **102**, 79–107 (1959)
72. A. Koldobsky, Operators preserving orthogonality are isometries. Proc. Roy Soc. Edinburgh Sect. A **123**, 835–837 (1993)
73. N. Komuro, K. Saito, R. Tanaka, A characterization of Radon planes using generalized Day-James spaces. Ann. Funct. Anal. **11**, 62–74 (2020)
74. D. Larman, On a conjecture of Lindenstrauss and Perles in at most 6 dimensions. Glasgow Math. J. **19**, 87–97 (1978)
75. C.K. Li, H. Schneider, Orthogonality of matrices. Linear Algebra Appl. **347**, 115–122 (2002)
76. Å. Lima, On extreme operators on finite-dimensional Banach spaces whose unit balls are polytopes. Ark. Mat. **19**(1), 97–116 (1981)
77. Å. Lima, Intersection properties of balls in spaces of compact operators. Ann. Inst. Fourier **28**, 35–65 (1978)
78. Å. Lima, G. Olsen, Extreme points in duals of complex operator spaces. Proc. Am. Math. Soc. **94**(3), 437–440 (1985)
79. B. Lin, T.S.S.R.K. Rao, Multismoothness in Banach Spaces. Int. J. Math. Math. Sc. **2007** (2007). Article ID 52382
80. J. Lindenstrauss, M.A. Perles, On extreme operators in finite-dimensional spaces. Duke Math. J. **36**, 301–314 (1969)
81. G. Lumer, Semi-inner-product spaces. Trans. Amer. Math. Soc. **100**, 29–43 (1961)
82. B. Magajna, On the distance to finite-dimensional subspaces in operator algebras. J. London Math. Soc. **47**(2), 516–532 (1993)
83. A. Mal, S. Dey, K. Paul, Characterization of $k-$smoothness of operators defined between infinite-dimensional spaces. Linear Multilinear Algebra **70**(18), 3477–3489 (2022)

84. A. Mal, K. Paul, Birkhoff-James orthogonality to a subspace of operators defined between Banach spaces. J. Operator Theory **85**(2), 463–474 (2021)
85. A. Mal, K. Paul, Characterization of k-smooth operators between Banach spaces. Linear Algebra Appl. **586**, 296–307 (2020)
86. A. Mal, K. Paul, S. Dey, Characterization of extreme contractions through k-smoothness of operators. Linear Multilinear Algebra **70**(20), 5301–5315 (2022)
87. A. Mal, K. Paul, T.S.S.R.K. Rao, D. Sain, Approximate Birkhoff-James orthogonality and smoothness in the space of bounded linear operators. Monatsh. Math. **190**, 549–558 (2019)
88. A. Mal, D. Sain, K. Paul, On some geometric properties of operator spaces. Banach J. Math. Anal. **13**(1), 174–191 (2019)
89. K. Mandal, D. Sain, A. Mal, K. Paul, Norm attainment set and symmetricity of operators on ℓ_p^2. Adv. Oper. Theory **7**(1) (2022) Article: 3
90. K. Mandal, D. Sain, K. Paul, A geometric characterization of polygonal Radon planes. J. Convex Anal. **26**(4), 1113–1123 (2019)
91. M. Martín', Norm-attaining compact operators. J. Funct. Anal. **267**, 1585–1592 (2014)
92. H. Martini, K.J. Swanepoel, Antinorms and Radon curves. Aequationes Math. **72**(1–2), 110–138 (2006)
93. H. Martini, K.J. Swanepoel, G. Weiß, The geometry of Minkowski spaces-a survey. I. Expo. Math. **19**(2), 97–142 (2001)
94. R.E. Megginson, *An Introduction to Banach Space Theory* (Springer, New York). ISBN 0-387-98431-3
95. P. Morris, R. Phelps, Theorems of Krein-Milman type for certain convex sets of operators. Trans. Amer. Math. Soc. **150**, 183–200 (1970)
96. M.A. Navarro, Some characterizations of finite-dimensional Hilbert spaces. J. Math. Anal. Appl. **223**, 364–365 (1998)
97. J.C. Navarro-Pascual, M.A. Navarro, Nice operators and surjective isometries. J. Math. Anal. Appl. **426**, 1130–1142 (2015)
98. K. Paul, Translatable radii of an operator in the direction of another operator II. Math. Slovaca **60**, 121–128 (2010)
99. K. Paul, Translatable radii of an operator in the direction of another operator. Sci. Math. **2**, 119–122 (1999)
100. K. Paul, S.M. Hossein, K.C. Das, Orthogonality on $B(H, H)$ and minimal-norm operator. J. Anal. Appl. **6**, 169–178 (2008)
101. K. Paul, A. Mal, Birkhoff-James orthogonality: its role in the geometry of Banach space. Math. Student **91**, 55–78 (2022)
102. K. Paul, A. Mal, P. Wójcik, Symmetry of Birkhoff-James orthogonality of operators defined between infinite dimensional Banach spaces. Linear Algebra Appl. **563**, 142–153 (2019)
103. K. Paul, D. Sain, Orthogonality of operators on $(\mathbb{R}^n, \|\ \|_\infty)$. Novi Sad J. Math. **43**, 121–129 (2013)
104. K. Paul, D. Sain, K. Jha, On strong orthogonality and strictly convex normed linear spaces. J. Inequal. Appl. **2013**, 242 (2013)
105. K. Paul, D. Sain, P. Ghosh, Birkhoff-James orthogonality and smoothness of bounded linear operators. Linear Algebra Appl. **506**, 551–563 (2016)
106. K. Paul, D. Sain, A. Mal, Approximate Birkhoff-James orthogonality in the space of bounded linear operators. Linear Algebra Appl. **537**, 348–357 (2018)
107. K. Paul, D. Sain, A. Mal, K. Mandal, Orthogonality of bounded linear operators on complex Banach spaces. Adv. Oper. Theory **3**(3), 699–709 (2018)
108. T.S.S.R.K. Rao, Operators Birkhoff-James orthogonal to spaces of operators. Numer. Funct. Anal. Optim. **42**(10), 1201–1208 (2021)
109. T.S.S.R.K. Rao, Smooth points in spaces of operators. Linear Algebra Appl. **517**, 129–133 (2017)
110. T.S.S.R.K. Rao, Adjoints of operators as smooth points in spaces of compact operators. Linear Multilinear Algebra **66**(4), 668–670 (2018)

111. A. Ray, S. Roy, S. Bagchi, D. Sain, Extreme contractions on finite-dimensional polygonal Banach spaces-II. J. Operator Theory **84**(1), 127–137 (2020)
112. M. Reed, B. Simon, *Methods of Modern Mathematical Physics*, vol. I (Academic Press, 1980)
113. S. Roy, S. Bagchi, Orthogonality of sesquilinear forms and spaces of operators. Linear Multilinear Algebra **70**(19), 4416–4424 (2022)
114. S. Roy, S. Bagchi, D. Sain, Birkhoff-James orthogonality in complex Banach spaces and Bhatia-Šemrl Theorem revisited. Mediterr. J. Math. **19** (2022). Article number: 58
115. S. Roy, D. Sain, Approximate orthogonality in complex normed spaces: a directional approach. Linear Multilinear Algebra **71**(5), 711–727 (2023)
116. S. Roy, T. Senapati, D. Sain, Orthogonality of bilinear forms and application to matrices. Linear Algebra Appl. **615**, 104–111 (2021)
117. W.M. Ruess, C.P. Stegall, Extreme points in duals of operator spaces. Math. Ann. **261**, 535–546 (1982)
118. D. Sain, P. Ghosh, K. Paul, On symmetry of Birkhoff-James orthogonality of linear operators on finite-dimensional real Banach spaces. Oper. Matrices **11**(4), 1087–1095 (2017)
119. D. Sain, A. Mal, K. Paul, Some remarks on Birkhoff-James orthogonality of linear operators. Expo. Math. **38**, 138–147 (2020)
120. D. Sain, K. Paul, Operator norm attainment and inner product spaces. Linear Algebra Appl. **439**, 2448–2452 (2013)
121. D. Sain, K. Paul, P. Bhunia, S. Bag, On the numerical Index of polyhedral Banach Spaces. Linear Algebra Appl. **577**, 121–133 (2019)
122. D. Sain, K. Paul, S. Hait, Operator norm attainment and Birkhoff-James orthogonality. Linear Algebra Appl. **476**, 85–97 (2015)
123. D. Sain, K. Paul, K. Jha, Strictly convex space: strong orthogonality and conjugate diameters. J. Convex Anal. **22**(4), 1215–1225 (2015)
124. D. Sain, K. Paul, A. Mal, On extreme contractions between real Banach spaces. Expo. Math. **39**, 33–47 (2021)
125. D. Sain, K. Paul, A. Mal, On approximate Birkhoff-James orthogonality and normal cones in a normed space. J. Convex Anal. **26**(1), 341–351 (2019)
126. D. Sain, K. Paul, A. Mal, A complete characterization of Birkhoff-James orthogonality in infinite dimensional normed space. J. Operator Theory **80**(2), 399–413 (2018)
127. D. Sain, K. Paul, K. Mandal, On two extremum problems related to the norm of a bounded linear operator. Oper. Matrices **13**, 421–432 (2019)
128. D. Sain, K. Paul, A. Mal, A. Ray, A complete characterization of smoothness in the space of bounded linear operators. Linear Multilinear Algebra **68**(12), 2484–2494 (2020)
129. D. Sain, A. Ray, K. Paul, Extreme contractions on finite-dimensional polygonal Banach spaces. J. Convex Anal. **26**(3), 877–885 (2019)
130. D. Sain, S. Roy, S. Bagchi, V. Balestro, A study of symmetric points in Banach spaces. Linear Multilinear Algebra **70**(5), 888–898 (2022)
131. D. Sain, On best approximations to compact operators. Proc. Amer. Math. Soc. **149**(10), 4273–4286 (2021)
132. D. Sain, On the norm attainment set of a bounded linear operator and semi-inner-products in normed spaces. Indian J. Pure Appl. Math. **51**(1), 179–186 (2020)
133. D. Sain, On extreme contractions and the norm attainment set of a bounded linear operator. Ann. Funct. Anal. **10**(1), 135–143 (2019)
134. D. Sain, On the norm attainment set of a bounded linear operator. J. Math. Anal. Appl. **457**, 67–76 (2018)
135. D. Sain, Birkhoff-James orthogonality of linear operators on finite dimensional Banach spaces. J. Math. Anal. Appl. **447**, 860–866 (2017)
136. J. Sen, A. Mal, K. Paul, Characterization of Approximate Birkhoff-James Orthogonality Sets in a Banach Space. J. Convex Anal. **27**(3), 881–892 (2020)
137. M. Sharir, A non-nice extreme operator. Israel J. Math. **26**(3–4), 306–312 (1977)
138. M. Sharir, A counterexample on extreme operators. Israel J. Math. **24**, 320–328 (1976)
139. M. Sharir, Extremal structure in operator spaces. Trans. Amer. Math. Soc. **186**, 91–111 (1973)

140. M. Sharir, Characterization and properties of extreme operators into $C(Y)$. Israel J. Math. **12**, 174–183 (1972)
141. I. Singer, Best approximation in normed linear spaces by elements of linear subspaces. Die Grundlehren der Mathematischen, vol.171 (Springer, Berlin, Heidelberg, New York, 1970)
142. J.G. Stampfli, The norm of a derivation. Pacific J. Math. **33**, 737–747 (1970)
143. A. Turnšek, A remark on orthogonality and symmetry of operators in $\mathcal{B}(\mathcal{H})$. Linear Algebra Appl. **535**, 141–150 (2017)
144. A. Turnšek, On operators preserving James' orthogonality. Linear Algebra Appl. **407**, 189–195 (2005)
145. D. Werner, Extreme points in spaces of operators and vector-valued measures. Proceedings of the 12th winter school on abstract analysis (Srní, 1984). Rend. Circ. Mat. Palermo **2**, Suppl. No. 5, 135–143 (1984)
146. W. Werner, Smooth points in some spaces of bounded operators. Integral Equ. Oper. Theory **15**, 496–502 (1992)
147. P. Wójcik, Approximate orthogonality in normed spaces and its applications II. Linear Algebra Appl. **632**, 258–267 (2022)
148. P. Wójcik, Extensions of linear operators from hyperplanes and strong uniqueness of best approximation in $\mathcal{L}(X, W)$. J. Approx. Theory **246**, 28–42 (2019)
149. P. Wójcik, k-smoothness: an answer to an open problem. Math. Scand. **123**, 85–90 (2018)
150. P. Wójcik, Birkhoff orthogonality in classical M-ideals. J. Aust. Math. Soc. **103**, 279–288 (2017)
151. P. Wójcik, Orthogonality of compact operators. Expo. Math. **35**(1), 86–94 (2017)
152. A. Zamani, M.S. Moslehian, Norm-parallelism in the geometry of Hilbert C^*-modules. Indag. Math. **27**, 266–281 (2016)
153. K. Ziętak, From the strict Chebyshev approximant of a vector to the strict spectral approximant of a matrix, Études opératorielles, 307–346, Banach Center Publ. 112, Polish Acad. Sci. Inst. Math., Warsaw (2017)

GPSR Compliance
The European Union's (EU) General Product Safety Regulation (GPSR) is a set of rules that requires consumer products to be safe and our obligations to ensure this.

If you have any concerns about our products, you can contact us on

ProductSafety@springernature.com

In case Publisher is established outside the EU, the EU authorized representative is:

Springer Nature Customer Service Center GmbH
Europaplatz 3
69115 Heidelberg, Germany

www.ingramcontent.com/pod-product-compliance
Ingram Content Group UK Ltd.
Pitfield, Milton Keynes, MK11 3LW, UK
UKHW022212051026
15624UKWH00001B/124
9789819971107